Troubleshooting Consumer Electronics Audio Circuits

Troubleshooting Consumer Electronics Audio Circuits

Homer L. Davidson

PROMPT© Publications is an imprint of Howard W. Sams & Company, A Bell Atlantic Company, 2647 Waterfront Parkway, E. Dr., Indianapolis, IN 46214-2041.

International Standard Book Number: 0-7906-1165-1
Library of Congress Catalog Card Number: 99-64767

Acquisitions Editor: Loretta Yates
Editor: J.B. Hall
Assistant Editor: Pat Brady
Typesetting: J.B. Hall
Indexing: J. B. Hall
Proofreader: Stacy Nolan
Cover Design: Christy Pierce
Graphics Conversion: Joe Kocha, Dick Raus, Lora Robinson, Bill Skinner, Terry Varvel
Illustrations and Other Materials: Courtesy of the Author

PRINTED IN THE UNITED STATES OF AMERICA

9 8 7 6 5 4 3 2 1

Contents

Chapter 7
Repairing SMD Audio Circuits 131

Chapter 8
Servicing High-Powered Audio Circuits 149

Chapter 9 Troubleshooting Inexpensive Electronics Circuits....171

Chapter 10 Servicing Special Consumer Electronics Audio Circuits....193

Chapter 12
Important Audio Tests and Adjustments 233

INTRODUCTION

90% of all consumer electronic products contain circuitry that you can troubleshoot. The audio circuits might start with a simple preamp circuit and zoom up to the high-powered semiconductor or tube audio output circuits. Servicing the electronic audio circuits can be accomplished by anyone interested in consumer electronics. Troubleshooting the mono and stereo audio circuits are easy to repair and sometimes one can listen to the music while working on the audio chassis. Besides, troubleshooting audio circuits is a lot of fun.

The purpose of this book is to show the various audio circuits found in the audio amplifier, cassette and tape player, TV, CD player, auto AM/FM/MPX receiver, home receiver, boombox, auto CD player, telephone answering machines, and the high-powered amplifier.

This is a practical book in troubleshooting consumer audio circuits. The book is filled with hints, kinks, and practical audio data. Throughout the many chapters there are actual case histories of the defective audio circuits, what tests were made, and how to replace the defective component. Locating the defective audio component takes up to 85 % of your service time, while removing and replacing the defective part requires the least amount of time in troubleshooting electronic circuit procedures.

The basic audio circuit tests are provided in detail within Chapter 1. Knowing how to take critical voltage, resistance, and semiconductor tests is required with the various test instruments. Learn how to locate the defective component when weak, distorted, and intermittent circuits are found in the basic circuits.

Knowing how to use and locate the defective audio component with the multi-tester, DMM, noise and tone generator, semiconductor tester, audio and function generator, and audio signal tracing test instruments is important also. The various load, test cassettes and tape head devices are found in Chapter 2. The most important list of test instruments are required for those technicians who want to become audio specialists within the audio consumer electronic field.

The most important circuits within the audio products are found in the low voltage power supply. Most technicians check the low voltage sources, first, in the audio circuits or chassis. Chapter 3 provides the many different power supply circuits found in the different electronic products in the consumer electronic field. The different service problems within the low voltage circuits are given in this chapter.

Troubleshooting the preamp, AF and power output circuits are found in Chapters 4 and 5. The different preamp and AF circuits with the many service problems are found in Chapter 4. Chapter 5 describes the different audio output circuits, the weak, distorted, intermittent, and noisy circuits within the audio output circuits. The special high-wattage and tube amplifier circuits are provided in this chapter.

Chapter 6 illustrates the different service problems found in the stereo audio circuits. This chapter shows how to troubleshoot, locate, and service the right and left stereo circuits found in the cassette and CD players, auto stereo receivers, and auto high-powered amplifiers. It also describes how to replace power output transistors and IC components in the stereo audio circuits.

Steps in repairing the new surface-mounted devices (SMD) are given in Chapter 7. Learn how to locate and know how to identify the SMD part and replace the defective component in the mini-cassette, portable CD player, TV, and stereo amplifiers. The different SMD parts are now found in SMD construction projects.

Information on servicing the high-powered audio circuits can be found throughout Chapter 8. How to repair the tough dog chassis, high-powered circuits, four channel amps, bridged power output circuits and several different case histories are listed in this chapter.

Information on how to service the simple and inexpensive consumer audio electronic circuits can be found in Chapter 9. The early and simple radio circuits, transistor and IC radio, compact disc, and phonograph circuits are found in this chapter. Also listed here, are the boom-box, portable CD, auto cassette, and the various headphone circuits.

The special audio circuits of the erase head, tape head, phono, remote control, mute, speaker relay, and tube bias circuits, to name a few, are found in Chapter 10. Information on how to repair the auto cassette reverse, receiver mute, center power amp, ac-dc tube radio, tube amplifier, and Dolby circuits are also found in this chapter.

The audio technician services many different audio products without a schematic, each day of the week. Chapter 11 shows the many different service problems and symptoms with the various audio circuits and how to repair them. Information on how to handle the hot, weak, intermittent, dead, and noisy audio circuits without a schematic is provided. Servicing the remote control circuits can be done by almost anyone with only a few hand tools and test instruments.

Information on how to connect, use, and make different tests with different test instruments in troubleshooting audio circuits is found in Chapter 12. Issues that are addressed include: What do the various scope waveforms indicate, how test instruments are tied into the audio circuits, and the many adjustments required within the cassette player and amplifiers for best recording and playback modes.

Besides providing the different electronic audio circuits and test data, many photos and schematics are found throughout the book. You will find many practical case histories in each chapter. The earlier and the latest audio circuits were taken from the Howard W. Sams Photofact® series. Now is the time to troubleshoot and repair that next audio circuit, found in the different audio electronic products. So lets begin.

Chapter 1

Basic Audio Circuit Tests

Troubleshooting consumer electronics audio circuits can be as simple as learning your ABC's or starting to take a step as a child. You begin with the basic audio circuits and how to use audio test equipment. Then, you learn how audio circuits react and perform, and finally, you learn how they are tied together. Besides learning how to service audio circuits, the electronic hobbyist or beginner learns how to solder, remove, and replace defective components. Even the electronic technician that only repairs TVs and VCRs might return to audio basics, and add another product to the service bench. Above all, you can receive many high and low rewards, sometimes even defeat, but still have fun while servicing electronic audio products.

The most popular audio circuits are found in many different consumer electronic products such as, the table radio, AM-FM MPX stereo receiver, phonograph, cassette and CD player, auto radio CD/cassette player, and the TV to name a few *(Figure 1-1)*. Of course, there are special audio circuits found in the telephone answering machine, audio test equipment, camcorder, high powered amplifier, and the everlasting high-priced tube amplifier.

Figure 1-1. The internal view of a Radio Shack 12-639 portable radio with a 5 1/2 inch speaker.

Audio circuits within the consumer electronic field produce a sound that is audible to the ear. The sound might be clean and clear, or weak and distorted. The audio frequency (AF) amplifier operates within a frequency range of 20 Hz to 20 kHz. Very few men, 50 years old, can hear above 10 kHz, while some women can hear above the 15 kHz range.

The sound we hear from the 1 inch to the large 15 inch speakers or a little old pair of headphones, is produced by a preamp, AF, driver and power output circuits within the consumer electronic product. Sometimes audio circuits appear simple and easy to repair, but then again, the intermittent sound problem can be down right mean to service.

THE BLOCK DIAGRAM

The block diagram is a simplified schematic of an electronic audio system with the various stages or circuits found in boxes. It is a simplified version of the various audio circuits. The block diagram shows how the various stages are tied together. Often, arrows point to the various boxes indicating how the signal path travels or power supply sources feed the other circuits. You can quickly isolate the defective sections in the electronic block circuits. Locating the circuits upon the electronic chassis is another hill to climb.

When checking the block diagram of a defective audio amp within the boom-box cassette player, you can quickly locate the defective circuit by various symptoms *(Figure 1-2)*. For instance, when the sound is distorted at the speaker, suspect the audio output circuits.

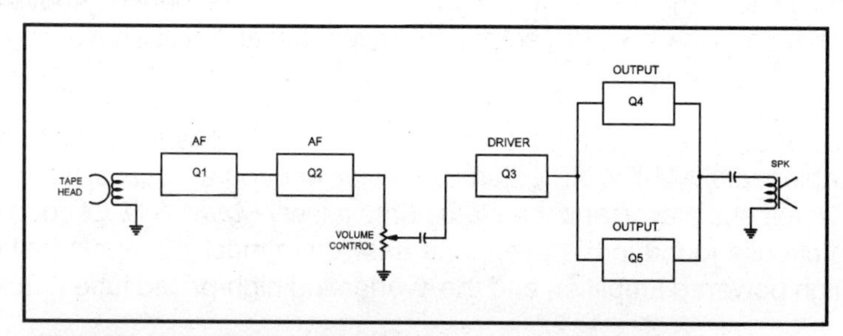

Figure 1-2. The block diagram of an early audio amplifier with an AF, driver and two transistors in push-pull operation.

Suspect a front-end or open tape head when a loud rushing noise is heard in the speaker. When a low frying noise is heard all the time and controlled by the volume control, suspect a defective component ahead of the volume control. Suspect a weak or open component when the stereo amp circuits will not balance up, and on it goes.

THE RIGHT ONE

The first peek at a schematic diagram of a clock radio output circuit might appear complicated to some, but if you break the schematic down into the various sections or circuits, servicing becomes much easier. For instance, if music from the radio speaker is quite weak and faint, go directly to the volume control and check the signal at that point in the audio circuit. Determine if the loss of signal is ahead or after the volume control. Proceed through the audio circuit with a signal tracer or scope to determine where the music becomes weak.

Many different manufacturers do not list the various voltages upon the schematic of transistor or IC components. Some list only the symbol and number of each part and no values. An electrolytic coupling capacitor might have a part listing with no value or working voltage. You might find the resistor's value and wattage missing from the circuit. The zener diode might have a part symbol with no zener voltage. Only one fixed diode symbol might be found in a low voltage power supply bridge circuit. When the various voltages, values and part numbers are missing, servicing the audio circuits can be more difficult.

A lot of the manufacturers have the complicated audio circuits of a high powered amplifier (above 100 watts) with the part symbol number and no voltage applied to the higher wattage transistors. You may find in the other stereo channel the same parts are marked with voltages or a separate voltage chart of each transistor is listed in a table format. When the operating voltages are listed in a chart or table, you must apply them to the transistor or IC in the circuit. It's best to mark the voltages on the schematic, from the chart, before attempting to service the audio chassis.

Some manufacturers have arrows or thick black lines indicating the signal path from stage to stage within the cassette-recorder. Here one can trace the recorded or playback signal very quickly through the various amplifier circuits. Some schematics have voltages listed in various colors. After locating a defective component, circle it and draw a line out to the side of the margin area to record the service problem. Now take correct voltage measurements and mark them on the transistors or IC components in the audio circuit.

VOLTAGE IN-CIRCUITS TESTS

The most common test instruments used in taking voltage measurements are the VOLT-OHM-METER (VOM), digital-multimeter DMM, and FET-VOLT-OHM-METER (FET-VOM) *(Figure 1-3)*. A quick method to locate a defective component in a given circuit is with voltage measurements. By taking voltage measurements on the transistor or IC terminals, you can determine if the circuit or component is defective. A quick forward bias voltage measurement between the emitter and base terminals of a transistor can indicate a defective transistor. Critical voltage measurements through the audio circuits can quickly locate a defective component or circuit.

The DMM is ideal in measuring critical voltages in the solid-state audio circuits. A normal silicon transistor may have a forward bias of 0.6 volts, while the normal germanium transistor has a forward bias of 0.3 volts dc. You can quickly determine if a transistor is open or leaky with a forward bias voltage reading between base and emitter terminals. Some electronic technicians take a quick forward bias measurement on all transistors in the electronic chassis to locate the defective component or circuit.

Figure 1-3. The VOM, DMM and FET-VOM test instruments are used in making voltage measurements in solid-state circuits.

Check the voltage measurements and compare them with those found on the schematic diagram. Rotate the function switch of DMM to the desired voltage range. If the overrange display comes on, turn to a higher voltage scale. The overrange symbol might be a 1 or L figure. One of the biggest advantages of the DMM is to indicate the correct voltage polarity without changing test leads. The DMM takes very accurate low voltage measurements found in solid-state audio circuits.

Take critical voltage measurements of all transistors within the audio input circuits *(Figure 1-4)*. Clip the negative (-) probe to chassis or common ground and place the positive (+) probe to the collector, base and emitter terminals in that order. Compare these voltages to those located on the schematic. Remember, the collector voltage is higher or more positive than the emitter or base terminal (NPN). A very low collector terminal voltage might indicate a leaky transistor or improper power source. A close voltage measurement on all three terminals may indicate a leaky or shorted transistor. No voltage measurement upon the emitter terminal might indicate an open transistor or emitter resistor. Higher than normal voltage found on one of the transistors in a push-pull output circuit might indicate an open transistor in the ground circuit.

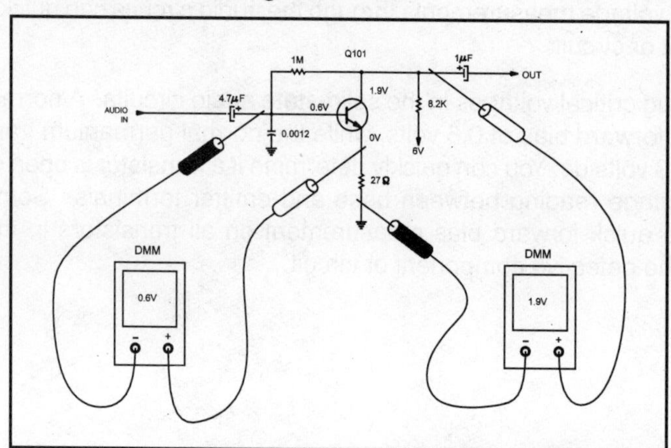

Figure 1-4. Taking voltage measurements upon the early PNP input amplifier stages with the DMM.

The VOLT-OHM-METER (VOM) was one of the first meters placed on the market to take voltage, resistance and current measurements. The VOM can load down the circuit producing inaccurate measurements. A VOM can make quick continuity tests of low resistance components, broken wires, and PC wiring. The VOM is not as accurate as the DMM or FET-VOM when taking critical voltage and resistance measurements.

The VOM is ideal when used as a meter monitor in tape head azimuth adjustments and other alignment procedures. The meter hand can be easily seen to raise or lower, where the DMM numbers change very rapidly, in ac alignment adjustments.

The FET-VOM (analog meter) is a brother to the vacuum tube voltmeter (VTVM). The FET-VOM has a very high impedance input and does not load down the circuit to be measured. The FET-VOM is a VOM meter employing a field-effect transistor amplifier circuit. This meter is more accurate than the regular VOM in taking voltage and resistance measurements. The FET-VOM is ideal when making alignment and adjustment procedures. Like the VOM, the FET meter must reverse test probes when the meter hand goes the wrong way. This meter is a great deal more accurate than the small VOM in voltage, resistance and current measurements. Of course, the FET-VOM meter costs three or four times the pocket VOM, but it's worth it. The DMM and FET-VOM meters are two dependable test instruments in servicing electronic audio circuits.

RESISTANCE MEASUREMENTS

A resistor offers opposition to current flow in the electronic circuit. The many different kinds of resistors can be accurately checked with the DMM or FET-VOM test instrument. Choose the DMM when taking low resistors under 10 ohms. The digital-multimeter can indicate a poor switching contact under a fraction of an ohm. Remove one end of a resistor to take accurate resistance measurement; this is especially so with 500k or higher ohm resistors in the electronic circuit. The amplifier circuits should be turned off when making resistance measurements.

The FET-analog multi tester might measure ac and dc voltages from 0 to 1000 volts. The resistance measurement might measure resistance from 0 to 10 megohms. The analog FET meter may measure current from 0 to 100 μA, 1 to 300 mA, and up to 10 amps.

Some FET-multi testers have a continuity buzzer that sounds when the resistance of the circuit being tested is below 300 ohms or less. Start with the highest voltage range or above the minimum battery or supply voltage since the analog meter hand will crash against the meter peg.

The analog meter probes must be reversed when the polarity is wrong and the meter hand goes backwards. Do not apply voltage to the test leads when the range selector is in ohm or current position. Do not attempt to measure RF voltage with the analog meter. Remember that the voltage and resistance measurements of the FET-VOM are made with the meter connected parallel to the component. Current measurements are made with the meter connected in series.

Accurate resistance tests can be made with the DMM in or out of the circuit. Measuring resistors in circuits containing transistors and diodes is quite accurate with the DMM. The digital multimeter quickly and accurately checks the critical tolerance of emitter resistors in transistor circuits. Comparable resistance measurements of stereo tape heads might determine if the tape head is defective or not. The impedance of an 8 ohm speaker voice coil will measure around 7.5 ohms with the DMM and 7 ohms upon the FET-VOM tester.

TRANSISTOR IN-CIRCUIT TESTS

The suspected transistor can be tested in or out of the circuit. The in-circuit transistor tester or the diode-function test of the DMM can quickly test transistors within the audio circuits. It is possible to receive an improper reading when a diode, low ohm resistor, or another transistor is directly attached or paralleled with the transistor to be tested. If a certain transistor has an improper reading, the transistor might be defective or have another low resistance component across it. Check the schematic to determine if the transistor is shunted with another transistor or fixed diode.

If in doubt, quickly remove the base terminal from the circuit and take another in-circuit test. Remove the base terminal from the PC wiring with solder-wick and soldering iron. Sometimes by just touching the transistor with a test probe the intermittent transistor will act up. If the suspected transistor tests open or intermittent while being tested, replace it. Replace any transistor that has a high-resistance junction test compared to the other elements.

A quick method to check a transistor in or out of the circuit is with the diode or transistor test of the digital multimeter (DMM). Remember, general purpose audio transistors have diodes back to back and can be tested with the diode test of the DMM. The early transistor audio circuits were made with PNP transistors. Today, most transistor audio circuits contain NPN types.

The unmarked or unknown transistor terminals can be checked with the transistor tester or junction-diode test of the DMM *(Figure 1-5).* Remember, the base terminal measurements are common to the collector and emitter terminals with the diode test of the digital-multimeter (DMM). The resistance measurement is always lower from base to emitter than base to the collector terminal. Place the positive (+) probe from the DMM on the base terminal with the NPN transistor. Likewise, place the negative probe of the DMM on the base terminal for a PNP transistor test.

When the transistor terminals are unknown within the audio circuits, start first with the NPN diode test. Today, most audio circuits have NPN type transistors. Place the red probe to one of the terminals and see if you can get a resistance measurement on one other terminal. Switch the bias test leads if there is no reading; you may have the wrong polarity or the transistor is a PNP type.

When you receive a transistor resistance measurement, leave the positive probe on that terminal and switch the negative probe to the other terminal. If the transistor is a NPN type and the red probe is at the base terminal, you should get a resistance measurement between the base and emitter terminals, and base and collector terminals.

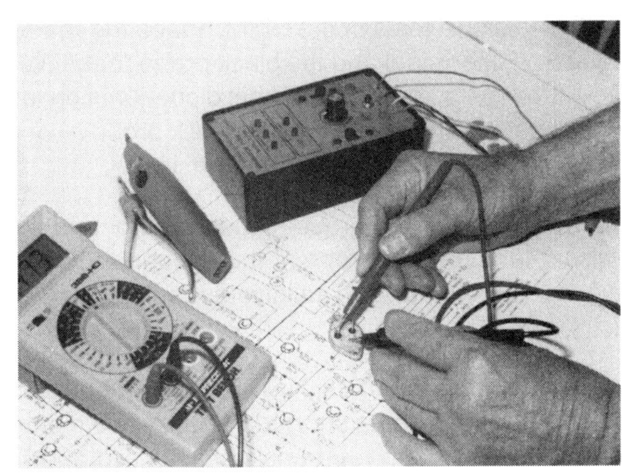

Figure 1-5. Checking for correct transistor test with the diode test of a DMM and high junction resistance measurement.

You have now identified the base terminal and the transistor is an NPN type, because the base terminal of an NPN transistor is always positive and is common to the emitter and collector terminals. Notice what terminal with the base terminal has the highest measurement. This is the collector terminal. The remaining terminal must be the emitter terminal. Double check the measurement between base and collector terminal with the highest resistance reading.

The defective transistor might be open, shorted, leaky, intermittent, or have a high-resistance junction. An open transistor might not show any measurement between base and collector or base and emitter terminals. A leaky transistor will show a low resistance between any three elements. The shorted transistor might have a leakage below 5 ohms. Most transistors become leaky between the collector and emitter terminals. A high-resistance diode-junction is generally above 1K ohms *(Figure 1-6)*.

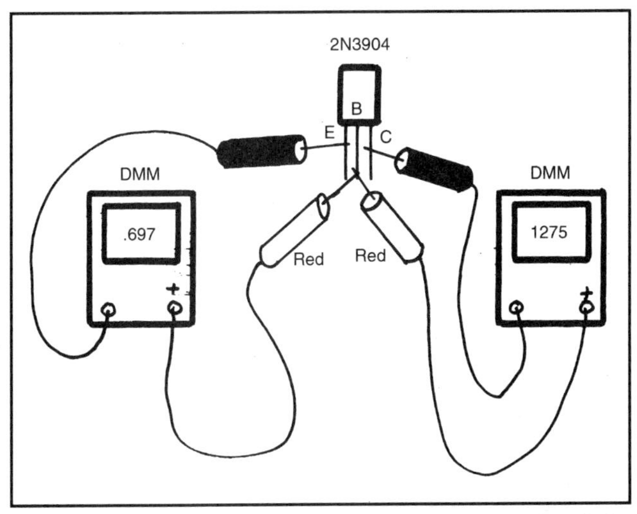

Figure 1-6. The normal DMM transistor test with the diode test of a DMM and high junction resistance measurement.

Test each transistor with the diode-tests of the DMM. Place the red probe (positive) of an NPN transistor at the base (B) terminal and the black probe (negative) at the collector (C) terminal of the suspected transistor. Note the normal diode-junction test resistance. Leave the red probe (+) at the base terminal and place the black probe (-) at the emitter terminal. Both readings should be quite close with a normal transistor.

Now reverse the procedure with the red probe at the collector and the black probe at the base terminal. Likewise, place the red probe at the emitter and black probe at the base terminal. Both measurements should have an infinite reading that indicates the transistor is good or normal.

If a low resistance measurement is found below 100 ohms, in both directions, the transistor is leaky *(Figure 1-7)*. The transistor is shorted between two elements if the measurement is below 5 ohms. A leaky or shorted transistor will have a low ohm measurement with reverse test leads in both directions.

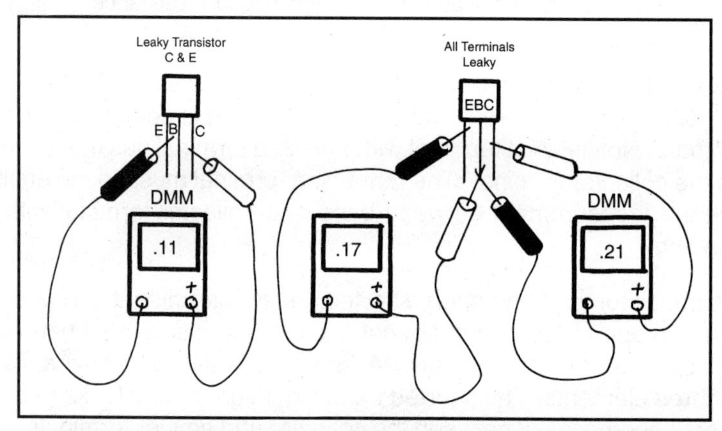

Figure 1-7. The transistor might be leaky between any two elements or all three terminals.

TRANSISTOR OUT OF CIRCUIT TESTS

Transistors can be tested very accurately when removed from the circuit. They can be tested in a commercial transistor tester, digital multimeter with a transistor gain hFE test, and the diode-test upon the DMM. The commercial transistor/FET tester might identify transistor leads, leakage paths, dynamic gain, a good or bad scale, Bi-polar transistor Beta, Dynamic Beta, Bi-polar leakage, and tests the field effect (FET) transistor with analog meter movement.

The lower priced transistor/diode tester with LED indicators, checks all types of diodes such as germanium silicon, power, LED's and zener. A LED-transistorized tester might indicate a NPN or PNP transistor, and checks all types of transistors as germanium, silicon power, RF, audio, switching, and FET's. The tester indicates a leaky or shorted transistor, open, and relative beta of two transistors (*Figure 1-8*).

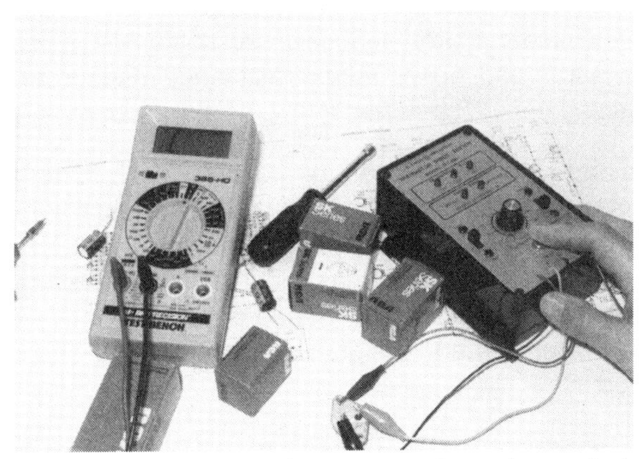

Figure 1-8. Checking the suspected transistor out of the circuit upon an LED transistor tester.

IC CIRCUIT TESTS

Integrated circuits (IC) are found in the preamp, AF and power output circuits of the audio stages. An IC might include the AF and power output circuits in one chip. In fact, one large IC might include both stereo output circuits in the low and high powered output circuits. One audio IC might include all the audio circuits within the portable cassette player.

The defective IC can be located with signal in and out tests, critical voltage and resistance measurements. Check the audio signal going into the IC chip terminal with an audio signal tracer, signal injection, and scope. If signal is found going into the IC and no sound at the output terminal, suspect a defective IC, defective components tied to the IC terminals, or improper supply voltage. The defective IC might become leaky or open.

Check the supply voltage pin terminal coming from the low voltage source. Take a peek at the schematic and see which pin has the highest applied voltage *(Figure 1-9)*. The terminal pin might have a Vcc mark on the supply voltage terminal. Measure the voltage applied to the supply pin to common ground. Suspect a leaky IC when the supply voltage is a lot lower than marked on the schematic.

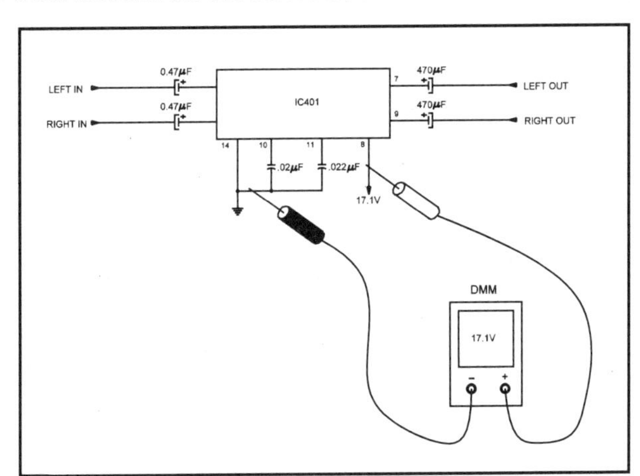

Figure 1-9. The highest voltage measured upon a normal IC is the supply pin terminal (Vcc).

Remove the terminal pin from the PC wiring. Remove the excess solder from around the pin with solder-wick and soldering iron or suck up the solder with a desoldering iron. Make sure the terminal pin is free and loose. Take a resistance measurement between pin and common ground. Replace any IC that has a low resistance to chassis ground.

Take critical voltage and resistance measurements for each IC terminal pin. Compare these measurements to the schematic. If one or two pin terminal voltages are low or high, take a resistance measurement from that pin to ground. Check for a possible leaky capacitor or a change in a resistor, if the pin has a low resistance measurement. Suspect an open IC when all voltages and resistance measurements are quite close to those found upon the schematic. When the audio signal goes into the IC and not out, remove and replace the suspected IC.

THE DOCTOR IS IN

Before the medical Doctor can decide what is wrong with your illness, he or she must know the symptoms. Likewise, the electronic technician must know the symptoms before tearing into the audio circuits. The audio output sound symptoms at the speaker might be dead, weak, distorted, erratic, or intermittent sound.

Now determine if the weak symptom is ahead or after the volume control. A weak stage might be caused by a defective coupling capacitor, transistor, IC, or improper voltages. Often, the distorted speaker symptom is produced in the audio output circuits. The erratic or intermittent sound might result from a defective coupling capacitor, transistor, IC, or poor board connections, anywhere in the audio circuits. Take the symptom from the speaker and apply it to the block diagram or schematic to locate the defective circuit.

I CAN'T HEAR YOU

The weak sound problem might be compared to the old electronic technician, who used to hear a TV, clock radio and boom box player operating at the same time; and at once he can pick out the product, when the sound became intermittent. Now, he cannot hear his wife within ten feet and the TV set must be turned up to full volume; the sound is weak and a hearing aid might solve the audio problem.

The coupling capacitor, bias resistors, transistors and IC components cause most of the weak sound problems within the audio circuits *(Figure 1-10)*. The dead symptom is much easier to locate than the weak or intermittent problem. Check the weak sound symptom by signal tracing the audio from stage to stage with an external audio amp or scope. Then, take critical voltage and resistance measurements where the sound stops, to locate the defective component. An open or leaky transistor or IC might cause a weak audio problem.

The weak sound and distorted symptom might be produced by a frozen speaker cone. A distorted and weak sound might result from a leaky or shorted output transistor or IC. The input and output terminal of an output IC can be signal-traced with the scope or external audio amp. Suspect the IC or improper voltages if the input terminal is normal and the output terminal is weak and distorted. Compare the normal stereo channel to the weak and distorted one with voltage and resistance measurements.

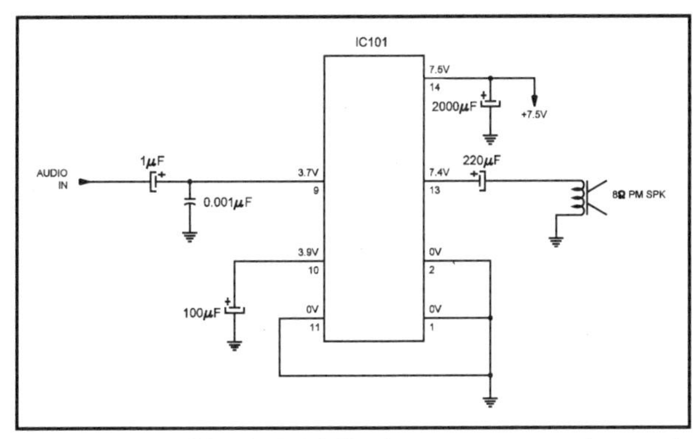

Figure 1-10. Check the following components for weak sound in the tape cassette player.

THAT SOUNDS TERRIBLE

Distorted music not only sounds terrible, but it is difficult to listen to. A very weak distorted symptom is difficult to locate. Transistors, ICs, speakers, resistors, capacitors, and sound alignment produce most distortion problems in the audio circuits. Extreme distortion is often found in the audio output circuits. Locate the distorted audio circuit with the external audio amp and scope.

A leaky coupling capacitor can cause some distortion in the sound circuits. Check for a leaky or shorted transistor in the driver, AF or output stages for a distortion symptom. Do not overlook burned, open or a change in resistance of bias resistors within the base and emitter circuits. Carefully inspect each bias resistor when locating a shorted or leaky transistor *(Figure 1-11)*. A change in resistance of a collector or emitter resistors can cause distortion. Check for leaky diodes in the base circuit of a transistor output circuit.

Suspect a defective output IC for audio distortion. If the audio input signal is normal and the output of the IC is distorted, suspect a defective IC. Take accurate voltage and resistance on the IC terminals to determine if the IC is open or leaky. Improper supply voltage might indicate a leaky IC or improper voltage source. One audio channel can be distorted and the other stereo channel normal with a dual-output IC. Check for leaky bypass capacitors on the IC terminals.

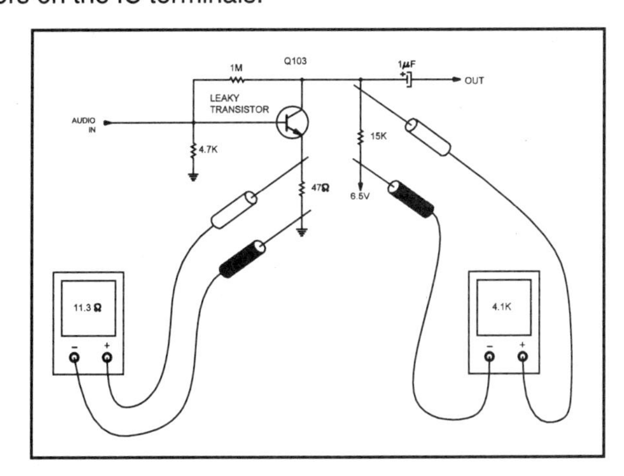

Figure 1-11. Check the bias and collector load resistors after locating a shorted transistor.

Sub another PM speaker for one that might cause distortion. A frozen voice coil or dropped cone can produce a distorted sound. Suspect the cone frozen against the PM magnet when the sound is tinny or mushy. A speaker blatting sound might be caused at higher volume with loose cone or speaker spider section. Loose dirt or particles in the cone area might cause a vibration noise. A blown voice coil is the result of too much power applied to the speakers.

Low distortion might result from leaky coupling and bypass capacitors, a change in resistance, and leaky AF and driver transistors. Often low signal distortion occurs in the preamp or AF circuits. Very low signs of distortion are very difficult to locate. Low distortion can be located with a square wave generator and scope, or a distortion meter.

Extreme distortion can be caused by a leaky output transistor or IC component. Check the output circuits for extreme audio distortion. Often leaky or shorted output transistors and IC's might have burned or open bias resistors. Check for a leaky driver or AF transistor that is coupled directly to an output transistor for distortion. Do not overlook a leaky diode or electrolytic coupling capacitor that might produce extreme distortion. A blown speaker voice coil can cause extreme distortion.

POPS IN AND OUT

Erratic sounds or sound that pops in and out can be caused by transistors, IC's, speakers, poor terminals and board connection. In the early table or auto radio chassis, when only one power output transistor was found in the output stage, these power transistors would produce a popping noise with intermittent audio. A poor flexible speaker connection at the voice coil can cause a loud off and on noise. Suspect a dual-power output IC component with popping noises in the sound.

Try to isolate the erratic sound with an external amp and signal trace the popping noise through the amplifier circuits. Replace the suspected semiconductor, if the popping noise is not heard at the base terminal of a power output transistor or IC and the noise is heard at the input terminal. Replacing the transistor or IC is the only answer as voltage and resistance measurements will not give any indications of a popping component or circuit. Often, the defective transistor or IC output component runs very warm.

HERE TODAY, GONE TOMORROW

The intermittent audio problem may act up at once or go for days before the music stops. The intermittent component is easy to locate if the music comes off and on at once. Usually the intermittent product should be left to operate in the home, when it takes hours or days to cut up and down.

Although, any electronic component can cause an intermittent condition, look for defective transistors, IC's, poor part connections, corroded terminals, and poor board connections. Determine if one or both channels in a stereo circuit are intermittent. Compare the normal channel with the intermittent channel in stereo sound circuits. It's possible to have one half of a dual-IC intermittent and the other channel normal *(Figure 1-12)*.

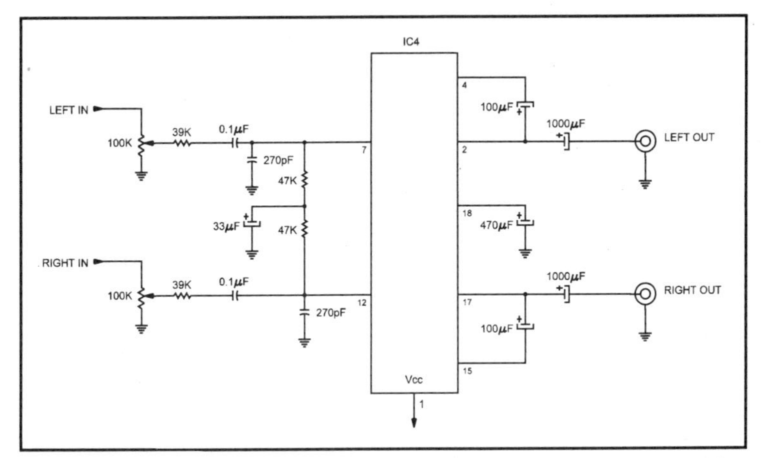

Figure 1-12. One channel of a dual-input IC might be distorted while the other channel is normal.

Monitor the intermittent audio circuit with external power amp or scope. Isolate the sound stages by starting at the volume control. It's best to cut the audio circuits in half to locate the intermittent component. Notice if the sound is intermittent ahead of the volume control or within the output sound circuits. Now go from stage to stage to monitor the intermittent audio. Check from base to collector on transistors and input terminal to output terminal of IC components. Check on both sides of a coupling capacitor to locate the intermittent capacitor.

Intermittent components, such as transistors and IC's can be sprayed with heat or coolant to make the intermittent to start up or stop. Poor board connections can be located by moving parts around upon the PCB. Push up and down upon the PC board with an insulated tool. Sometimes poorly-tinned part terminals might produce an intermittent connection under a large blob of solder. Check for poor socket and wire terminals for intermittent connections. Do not overlook an open or poor speaker voice coil connection for intermittent sound.

VACCINATION TIME

A dead or weak audio circuit can be signal traced by injecting a signal from the audio or noise generator. Inject an audio signal (1 kHz) at the center volume control terminal to signal trace the audio output circuits. If the signal can be heard in the speaker, proceed to the front-end audio circuits. The signal can be injected from a noise generator to locate a defective RF or AF stage.

Inject the audio at the tape head terminal of the cassette player or preamp-amp circuit of an AM-FM receiver. Check the schematic for the input and output terminal of a suspected preamp IC. Inject the audio signal at the input and then the output terminal until a sound is heard in the speaker. Proceed through the circuit until the sound is heard.

Likewise, inject the audio signal at the base or coupling capacitor of the preamp sound circuit *(Figure 1-13)*. Inject the signal on both sides of the coupling capacitor to determine if the capacitor is open. Go next to the collector terminal of the preamp transistor. Proceed through the audio circuits until the 1 kHz tone can be heard.

For instance, if sound can be heard with the signal injected at the collector terminal and not on the base, you have located the defective stage. Test the suspected transistor for open or leaky conditions. Then take critical voltage and resistance measurements.

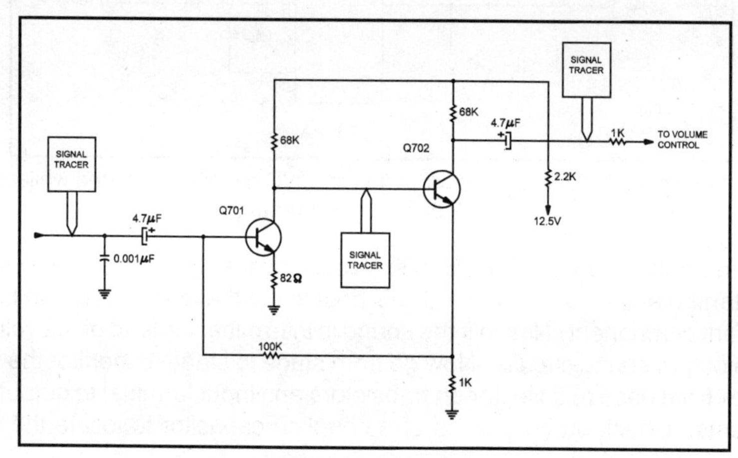

Figure 1-13. Inject a noise or tone generator signal at the base of coupling capacitor to signal trace a preamp sound circuit.

A dead audio circuit is found with an open coupling capacitor, leaky or shorted transistor or IC. Often a weak sound is heard with a shorted or leaky component. Sometimes a weak sound condition might bleed through with an open coupling capacitor. Check for poor soldered terminals for a no sound symptom. The open speaker voice coil or poor soldered terminals can cause a no sound problem.

CHASING THE DOG

Signal trace the audio circuits with the audio signal generator and external audio amplifier. The external amp, speaker or scope can be used as indicators or monitors. Insert a test or music recording cassette to signal trace the audio through the cassette or boom-box player. A CD disc can be played to test the music from stage to stage in the amplifier of a CD player. The tough dog sound problem can be signal-traced stage by stage with the audio sine or square wave generator and scope *(Figure 1-14)*.

Inject the audio signal at the volume control and if the audio is normal at the speaker, the defective stage is ahead of the control. Start at the volume control and signal trace the audio output if the preamp stages are normal. Sometimes the electronic technician might start at the speaker with the external amp and work backwards through the sound circuits.

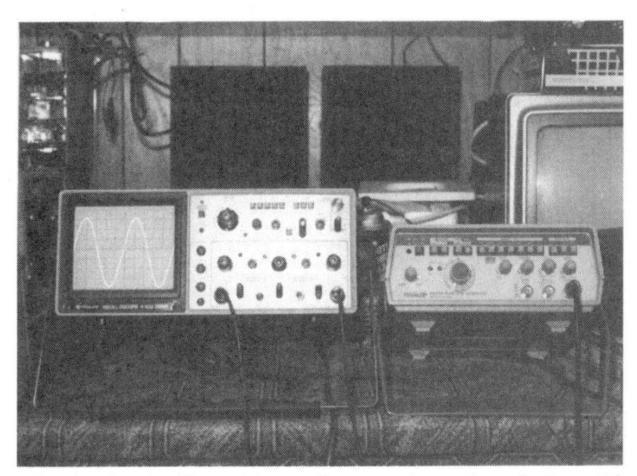

Figure 1-14. Signal trace the weak or distorted audio problem with the square wave generator and scope.

The external audio amp can be used as indicator to test out the audio circuits. Start at the volume control and go from base to base of each AF or output transistor with the audio amp. When one stereo channel is dead, weak or intermittent, play a cassette or inject audio signal into both channels. Now compare the audio at each stage with the good stereo channel. Likewise, inject the audio signal at the input terminal of a suspected preamp or output IC and signal trace the audio at the output terminal.

When one large dual-IC is found in the stereo audio circuits and the left channel is weak and distorted, signal-trace the distortion through the audio circuits with the scope or external amp. Keep the volume control on the external amp as low as possible to not add distortion to the original audio signal. Check the signal on each side of an electrolytic coupling capacitor to the IC input circuit. Make sure the supply voltage source is measured at the IC or collector terminal of each audio transistor. Compare the audio signal with the normal channel.

THE ORIGINAL, THE SUB, THE UNIVERSAL

Most audio transistors and IC components can be replaced with the original or universal replacements. Replace all high-powered and MOSFET audio transistors with the originals. Large and special IC components should be replaced with the original part number *(Figure 1-15)*. Some original foreign and Japanese semiconductors are now available from wholesale and mail-order firms. The small signal audio transistors can be replaced with universal types without any problems.

Special volume, treble, bass, dual, and balance controls must be replaced from the manufacturer. The many position function switches found in cassette players and receivers should be replace with originals. Replace large high-wattage column speakers with the right part number. Special gears, idlers, and pressure rollers must be replaced with the manufactured part number. Replace selector matching knobs and dial assemblies with the exact part number.

Figure 1-15. Replace special and high-powered transistor and IC's with originals.

Small electronic parts such as resistors, capacitors, fuses, and some switches can be replaced with universal replacements. The original filter capacitor might mount in a small space and can be replaced with universal capacitors. Filter capacitors can be added or paralleled when the original is not handy. You can replace a 1000 µF electrolytic capacitor by paralleling two 500 µF filter capacitors. A can-type filter capacitor with several capacitors in one section can be subbed with universal capacitors. Make sure the capacitor and working voltages are the same or higher. The defective 2000 µF 35 volt electrolytic capacitor can be replaced with a 3000 µF 50 volt filter capacitor.

The leaky or shorted power supply diodes can be easily replace with universal units. A 2.5 amp 1000 volt diode can replace the 1 amp 500 volt fixed diode. When a special bridge diode is not available, make up a bridge-diode circuit out of four fixed diodes with the same or higher amperage and voltage. Make sure the correct polarity of each diode is connected into the bridge circuit.

THOSE TINY CREATURES

The surface-mounted device (SMD) is a miniature component that takes up very little mounting space. The SMD part is found in today's TV set, VCR, camcorder, compact disc and cassette player. Also, they are found in hand-held portable radios, telephone, wireless telephones, cellular, pagers, scanners, and receivers of all kinds. You must have a steady hand, good eyesight, and a lot of patience to locate and replace SMD components *(Figure 1-16)*. The SMD part is mounted directly on the PC wiring.

Because many surface-mounted parts have similar shapes and sizes, sometimes they are difficult to identify on the PC wiring. Some of these parts look like black, brown or white specks on the PCB. The commercial resistors might appear as round, flat, leadless devices. The ceramic capacitor is a flat, solid part with the terminal connections at each end. These connections are tinned with solder for easy circuit mounting. The fixed resistor might have several white numbers stamped on top of a black body to identify the resistance value. A fixed SMD ceramic capacitor might have a line at the top with a letter of the alphabet and numbers. Sometimes universal ceramic capacitors have no value marked on top *(Figure 1-17)*.

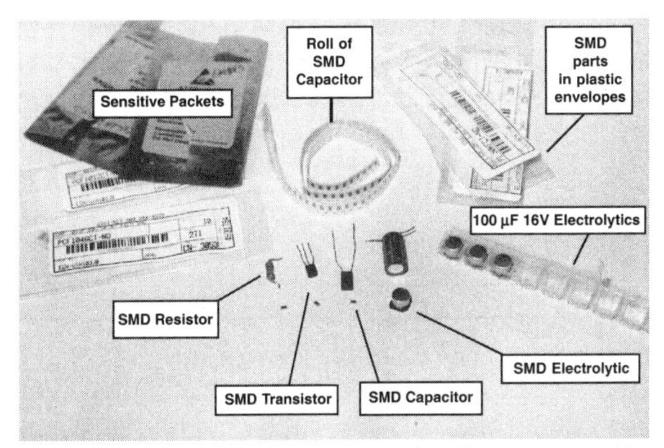

Figure 1-16. SMD components are tiny compared to standard electronic parts.

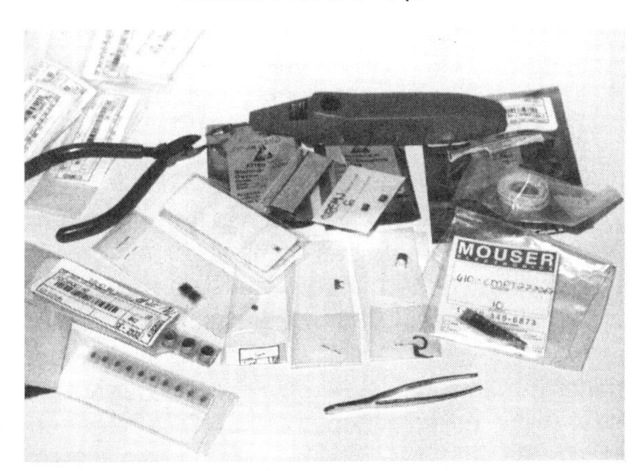

Figure 1-17. Since many SMD parts do not have any markings, keep them in packages they were shipped in.

Transistors and diodes are often identified with two letters or a letter and number. The SMD component terminals are found at each end, except transistors and IC chips. Of course, the SMD IC has many gullwing terminals, while the transistor has only three. The commercial surface-mounted transistor (SMT) might appear in a chip form with flat contacts at one side, top and bottom, or on both sides. You might find more than one transistor in one chip. The same applies to fixed diodes and LED's; two or more diodes might be found in one chip. You can test these SMD semiconductors like the standard component.

These SMD components are miniature in size and must be handled with care. Since these parts are so tiny, they can easily be lost or flipped out of sight. The special SMD component should be replaced with the original part number from the manufacturer. Of course, some of these parts can be replaced with universal SMD parts such as fixed ceramic capacitors, resistors, diodes, LEDs, and some universal transistors.

Identify the SMD part that is soldered directly on the PC wiring, while larger components are mounted on top of the board. The chip resistor value can be identified with numbers stamped on top. A ceramic chip capacitor might have a letter with a number along side *(Figure 1-18)*. The electrolytic chip capacitor can be identified with a white line at one end, indicating the positive terminal with the value and voltage marked on top of the chip. Remember, the standard electrolytic capacitor has a black line that indicates the ground side. The electrolytic ceramic chip capacitor has the opposite polarity sign (+).

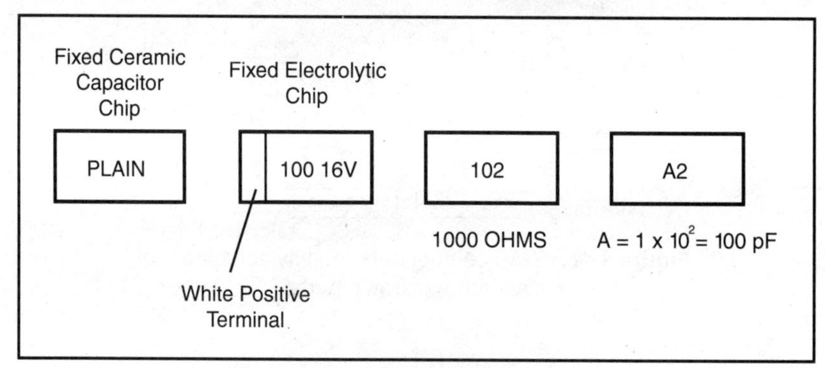

Figure 1-18. The ceramic chip capacitor and resistor might be marked on top of part with numbers and letters.

The SMD transistor has three terminals with two on one side and the other on the top side. The terminals might be marked 1, 2 and 3. Sometimes the chip might have two transistors in one component. The ceramic IC chip has many terminals on each side, while some microprocessors have gullwing-type terminals. The SMD transistor or IC might have the part marked on top or no markings at all. Some transistors are marked with a number and letter on the top side *(Figure 1-19)*. The digital transistor might have a resistor in series with the base and emitter terminal inside the chip. You must have a magnifying glass to identify the small numbers and letters on top of the SMD component.

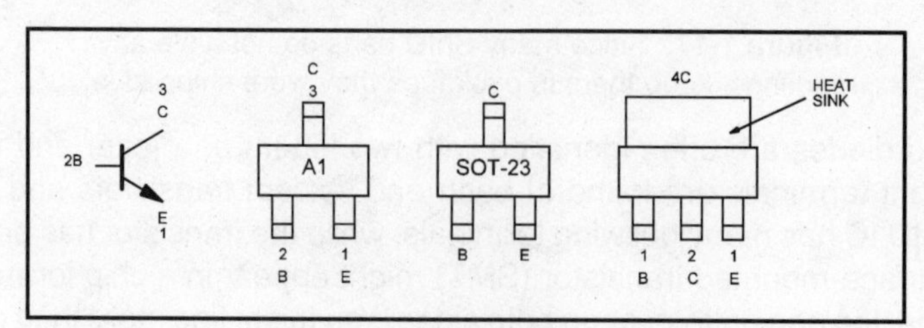

Figure 1-19. Check the three-legged transistor chip with a number and letter on top.

Remember the ceramic chip capacitor is a non-polarized capacitor. Likewise, the fixed chip resistor is the same. You can solder any end into the PC wiring without any problems. The ceramic chip electrolytic capacitor is polarized and the line at the end must be mounted to the positive terminal or connection. A universal SMD aluminum electrolytic capacitor is polarized with a black edge or line (-) on top of the capacitor. The negative terminal is mounted at ground potential. The capacity and working voltage might be stamped on the top side. Remember, the electrolytic capacitor must be installed correctly or if in backwards, it can overheat, blow fuses, and might blow up in your face.

IC chips are not heatproof or shockproof. These devices are made of ceramic or plastic molding, and they should not be subject to direct shock. Do not apply extra stress to the IC chip. Handle SMD semiconductors with extreme care. These sensitive semiconductors might appear in dark protective envelopes *(Figure 1-20)*. Keep all SMD components in the original package until they are ready to be mounted.

Figure 1-20. The sensitive semiconductors might appear in dark static-free envelopes.

The IC terminals are identified by a small circle indented on top of the IC. The dot indicates terminal 1. You can count off the number of terminals to take voltage and resistance measurements. Keep those test probes sharp so they will not short out any two gullwing terminals. Sometimes the IC part is marked on top with white letters. The surface-mounted IC processor has many gullwing terminals and is very difficult to remove and replace *(Figure 1-21)*. Check the soldered connection of the SMD component for intermittent sound and cracked or poor soldered connections.

LOSS OF HIGH FREQUENCY

The defective tape head might cause distortion, one dead channel, intermittent playing or recording, and a howling noise. Weak and distorted music may result from excess oxide packed on the tape head. In fact, the small gap might be full of oxide dust resulting in no music from the tape head. Clean the tape head with alcohol and a cleaning stick before taking tests or checking the tape head.

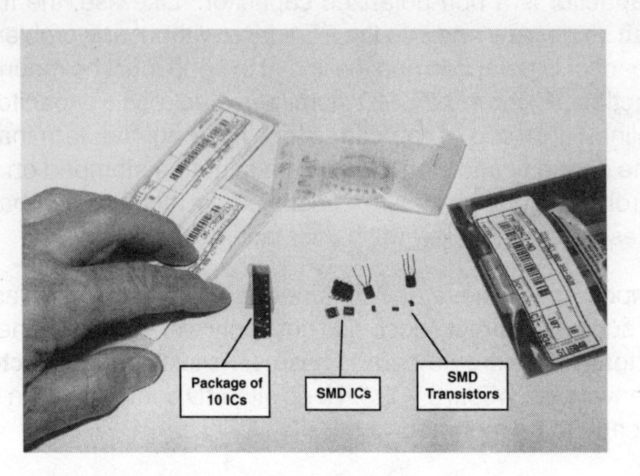

Figure 1-21. The suface-mounted IC's are tiny compared to the standard integrated circuit.

Inspect the tape head for excessive worn marks when a weak and tinny noise is heard in either channel. A loss of high frequency might be caused by a worn tape head. Often, the sound of music becomes weak and tinny compared to the normal channel. Replace the tape head if worn marks are found on the tape surface or tape head. Demagnetize the tape head after repairs are made to eliminate any background noise.

DON'T SCREW UP

Simply stay out of critical and difficult audio circuits if you don't have the correct test equipment, schematics or knowledge. Do not try to touch-up alignment or adjustment screws to increase the sound level if the correct tools and test equipment are not available. Trying to adjust bias controls within high powered amplifiers might make matters worse without the correct instructions. Leave these repairs to the electronic technician who is a specialist in servicing special sound circuits.

Chapter 2

Important Audio Test Equipment

The seven most used pieces of test equipment on the audio service bench are the DMM, FET-VOM, dual-trace scope, semiconductor and capacitance tester, audio generator, and isolation transformer. The DMM and FET-VOM take critical voltage, resistance and current tests, while the scope might be used as waveform and signal indicators. A semiconductor tester checks transistors, diodes, zener diodes, and signal diodes in and out of the circuits. The audio signal or function generator injects waveforms and signals into the defective audio circuits for signal tracing procedures *(Figure 2-1)*. The audio chassis is plugged into the isolation transformer, so as not to damage test equipment or the electronic product.

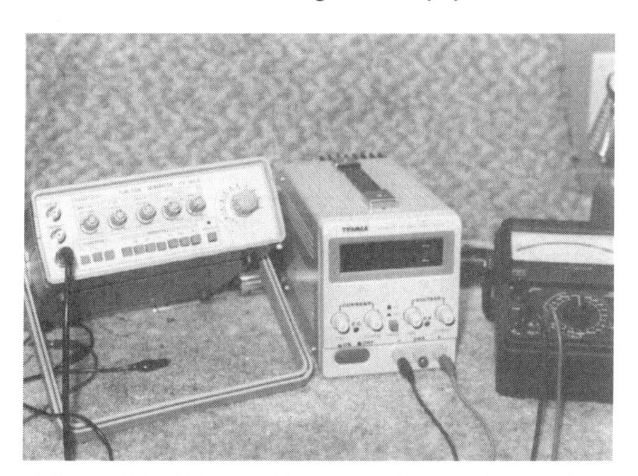

Figure 2-1. The function generator provides a sine and square wave waveform in making audio tests of the amplifier.

Of course, some electronic technicians trace out audio circuits with an external audio amp and power supply. The more experienced electronic technicians might require only a few audio test instruments. Several small tools, test equipment and test parts such as speaker load resistors, test speakers, test clip-leads, test discs and cassettes are used to round-up the required audio test equipment. The audio oscillator and frequency counter, distortion, wow and flutter meter might be found on the audio specialist bench.

THE ANALOG MULTITESTER

The new analog volt ohmmeter (VOM) should have a 20,000 to 50,000 ohms/volt DC sensitivity range. Some large VOM's have 25 or more ranges of AC/DC volts, dc current, and resistance measurements. The deluxe VOM might have auto-ranging where you select the function and the meter sets the correct range. Other features might include LED range indicators, auto polarity, continuity buzzer, milliamperes and micro amp ranges. The continuity buzzer lets you quickly spot problems like damaged connectors, broken wires, and blown fuses.

In the early days, the VOM was the standby test equipment of the electronic industry *(Figure 2-2)*. The Simpson 260 was the work horse of yesterday. Remember, the VOM can load down the circuit being tested. The VOM can still make quick continuity tests and remain the audio indicator in alignment and adjustment procedures. If you do not have a good VOM, choose the FET-VOM and DMM for greater accuracy.

Figure 2-2. The analog (VOM) meter is handy in taking voltage and continuity test in audio circuits.

HIGH IMPEDANCE MULTIMETER

Many of the early audio tests and adjustments were made with the vacuum-tube-voltmeter (VTVM). Although, the VTVM is no longer manufactured, the FET-VOM can take its place. The FET-VOM provides high input impedance with a dual field-effect transistor (FET) circuit. The analog meter hand is easier to read than the digital multimeter in alignment procedures.

The analog FET-multitester might measure up to 1000 volts (AC/DC), 10 amps (DC) and 100 megohms resistance with accuracy and ease. It can also measure down to 20 millivolts (DC) and a few micro amps (DC). The input impedance might be above 10 megohms. Some deluxe models have a special audible continuity check function with a built-in buzzer. The buzzer will sound when the circuit continuity is approximately 300 ohms or less *(Figure 2-3)*.

Figure 2-3. The FET-VOM is a brother to the famous VTVM with high impedance input and much more accurate measurements than the portable VOM.

The FET-VOM might have other features such as a mirrored scale for accurate meter readings, overload protection for both meter and internal circuitry, zero center scale, and polarity reversing switch. Other audio tests might include frequency response (4.5kHz to 10kHz) and decibel (dB) measurements. Set the range selector to one of the ACV positions and it is best to start at the highest range selection. For absolute dB measurements, the circuit impedance must be 600 ohms with a -10dB to +63dB on some FET-meters.

The FET-VOM can make very accurate resistance, voltage and current tests within the audio circuits. Do not apply voltage to the test leads when the range selector is in the ohms position. When attempting to identify cathode and anode ends of semiconductor junctions, or the type of transistor (PNP or NPN), the actual polarity of the tester's voltage is opposite of the tests lead colors. The red lead is the negative source and the black lead is positive. The FET-VOM and DMM test leads upon a diode are just the opposite of the VOM *(Figure 2-4)*. The FET-VOM is ideal in making audio adjustments and alignment procedures.

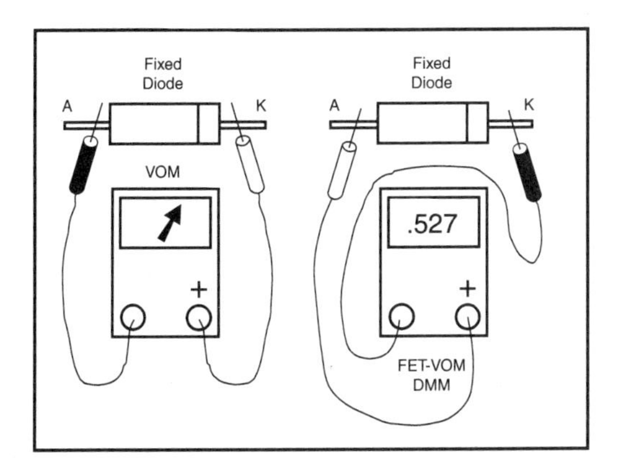

Figure 2-4. Notice the different polarity of test leads when taking a diode measurement with VOM or the FET-VOM and DMM.

TODAY'S CHOICE

The portable and bench-top digital multimeter is the most versatile, accurate and used test instrument upon the electronic service bench. The high-tech DMM has just the right combination of features for service technicians, hobbyists and students. The meter might have auto-ranging that lets you select DC volts, AC volts, resistance, DC current and the meter sets the correct range automatically.

A portable DMM can quickly check diodes and transistors on a diode-transistor test. Auto polarity gives you valid readings if you connect the test leads in reverse. These small DMM's might have manual override, continuity buzzer, data hold to freeze display and audible overload-alert features. A continuity buzzer lets you quick-check fuses, cords, and connections; its audible indication frees your eyes for faster testing.

Today the digital multimeter (DMM) has actually replaced the VOM *(Figure 2-5)*. The B & K TEST-BENCH DMM, model 388-HD, can make most of the electronic tests in servicing audio products. This meter can check transistors, diodes, frequency, current, logic, capacitance, current, and voltage. The diode test can accurately test diodes for open, leaky or shorted conditions. Also, the diode test can check transistor junctions for open, leaky, shorted and high resistance measurements. A separate transistor socket is located above the positive test probe to plug the PNP or NPN transistor for a hFE test.

Figure 2-5. The digital multimeter is today's workhorse and can take accurate voltage, resistance and current measurements in audio circuits.

This DMM has three different frequency ranges: 2 kHz, 20 kHz and 200 kHz. The current range is from 200 μA to 20A in five different ranges. Small capacitors can be checked from 2 NF to 20 μF with a separate CX socket. The ac voltages vary from 200 mV up to 1000 volts. Seven resistance ranges vary from 200 ohms to 2000 megohms. The ac voltage range measures up to 750 volts.

The portable DMM is ideal when taking the different tests upon audio components *(Figure 2-6)*. Besides critical voltage measurements upon diodes and transistors, transistors can be tested with the diode test or in a separate transistor socket. The frequency range can be used to indicate the audio amp frequency range and cassette speed problems with the frequency-counter test.

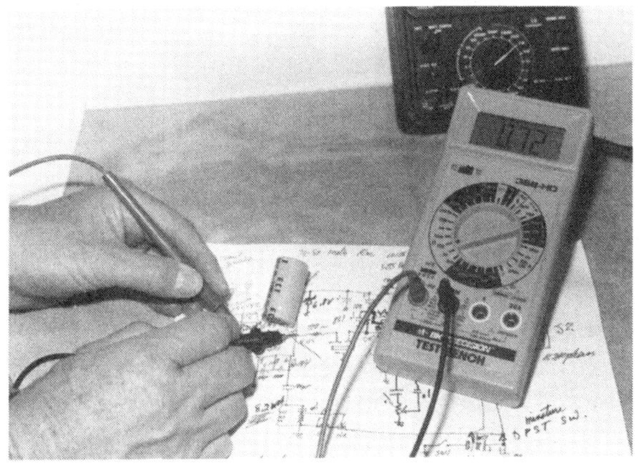

Figure 2-6. The digital multimeter (DMM) can check capacitors, diodes and transistors besides the regular voltage measurements.

The cassette tape head azimuth adjustments can be made with the low ac voltage range or frequency-counter tests. Remove one terminal from the circuit when testing unknown capacitors, high-ohm resistors and diodes.

THE DUAL-TRACE SCOPE

The dual-trace scope is ideal in taking waveforms or alignment in audio stereo channels. The scope can also check the stability of the amplifier and hum in the ac power supply. Select an oscilloscope with at least a 40 to 100 MHz bandwidth for audio signal measurements. Some of the features might be dual channel, dual trace, dual time base, delayed sweep, auto triggering, x-y operation, scale illumination, and intensity control *(Figure 2-7)*.

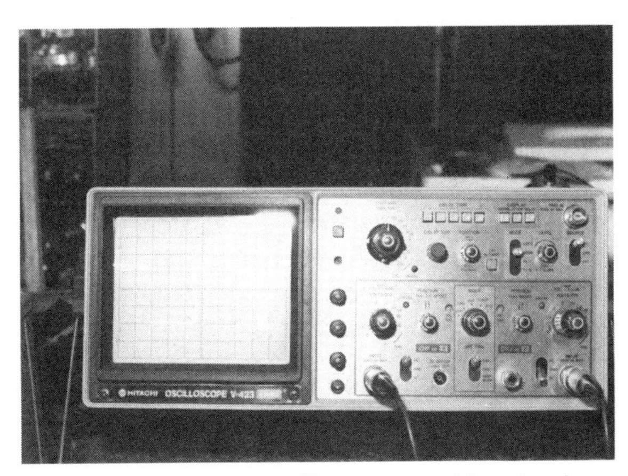

Figure 2-7. The oscilloscope provides visual waveforms of weak and distorted circuits.

The vertical features might include a sensitivity of 2mV/div-5V/div in 11 calibrated steps of 1-2-5 sequence. A bandwidth from DC to 40 Hz or 100 MHz (-3 dB), AC to 10 Hz to 100 MHz (-3 dB) with a rise time less than 3.5 ns. The maximum input voltage between 250 and 400 volts p-p (DC to peak AC) at 1 kHz or less. The input impedance should be 1 megohms with polarity inversion. The different modes might include CH-1, CH-2, ALT, CHOP and ADD.

The horizontal deflection features might include display modes of A, ALT, B, B TRIG'D, and X-Y. Delay line jitter better than 1:10,000 with sweep magnification of 10 times. The sweep time base should be at least .1 m sec/div to .5 sec/div in 20 calibrated steps. A maximum input voltage should be from 30 to 250V DC to peak AC.

The sine/square wave generator and oscilloscope can locate weak and distorted stages of the electronic audio circuits. The scope is also ideal when making alignment and adjustments procedures. A dual-trace scope can be used to check out the stereo circuits at the same time. Inject an audio signal from generator or test cassette into both stereo channels *(Figure 2-8)*. Go from stage to stage and locate the weak sound or distorted circuit. The dual-trace scope can be used in AM/FM/MPX, compact-disc signal tracing, and offset adjustments.

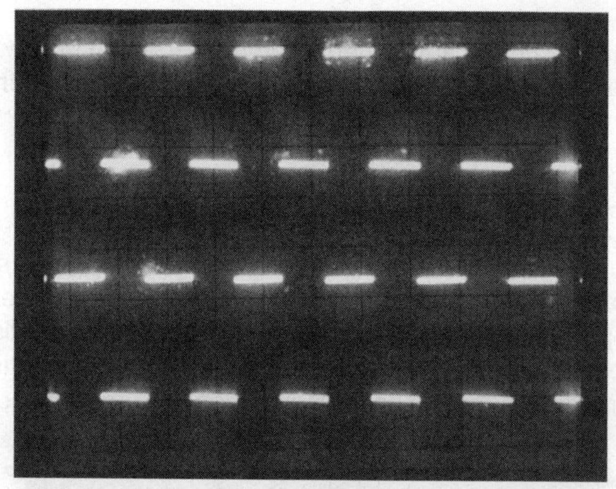

Figure 2-8. Both square waveforms are quite normal in each stereo channel of a suspected audio circuit.

ISOLATION TRANSFORMER

The isolation transformer provides isolation from the AC power line when servicing a "hot" chassis. The transformer eliminates a shock hazard and prevents damage to the test equipment, technician, and electronic chassis. An isolating transformer might have a 1:1 turn ratio for isolating equipment from direct connection to the power line.

The variable-isolation transformer can make the intermittent chassis act up by applying a variable power line voltage. Some variable transformers provide a variable AC voltage from 90 to 140 volts. Plug the intermittent audio chassis into the variable transformer and lower the ac line voltage. Often the intermittent will act up on either low or higher then normal-applied AC voltage.

All audio chassis that operates from the power line should be plugged into the isolation transformer before attaching any test equipment *(Figure 2-9)*. If not, the fuse might blow with extensive damage to the silicon diodes within the electronic power supply. Sometimes expensive test equipment is damaged and the technician can receive a power line shock between the audio chassis and test equipment; it's time to think. Be very careful out there.

Figure 2-9. The electronic chassis should be plugged into the isolation transformer before attaching test equipment.

NOISE GENERATOR

The noise generator is just as effective in RF front-end circuits as the sound circuits. A noise generator produces a noise signal for signal tracing the audio circuits. The handheld noise generator in the early radio days was to go from the RF to the audio stages to locate a defective circuit or component. The noise generator can quickly locate a defective stage within the AM/FM/MPX receiver circuits.

TONE GENERATOR

The tone or 1 kHz audio generator can be used like the noise generator to quickly locate a dead or weak stage within the audio circuits. The weak audio circuit in a cassette player might result from a packed-oxide tape head, leaky or open transistors and IC, weak battery and defective electrolytic coupling capacitor. A dead stereo channel can be caused by defective transistor or IC, open coupling capacitor, and improper voltage source. A tone generator can be purchased or easily constructed for less than a ten dollar bill.

The tone generator circuit is shown in *Figure 2-10*. The tone generator is constructed around a low voltage LM386 IC. The frequency can be changed from 1 kHz to 10 kHz by switching R1 to 4.7K ohms. The output frequency is lowered (kHz) with R1 at 56K ohms. Although, the frequency might not be on the nose at 1 kHz frequency, it is close enough for signal injection of sound circuits. The tone generator can be housed in a small metal box.

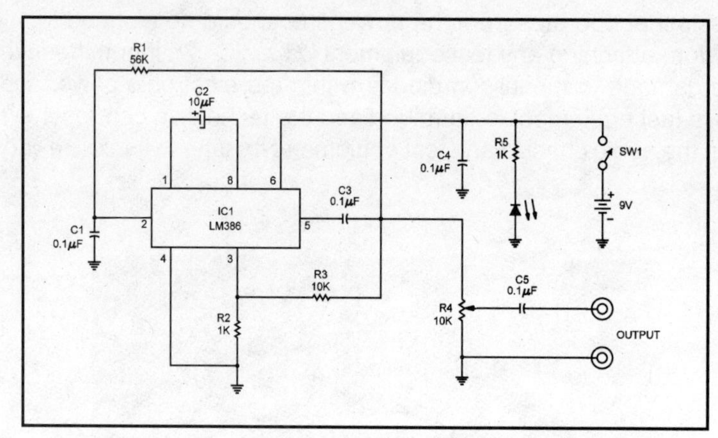

Figure 2-10. The circuits of a 1 kHz portable tone generator.

Select a general-purpose IC PC board so the IC socket can straddle the PC board terminals. Tie each part of the circuit into the correct board holes and solder underneath the board. All parts are mounted on top and hand wired with wire terminals and hookup wire beneath the PC board. LED1 is a pilot light indicator to indicate when the generator is operating. C5 should have a 1000 volt rating to prevent damage to the tone generator, if higher voltages are encountered.

PART LIST

SEMICONDUCTORS

IC1 - LM 386 low voltage audio amp IC
LED1 - 2 volt chassis mount LED

CAPACITORS

C1, C3, C4 - 0.1 µF 50 volt ceramic capacitors
C2 - 10 µF 15 volt ceramic capacitor
C5 - 0.1 µF 1000 volt ceramic capacitor

RESISTORS

R1 - 56K ohm 1/2 watt fixed resistor
R2, R5 - 1K ohm 1/2 watt fixed resistor
R3 - 10K ohm 1/2 watt fixed resistor
R4 - 10K ohm linear control with SPST switch
SW1 - SPST switch on rear of R4
misc - 276-150A Radio Shack general purpose IC board
    8 pin IC socket, red and black banana jacks,
    9 volt battery, 9 volt battery clip, knobs,
    cabinet base, etc.

SEMICONDUCTOR TESTER

The semiconductor tester should automatically identify transistor leads, NPN or PNP types, leakage test, dynamic gain test, and checks Bi-Polar and FET transistors. A good semiconduction tester should check the transistor in or out of the sound circuits. Some expensive semiconduction testers have an in-circuit "go-no-go" and a good and bad scale. The low-priced transistor-diode tester might not have an analog meter.

The LED transistor and diode tester might have LED's that indicate a normal, open or leaky transistor or diode. Some LED testers will test transistors in-circuit provided the base biasing resistance is greater than 100 ohms *(Figure 2-11)*. Pull the power cord from the electronic audio product before testing transistors or diodes in the circuit. Shut off the battery power before taking the transistor tests. Remove one end of suspected diode from the board to make an accurate diode test.

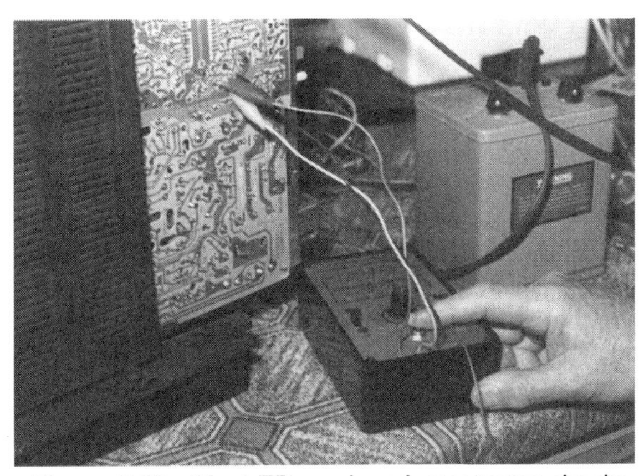

Figure 2-11. The LED semiconductor tester checks transistors in and out of the circuit.

CHECKING CAPACITORS

The standard digital-multimeter might include a capacity range from 2 NF to 20 NF in 4 or 5 steps. Choose a capacitance meter that accurately measured the value of any capacitor between 0.1 PF to 2000 μF. The extended range capacity tester might test capacitors from 0.1 pF to 20,000 μF in a different range. Some of these capacity meters test the suspected capacitor in or out of the circuit. Always discharge the electrolytic capacitor before testing. The LCR meter can accurately test capacitors in and out of the circuit *(Figure 2-12)*.

The ohmmeter range can check the fixed capacitor for leakage and charging qualities. Check the capacitor out of the circuit for leakage. Place the two ohmmeter probes on the electrolytic and watch the numbers charge up and discharge. Reverse the test probes and again the good electrolytic will charge up. The numbers will increase until the over range sign appears in the LCD. Rotate the ohmmeter range to 200K ohms when testing a low capacitance of 4.7 or 10 μF to acquire the charging effect with the DMM. The analog FET-VOM meter hand will charge up and down with the same tests.

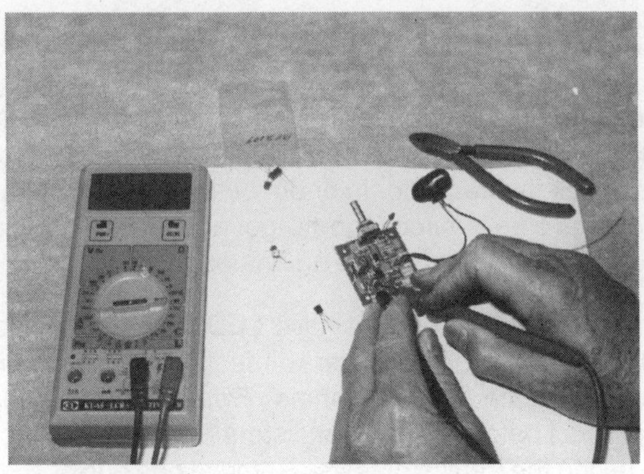

Figure 2-12. The LCR meter can check small capacitors and transistors in the circuit.

A new capacitor tester called the Capacitor Wizard is an extremely fast and reliable tester that measures the equivalent series resistance (ESR) on capacitors of 1 uF and larger. These accurate tests are made in-circuit, eliminating the need to remove the capacitor for accurate tests. Standard capacitor meters cannot detect any change in ESR, therefore they miss defective capacitors leading to time consuming "tough dog" repairs. This capacity tester is made by:

Howard Electronic Instruments, Inc.
6222 N. Oliver
Kechi, KS 67067
1-800-394-1984

THE SIGNAL CHASER

The audio signal tracer is nothing more than an external amplifier that picks up the signal at any spot in the defective amplifier. The audio signal tracer troubleshoots audio circuits from input to the speaker. The commercial signal tracer might include a 1 kHz injection signal. The audio signal chaser might have earphones or a speaker as indicator. The commercial amplifier speakers might be used as a signal chaser.

Check the audio circuits by the number. Start at the volume control and if audio signal is present, proceed towards the speaker. Move to the front end of the audio circuits by checking the output at preamp IC or transistor. The audio signal within a cassette player can be traced with the external amplifier from the tape head to the output speaker *(Figure 2-13)*. Compare the audio signal at any point with the normal audio channel in stereo circuits.

The external amplifier tracer can be made from an audio amplifier kit, external amp or build one yourself. The amplifier should be able to drive a small 8 ohm speaker. An amplifier with 1 to 5 watts is preferred. This amp should be powered by batteries or if operated from the power line should be isolated with a power transformer. A separate volume control should be included to control the increased audio as it is traced through the circuit.

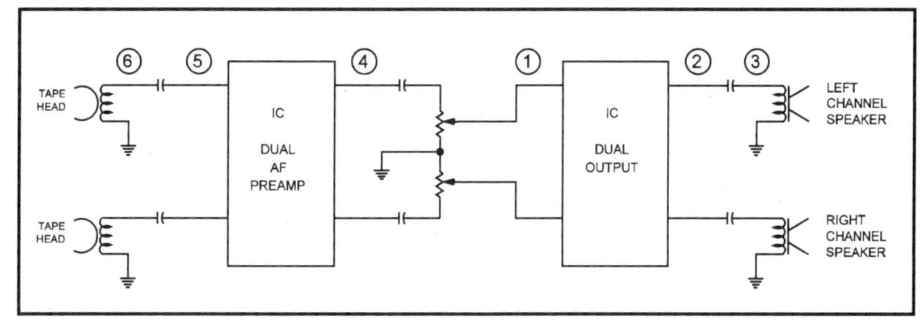

Figure 2-13. Signal trace the audio circuits by the number with the external audio amp.

AUDIO GENERATOR

The hand-held audio generator is ideal for field and bench service of audio equipment, stereo decks, car stereo and cassette players. The portable generator has a wide range of frequency, sine and squarewave output, sync output, and a continuously variable 20 dB fixed output. The frequency range varies from 20 Hz to 150 kHz in 40 to 50 steps. The sine wave output voltage of 1.2V and a squarewave of 8V maximum under no load, with a 600 ohm output impedance.

The bench audio generator operates from the ac power line which generates sine and squarewave forms. The frequency range might vary from 10 Hz to 1 MHz in five ranges. The sine wave frequency varies in X1, X10, X100, X1K, and X10K ranges, with a frequency response of 10 Hz to 1-mHz and a + or - .5 dB distortion. The output voltage is around 20 volt peak to peak with no load. The squarewave signal is the same with 10 volt p-p output and no load. These audio signal generators might have external synchronization with an output impedance of 600 ohms.

The audio oscillator test instrument might have a frequency counter in the same cabinet. The features might include low distortion, TTL output, sine and square wave output, 1MHz frequency counter and external sync input. The frequency range might vary from 10 Hz to 200 kHz in 7 or 8 steps with a standard 600 ohm output impedance. The enclosed frequency counter might have a display of four digits with a reading accuracy of .01% + or - 1 count.

FUNCTION GENERATOR

The function generator generates several different waveforms at a variable frequency *(Figure 2-14)*. The multi wave generator might appear in a 2 MHz, 5 MHz, and 20 MHz sweep function requiring either a standard function generator or a sweep generator. Some sweep generators include a 10 MHz frequency counter. Most sweep generators have a six digit LED display. A function generator might produce a square, triangle, sine, TTL, CMOS, and also pulse, ramp and sweep outputs. The 2-MHz sweep function generator might have a frequency range from .2 Hz to 2 MHz in seven ranges. The frequency ranges are: 2 Hz to 2 Hz at 1, 2 Hz to 20 Hz at 10, 20 Hz to 200 Hz at 100, .2 kHz to 20 kHz at 1K, 2 kHz to 20 kHz at 10K, 20 kHz to 200 kHz at 100K, .2 MHz to 2 MHz at 1 megohm.

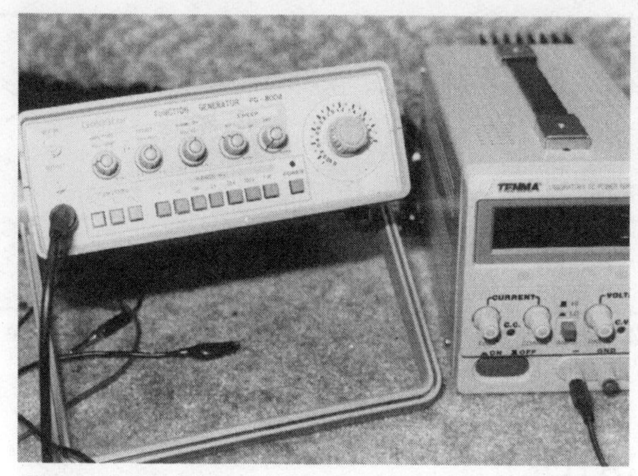

Figure 2-14. The audio or function generator injects sine and square waveforms into the defective audio circuits.

The frequency counter included within the function sweep generator might have a time base oscillation frequency of 10 MHz. The gate time: 10sec, 1sec, .1sec, and .01sec. A frequency range from 5 Hz to 10 MHz, a maximum input voltage of 42V peak, and an input impedance of 1 megohms (-20 dB), 500K ohm (0dB). The function generator can be used to inject different waveforms within the audio circuits to locate weak or distorted circuits and for audio test procedures.

EXTERNAL POWER SOURCES

The external power supply might provide DC voltage to the auto CD and cassette player, high wattage amplifier and for injecting voltage to eliminate intermittent problems within the electronic chassis. The 10A and 15A regulated power supply might provide voltage to the car radio and cassette player, while a 30 or 40 amp power supply might be needed to operate those high-powered amp circuits. The 30 amp power supply is ideal for servicing or powering car stereo products.

The 10A regulated power supply has a 13.8 VDC regulated output for most auto radios. The higher amp power supply might have a variable voltage from 1 to 15V DC with a voltage fluctuation of less than 1%. The 40 amp power supply might have a meter to read the output voltage and current drawn by the electronic product.

The variable single, double or triple output DC power supply might have an output voltage from 0 - 18V DC, 0 - 30V DC, 0 - 60V DC. A triple output dc supply might have a fixed 5 volts, 0 - 24 volts, or two 0 -30V DC supplies rated at 3 amps. The laboratory VDC power supply has an adjustable output voltage with separate voltage and current meters. This power supply might also have fine and coarse voltage adjustments for accurate voltage settings *(Figure 2-15)*. Connect the external voltage source for a steady and accurate voltage in making alignment and adjustments upon battery operated products.

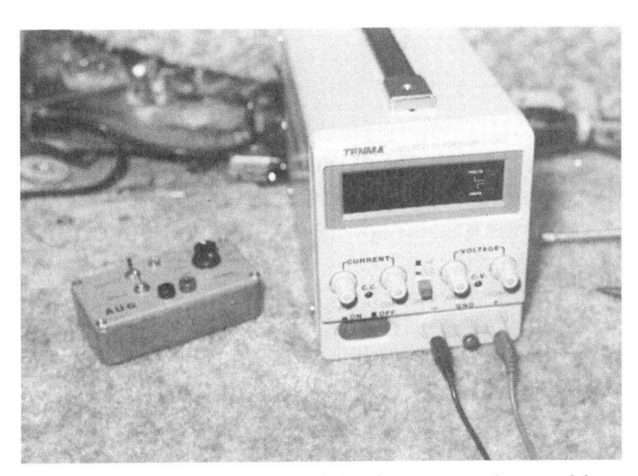

Figure 2-15. The external dc power supply provides a steady voltage to amplifiers for the auto and battery operated products upon the service bench.

TEST SPEAKERS

The test speaker should have the correct output impedance connected to the amplifier for a perfect match. Since the common speaker impedance is 8 ohms, just about any size will do. The test speakers should be mounted into the service bench or speaker enclosures. A pair of compact-shelf speakers are ideal where they can be moved and clipped to the audio chassis. A 10 or 12 inch pair of woofer speakers that have an average power of 50/100 watts might be required when servicing high-powered stereo amplifiers.

Most speaker enclosures have RCA-type plugs to plug into the amplifier. Some audio chassis have lug to lug, pin connectors, or tinned wire ends *(Figure 2-16)*. A pair of alligator clips on each speaker might make the quick connections. Make sure the connections are good and tight. Check the volume control setting before firing up the amplifier. Do not accidentally apply too much power to the test speakers so the voice coils are damaged at once.

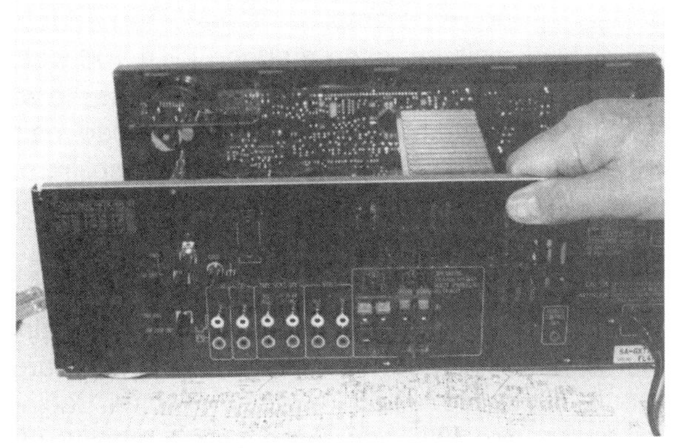

Figure 2-16. The output of the high-powered AM/FM/MPX receiver might have lug, plug or tinned wire speaker connectors.

LOAD IT DOWN

Attach the speaker before attempting to service any audio amplifier. The amplifier must have a load attached before the volume control is rotated or the amplifier output circuits can be damaged. High-powered dummy loads are required for servicing PA systems and high-powered auto or car amps. Sometimes the audio output circuits become defective and place a DC voltage upon the speakers voice coil destroying the speaker. Always check the speaker terminals or jacks for a DC voltage before attaching any PM speaker.

Since many of the latest audio amplifiers have high wattage output systems, a test-speaker dummy load must be able to withstand up to 100 watts or more. Several 50 or 100 watt resistors can be placed in series or paralleled to acquire the correct resistance and wattage. For instance, two 50 watt 8 ohm resistors can be paralleled to achieve a 100 watt resistance to match a 4 ohm impedance. Simply connect two 150 watt at 4 ohms in series to match an 8 ohm 150 watt dummy load *(Figure 2-17)*. A variable load resistor is designed for testing non-inductive load for bench testing amplifiers from 0-8 ohm at 90 watts and 0-18 ohms at 225 watts. Adjust the wiper ring for correct resistance upon the high-wattageload resistor.

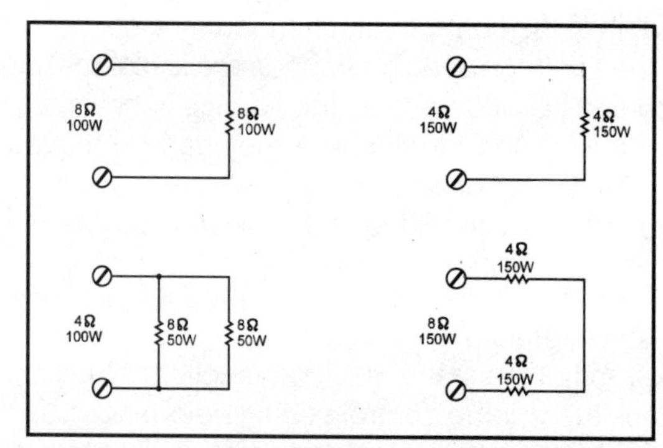

Figure 2-17. The various connections of power resistors for correct wattage and impedance dummy load resistors.

Since the full wattage of a high-powered amplifier is never turned full on while being repaired, a bank of 20 or 25 watt resistors can be added for a 100 watt output. Often the electronic technician keeps the volume turned down low as possible while trying to locate the weak or distorted circuit. A 100 to 250 watt 8 ohm dummy load resistor on each stereo channel can provide adequate loading of the most powerful amplifiers.

TEST CASSETTES & DISCS

There are several different cassette alignment tapes available for azimuth, frequency response, and speed adjustments for the micro cassette and standard cassette players. Besides these adjustments, the cassettes can be used for signal tracing the various audio circuits. A 15 minute micro cassette alignment test cassette consists of a 1 kHz audio tone recorded at a level of -4dB can be used for azimuth tape head adjustment. The 30 minute alignment cassette tapes include a full 20-20,000 Hz sweeps, as well as 14 other stereo tones recorded at -10dV. These alignment cassettes might include three different cassettes at normal bias, high bias, and metal bias types.

The deluxe BFC41 reference and frequency response cassette combines in one cassette all reference levels needed to set up and balance cassette recorders and decks. Part 1 provides a reference level of 315 Hz (0dB), and a 315 Hz (DE10dB). Part 2 provides the azimuth alignment of 6.3 kHz, 8 kHz, and 10 kHz (DE10dB). Part B is the speed adjustment cassette with frequency at 50 Hz, 3.15 kHz, and 3 kHz (DE15dB). Part 4 provides a frequency response test of 315 Hz, 31.5 Hz, 40 Hz, 63 Hz, 125 Hz, 250Hz, 500 Hz, 1 kHz, 4 kHz, 6.3 kHz, 8 kHz, 10 kHz, 12.5 kHz, 14 kHz, 16 kHz, 18 kHz, 315 Hz, (DE20dB) at fifteen second recording of each frequency. Part 1, Part 2, and Part 3 have a 30, 90, 120 second recording, respectively.

The BSC21 tape path mirror cassette makes it quick and easy to reliably check the way the tape moves through the recorder. Part 1 is a see-through lead-in path for optical check-up of heads and tape guiding parts. Part 2 is a normal FE tape running for approximately 10 minutes.

The DMC100 torque meter cassette, provides accurate torque meter measurement. The special feature of this cassette is a 5-sided display. By this method it is possible to read off either side of the cassette without having to remove cassette. This is particularly useful with the car cassette recorders where only one side of the cassette can be seen. This cassette measures the torque at normal, fast forward and rewind speed. Calibration is in gcm from 0 to 100.

The WFC1 wow and flutter cassette is recorded on both sides of the tape with 3000 and 3150 Hz respectively and so is fully compatible with all wow and flutter meters. Although these cassette alignment tapes are quite expensive compared to the 1 kHz at $19.95 and the audio alignment cassettes at $24.95, they are required if you are servicing many different cassette players. These alignment cassettes can be purchased at:

MCM Electronics
650 Congress Park Dr.
Centerville, OH 45459-4072
1-800-543-4330

TAPE HEAD CLEANERS

There are many different kinds of cassette tape head cleaners on the market. The automatic cassette cleaner has a gear driven cleaner with one wet and one dry pad that thoroughly cleans the tape head. The Recoton® deluxe head cleaning system uses a wet process cleaning system to maximize audio quality. A dishwasher System 11™ head cleaner cleans heads and prevents "eaten" tapes. The cleaner removes sticky residue from capstans and pinch rollers. A total cassette care kit might include a combination demagnetizer and wet process head cleaner for restoring audio-quality playback and recording.

DEMAGNETIZE THAT TAPE HEAD

The demagnetizer and head cleaner might operate from a 1.5 drycell (SR-44, S-76, RU-76, MS-76). After long hours of tape playing the audio heads can become magnetized causing a loss of high-end frequency response and increased distortion, resulting in overall poor performance *(Figure 2-18)*. Insert cassette demagnetizer with front face up. Place machine in play mode. After a few seconds, eject demagnetizer. These head cleaners and demagnetizers can be purchased at many different hardware and mall stores.

Figure 2-18. A self-powered tape head cleaner and demagnetizer cassette.

The electronic technician often cleans up the cassette tape head before attempting to service the cassette player. Use either a tape head cleaning kit or felt-tip cleaning sticks and alcohol. An old toothbrush with alcohol can remove the oxide packed tape head. Then, clean up with cleaning pads or sticks. Clean up all tape guides and paths with the top cover removed. Before buttoning up the cassette player, demagnetize the tape head with cassette or head demagnetizer with a curved tip to get at those hard to reach tape heads.

ADDITIONAL TEST EQUIPMENT

KEEP COUNTING

The frequency counter might be included in a hand-held battery operated test instrument or within the digital multimeter. The B&k 388-HD DMM has a 2 kHz, 20 kHz and 200 kHz frequency range. A frequency counter might be included within a function generator or audio oscillator test instrument. The frequency counter counts signal cycles or pulses in a frequency measurement. The frequency counter can check frequency, speed, head alignment, and overall frequency response of cassette players and amplifiers *(Figure 2-19)*.

The 100 MHz frequency counter might have selectable attenuation, low pass filter, selectable kHz/MHz, terminated BNC input connector, gate indicator light, and eight-digit LED display. A counter might have a frequency range from 5 Hz to 100 MHz with a + or - time base accuracy + or - 1 count. The time base is a crystal controlled oscillator. The frequency counter might have a low pass filter of 100 kHz and a 1 megohms impedance.

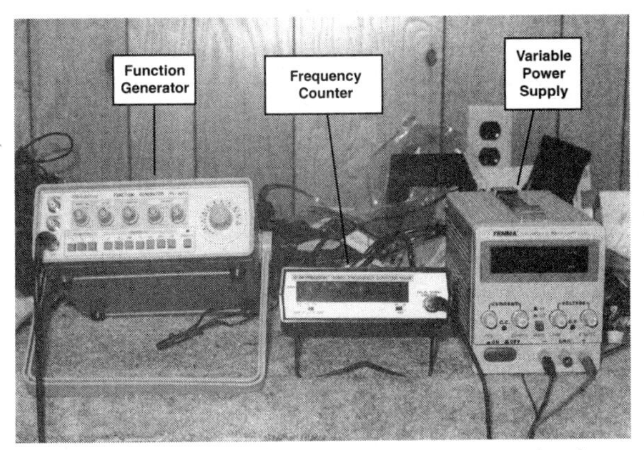

Figure 2-19. The frequency counter can check speed, head alignment and frequency response in the cassette player.

DISTORTION METER

The distortion meter is designed to measure total distortion at any frequency between 20 Hz and 20 kHz. The meter might also act as a level meter allowing simultaneous measurements of both signal levels and distortion. The distortion meter is usually connected to the line or amplifier output of the amplifier *(Figure 2-20)*.

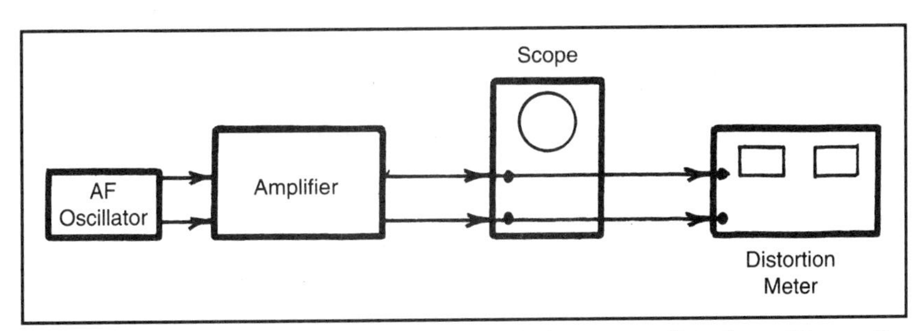

Figure 2-20. The distortion meter can check the output distortion of the audio amplifier circuits.

The distortion meter might have auto ranging signal level and a distortion measurement from 0.1% to 100%. The meter may measure signal levels from 1 mV RMS (root measure square) to 300V in 12 ranges. X and Y outputs allow the technician to observe the waveforms of the input signal and the total harmonics on an oscilloscope. The distortion measurement has an input level of 100 mV RMS to 300V RMS with residual distortion, including hum/noise less than .03%. The signal voltage frequency response of 20 Hz to 200 kHz with an output impedance of 600 ohms.

WOW AND FLUTTER METER

The wow and flutter meter is ideal when servicing any record/playback equipment such as cassette players, reel to reel, VCR's and VTR's. The meter should be compatible with JIS, NAD, CCIR, and DIN standards. The wow and flutter meter might have a four digit frequency counter with a center frequency of 3 kHz and 3.5 kHz. The tester should have a scope and recorder output for observation and recording. The input level should be from 5 mV to 10 Vrms.

The wow and flutter center frequency should be 3 kHz (CCIR, NAB, JIS) with a 3.15 kHz (DIN). The wow and flutter tape speed range is from 2.7 kHz to 3.5 kHz. The frequency range of the frequency counter might be from 10 Hz to 999.9 kHz. The distortion factor of output drift and outputs to the scope and recorder should be less than 2%.

A FINAL NOTE

The test instrument is no better than the electronic technician that takes the critical measurements. Some electronic technicians with a great deal of knowledge and experience use only a few test instruments everyday. Some repairmen only use one or two per day. The most often used test equipment is the DMM, semiconductor, and oscilloscope. The required test equipment depends upon what part of the electronic entertainment field the technician wants to specialize in. Although, the wow and flutter meter and distortion test instruments are quite expensive, the sound specialist technician might be required to use them in servicing high-powered and expensive audio equipment.

Chapter 3

Servicing The Low Voltage Power Supply

Most electronic products found in today's electronic entertainment field are powered by batteries or a low voltage power supply. The portable or shortwave radio might operate from several batteries wired in series or an AC adapter. A portable TV may operate from a battery pack, several batteries in series, or a self-contained power supply. The camcorder operates from a rechargeable battery while the large-screen TV contains several power sources, providing different voltages for the various circuits.

Often the batteries are switched out of the circuit when the electronic product is plugged into the AC power line receptacle. When the AC adapter is plugged into the portable radio or CD player, the batteries are switched out of the circuit *(Figure 3-1)*. The AC/DC adapter might provide a 6V, 7.5V, 9V 10V 12V and 18 volt source. The universal AC/DC adapter might have an output of 3V, 4.5V, 6V, 7.5V, 9V and 12 volts with several different male or female plugs. The AD/DC adaptor might provide a dc voltage to the portable radio, CD player, cassette player, power tools and Nintendo games.

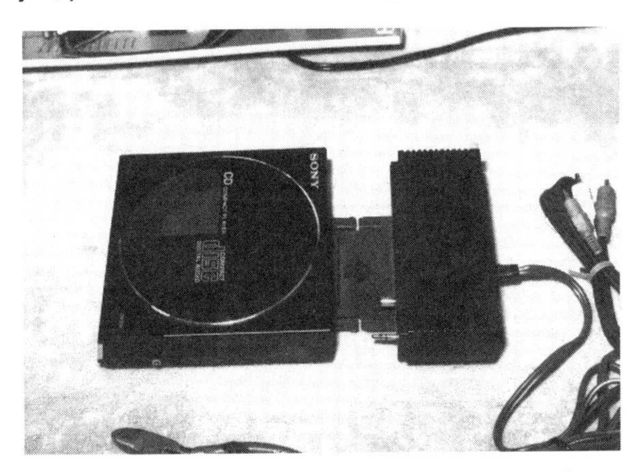

Figure 3-1. The Sony portable CD player can be operated from the power line with an AC adaptor power supply.

The portable cassette player or radio might operate from batteries, external power, or the AC power line. When the AC cord is plugged into the player, SW1 switches the battery and external power socket out of the circuit *(Figure 3-2)*. Now, AC is applied to preamp winding of transformer (T1) to a fullwave rectifier circuit. Likewise, when external power is plugged into P1, the power transformer and "C" cells are removed from the power circuits.

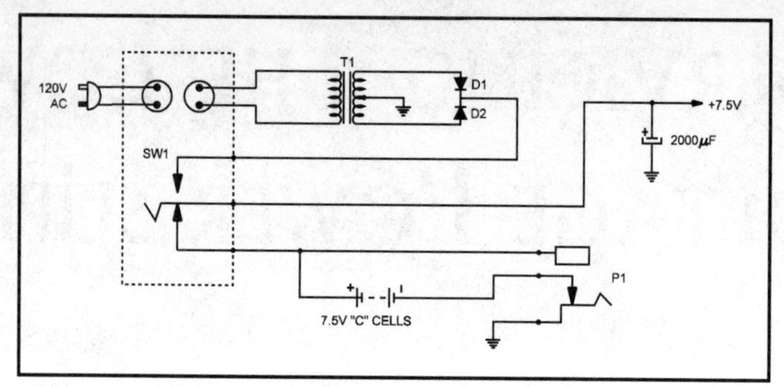

Figure 3-2. SWI switches the batteries or ac power supply circuits in and out of the portable cassette player.

The step-down transformer, (T1) supplies a low AC voltage to the anode terminals of rectifiers D1 and D2. D1 and D2 provide fullwave rectification that is switched by SW1 to a filter capacitor (C1). The 7.5 volt source is applied to the cassette motor and IC circuits of the cassette player.

HALFWAVE POWER CIRCUITS

There are two kinds of power rectifier circuits found in the AC power supplies that consists of a half wave and fullwave rectifier circuits. Halfwave rectification is used where small amount of power is required. In the halfwave circuit only one-half of the waveform is used, where the fullwave rectifier employs both halves *(Figure 3-3)*. Although, the halfwave circuits were found in the early tube radios and TV chassis, you might still see a halfwave circuit found in today's multiple power sources. You will find larger filter capacitors with higher capacitance in half wave rectifier circuits, to help smooth out direct current ripple effect. The halfwave rectifier might provide voltage to a tape motor, meter or indicator circuit.

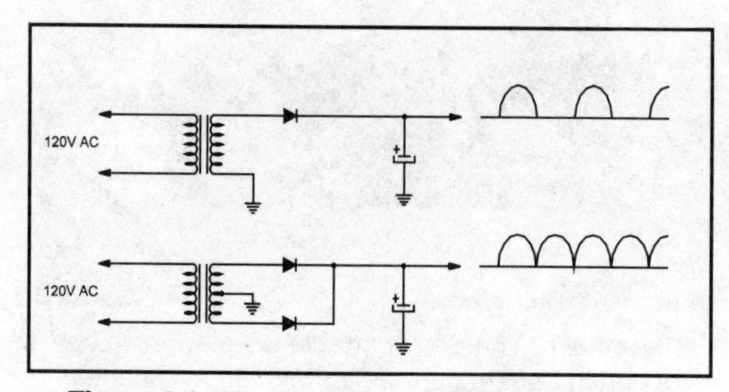

Figure 3-3. The half-wave rectifier contains one silicon diode while the fullwave circuit has two or four diodes.

The halfwave rectifier circuit might operate directly from the AC power line without a power transformer, from a tap off of a fullwave rectifier or from a separate winding of the power transformer. In the early AM/FM/MPX radios and cassette decks, the DC tape motor operated from a halfwave rectifier circuit *(Figure 3-4)*. A separate winding provided AC voltage to D102 and filter network (C102). The radios and amplifier circuits were powered from a bridge rectifier (fullwave) circuit.

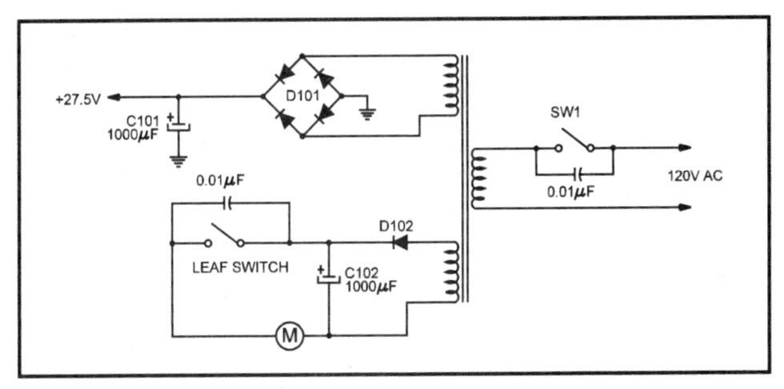

Figure 3-4. Notice the cassette motor operates from a halfwave diode while the amp and radio circuits have a fullwave bridge rectifier circuit.

THE FULLWAVE POWER CIRCUITS

The fullwave rectifier circuits might consist of two or four separate diodes. You will find only two silicon diodes in a fullwave rectifier circuit within the lower priced electronic products. Two silicon diodes might be found in the clock radio, cassette player, stereo recorder, and CD player. The fullwave or bridge rectifier circuits consist of four diodes in a bridge configuration *(Figure 3-5)*. You might find more than one bridge circuit in high-powered supply circuits. The bridge rectifier might have four diodes molded in one electronic component.

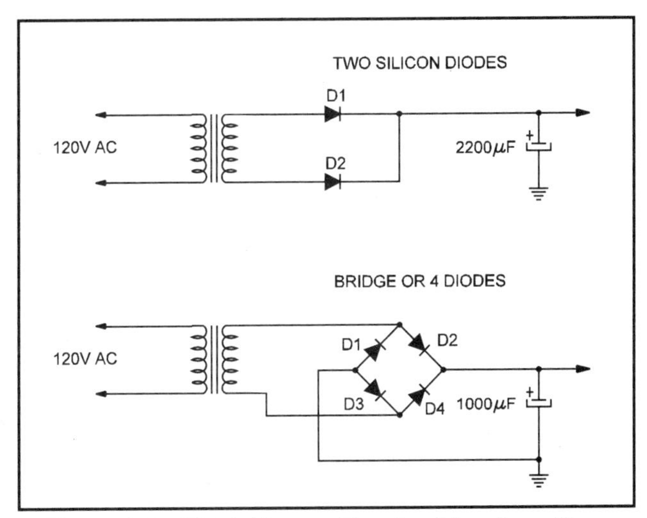

Figure 3-5. The bridge rectifier circuit has four silicon diodes.

The fullwave bridge rectifier is found in today's AM/FM/MPX receivers, cassette players, phonographs, amplifiers, CD players, and TV sets. The bridge circuit might have a single bridge component or four separate silicon diodes *(Figure 3-6)*. When a bridge component is not available, four separate diodes can be wired into a bridge circuit.

Figure 3-6. Here a Technics high-powered receiver has four diodes in a bridge rectifier circuit on a separate PCB.

The integrated stereo component system that includes an AM/FM/MPX tuner, tape deck, CD section, and high powered amplifier might have two or more bridge rectifier circuits. You might find more than one power transformer or two or more secondary windings feeding the bridge circuits. The bridge output circuits might have transistor, zener diode and IC voltage regulators *(Figure 3-7)*.

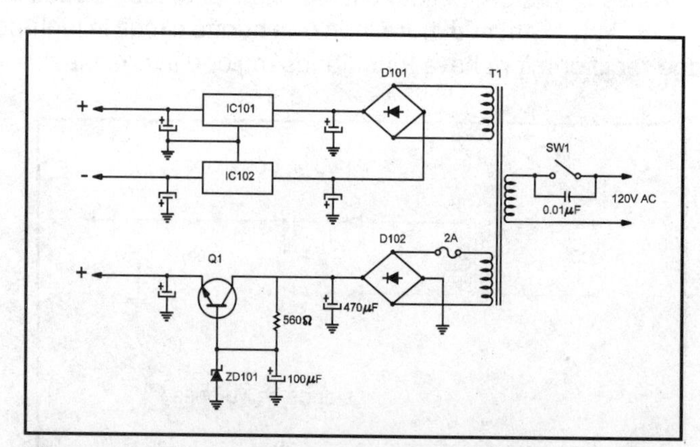

Figure 3-7. The regulator circuits might consist of transistor, zener diode and IC components or all three.

IC101 of the bridge rectifier provides a positive voltage, while IC102 has a negative voltage output to the high powered amplifier circuits. Q101 and ZD101 provides a voltage regulated source for the other different players. Notice the bridge symbol is provided with only one diode pointing in the positive direction. This diode symbol is found in many Japanese power supply circuits.

THE RECTIFIERS

After the tube rectifier, the germanium and selenium rectifiers were found in the early battery-tube circuits. The selenium rectifier consisted of large selenium plates stacked together to rectify the AC voltage which provided dc voltage to the tube filaments and B+ circuits. Very soon the selenium rectifier was replaced with the solid state silicon diode, The silicon diode is small in size, has high amperage characteristics, is very cheap to manufacture, and is very dependable.

The silicon diode is used today in half wave, fullwave and bridge rectifier circuits. Only one silicon diode is found in the half wave rectifier circuit, while two diodes are located in the fullwave, and four separate diodes work in the fullwave bridge circuit. Most low voltage power supply rectifiers are 1 and 3 amp silicon diodes. The electronic technician might replace all silicon diodes with a 2.5 amp type, except the power supply which requires the 3 amp diode. The defective diode can become leaky, shorted or appear open.

THE FILTER CIRCUITS

After the AC voltage has been stepped-down and rectified by the silicon diodes, the filter components must smooth out the remaining pulsating and dc ripple. The early radio and TV circuits included input-choke and capacitors. In fact, the early radio chassis used a magnetic field-coil as a choke to help eliminate the dc ripple in the low voltage power supply. The choke coil provides high impedance to AC current while practically no opposition to direct current.

The input-choke circuit provides better regulation than the capacitor input filter with a lower output voltage. A capacitor-input filtering network provides a higher output voltage and less filtering action *(Figure 3-8)*. Although the choke-input filtering is no longer used in many of the present day electronic products, capacitor input filter action is still used with transistors, zener diodes, and IC's as regulators.

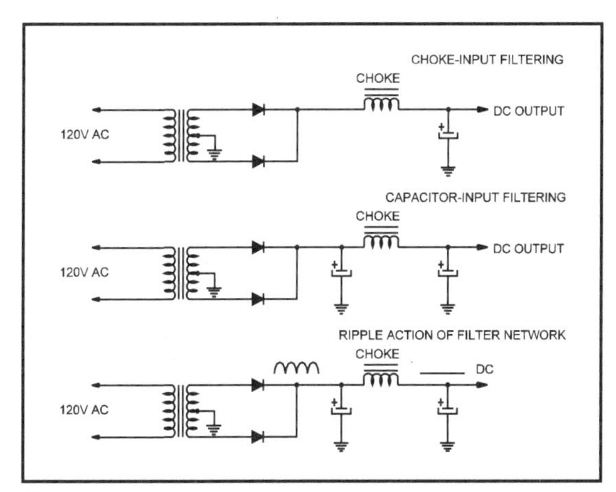

Figure 3-8. The early rectifier circuits had choke or capacitor input filter circuits.

Very large and higher wattage resistors were found as voltage dividers and bleeders in the early power supplies. Some high-wattage resistors had sliding taps to provide a certain working voltage. Now, the voltage regulators, resistor and decoupling electrolytic capacitors form the different voltage circuits. The electrolytic capacitor-input circuits are found throughout the electronic field of entertainment products.

You can spot the electrolytic filter capacitors on the electronic chassis since they are very tall and large compared to other radial type capacitors *(Figure 3-9)*. The defective filter capacitor might show signs of bulging sides, black and white substance oozing out of the bottom terminal area. A defective filter capacitor might become leaky, open, or shorted, damaging the silicon rectifiers and power transformer. The dried-up electrolytic can produce a very low voltage source. Most electronic technicians check the voltage across the large filter capacitors first to determine if the power supply is normal or provides improper voltages.

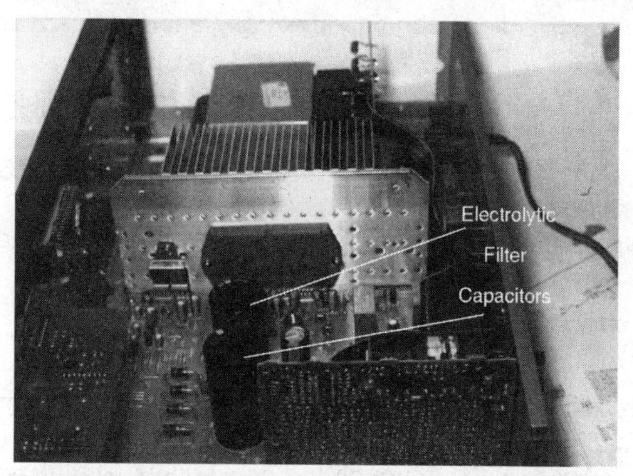

Figure 3-9. The filter capacitors are the largest capacitors found in the power supply circuits.

VOLTAGE REGULATOR CIRCUITS

The voltage regulator circuits might consist of either a transistor, zener diode, and IC or all three in the large power supplies. A zener diode might supply a positive or negative fixed voltage to the cassette motor, display tube, or operate in a transistor regulated circuit. The zener diode provides a simple voltage regulator whose output is constant and fixed. The zener diode regulator circuit consists of a zener diode and limiting resistor in series with the unregulated power source. The defective zener diode might appear open, leaky, shorted or overheated with burn marks.

The zener diode might be found in transistor regulator circuits. The diode is usually located in the base circuit of the transistor regulator. The transistor-diode regulator combination provides a low priced, dependable, and high current regulator circuit *(Figure 3-10)*. An open transistor regulator indicates no output voltage while a leaky or shorted transistor regulator might produce a low or higher dc voltage source. Often the zener diode operates quite warm and might be leaky when the transistor regulator becomes leaky or shorted.

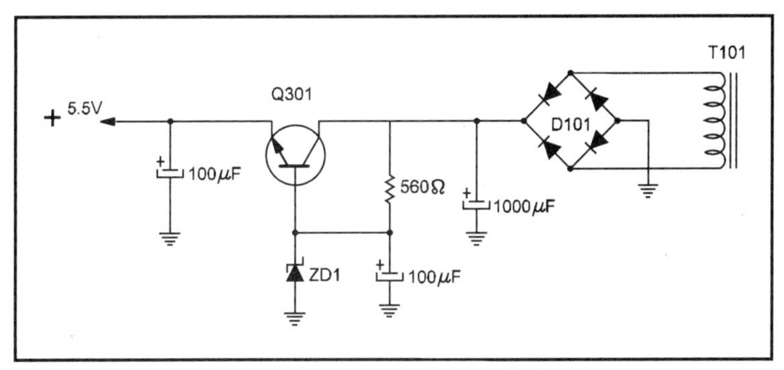

Figure 3-10. The transistor and zener diode regulators are dependable and have higher current capabilities.

The IC regulator might be found throughout the high-powered receiver, CD player, camcorder, amplifier, VCR's and TV set. In the TV chassis, a large fixed high-voltage regulator is found after the bridge rectifier circuit. The bridge rectifiers and IC regulator operate directly from the AC power line. The TV-IC regulator might regulate the output voltage from +115 to +135 volts; this regulated line-voltage is fed to the horizontal output transistor.

You might find several different IC regulators within the AM/FM/MPX receiver, camcorder, VCR and CD player. The IC regulator might provide a positive and negative voltage source. Several motors within the camcorder or CD player might operate from an IC regulator. The IC regulator provides a fixed voltage source to critical circuits within the low voltage power supply *(Figure 3-11)*. The IC voltage regulators might appear in low voltage power supplies that also contain a transistor-zener diode voltage source. The defective IC regulator might appear leaky or open, providing improper voltages.

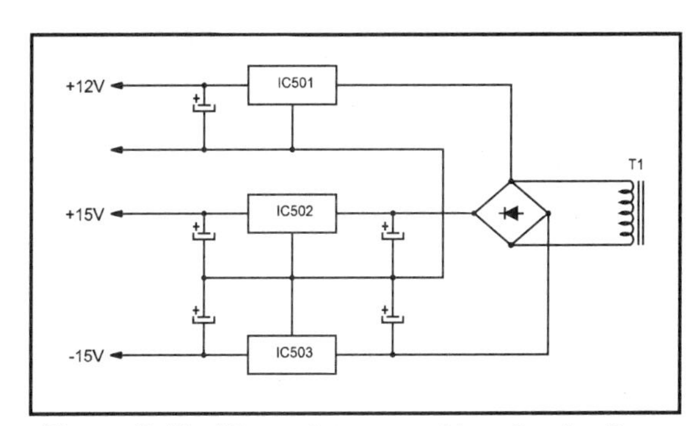

Figure 3-11. IC regulators provide a fixed voltage source in the power supply circuits.

SEVERAL DIFFERENT VOLTAGE SOURCES

Several different voltages sources are found in today's electronic products. Although the camcorder operates from a NI-CAD battery, the dc to dc converter provides several different voltage sources to feed the different capstan, drum, and loading motor circuits. The compact disc player power supply sources feed different voltages to the various electronic circuits, loading, disc, turntable and up and down motors.

The TV chassis might have a line-operated regulated power supply that supplies voltage to the horizontal output circuits and in turn produces many different voltage sources from the flyback or horizontal output transformer circuits. In the RCA CTC157 TV chassis, the low voltage power supply provides 129 volts to the horizontal output circuits and the flyback supplies several voltages to operate the different circuits *(Figure 3-12)*. The 200 volt source feeds the color output transistor, CRT and boost circuits. The 44 volt source supplies voltage to the beam limiter transistor, vertical reset and vertical sawtooth circuits.

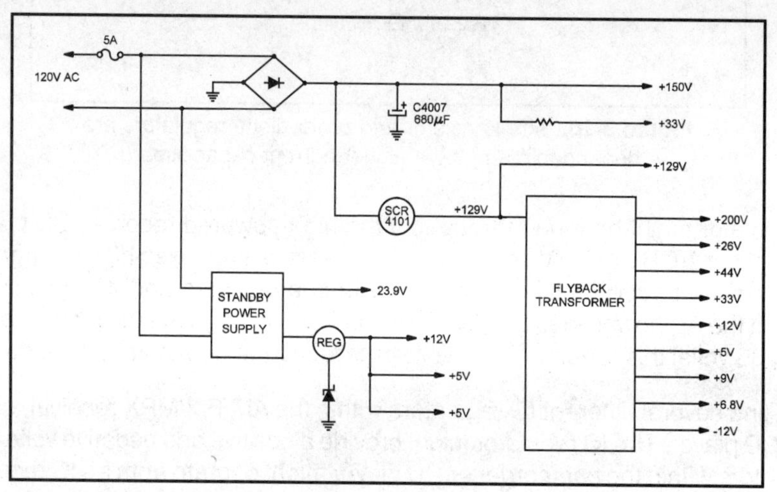

Figure 3-12. There are three different power supply circuits found in the RCA CTC157 TV.

The audio and vertical output circuits are fed from the scan-derived 26 volt source. The clamp and pin cushion circuits are also fed from the 26 volt source. Besides these secondary flyback voltages, there are 33V, 12V, -12V, 9V, and 5 volt sources supplied from the scan-derived horizontal output transformer. A 33V, 129V and 150 volt source is powered directly from the power line voltage supply. A separate standby power transformer supplies a 23.9V, 12V, and 5 volt source. You will find many different voltage sources in today's electronic products that can break down and provide no or improper voltages.

FILTER CAPACITOR PROBLEMS

When a loud hum is heard from the speakers with the volume turned down, suspect a defective electrolytic capacitor in the power supply. A shorted or leaky filter capacitor can destroy silicon diodes and bridge rectifiers. A dried-up or open filter capacitor in the TV power supply can produce black bars on the picture tube. If the unit is left on too long with a leaky capacitor the power transformer can be damaged. Often the primary winding of the step-down transformer will open up with an overloaded power supply. This can occur at once within the electronic chassis that has no protection fuse in the primary winding.

The defective filter capacitor can dry up, appear open, leaky and shorted. Electrolytic capacitors, after several years old, might dry up the electrolytic paste between the capacitor elements and lose capacitance. The capacitor terminal leads might break inside the capacitor or come loose from the aluminum foil and appear open. An open or dried-up decoupling filter capacitor can cause a lower voltage source. The shorted or leaky electrolytic capacitor can destroy resistors, transistors and zener diodes in the voltage sources. Shunt the suspected filter capacitor with the same or greater capacity and working voltage, to determine if the hum disappears and the voltage source increases.

A quick voltage test across the main filter capacitor terminals and compared to the schematic can indicate if the power supply is normal. Suspect a leaky diode, burned isolation resistor or voltage regulator when the voltage is quite low across the capacitor terminals. Discharge the capacitor. Measure the resistance across the filter capacitor terminals. A low ohm measurement might indicate a leaky capacitor or a leaky component tied to the voltage source. Remove the positive lead from the electrolytic capacitor.

Check the suspected capacitor in and out of the circuit with a capacitor tester *(Figure 3-13)*. A good electrolytic capacitor should have about the same capacity as marked on the capacitor or a higher measurement in microfarads. Replace the suspected capacitor with very little capacity or with internal leakage. A resistance measurement on the 20K ohm scale can indicate a good capacitor that charges and discharges with the meter hand. Reverse the test leads and the electrolytic capacitor will charge up again. Always observe correct polarity when replacing the defective electrolytic capacitor.

Figure 3-13. Check the suspected or replacement electrolytic with a capacity tester.

DECOUPLING VOLTAGE CIRCUITS

The decoupling capacitor provides a low-impedance path to the ground to prevent common coupling of the various electronic circuits. A decoupling capacitor within the low voltage sources follows a voltage dropping resistor. Often, the decoupling capacitor has a low capacity value compared to the main filter capacitor. Suspect a leaky or open decoupling capacitor when one stage is weak or dead. A decoupling electrolytic capacitor might isolate two different voltage circuits.

In the car radio, the AM and FM circuits might be isolated from the amplifier circuits with a voltage dropping resistor and decoupling electrolytic capacitor. The AF stage within the auto radio amplifier circuits might be isolated from the same voltage that feeds the power output IC *(Figure 3-14)*. R707 (680 ohm) resistor provides a lower voltage source and C705 (220 µF) is the decoupling capacitor. R703 (820 ohm) resistor is the collector load resistor of Q703. Suspect a leaky C705 when real low voltage is measured at the collector terminal of Q703. R707 might become burned and change resistance with a shorted C705.

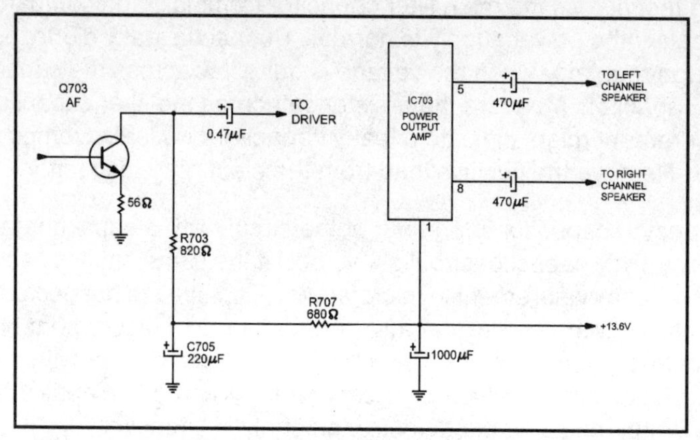

Figure 3-14. C705 and R707 provide a decoupling voltage circuit in the audio amplifier.

Check the low voltage source feeding the dead or weak amplifier stage. Improper voltage on the amplifier transistor or IC might result from a leaky decoupling capacitor. Suspect real low voltage applied to the amplifier stage when the decoupling capacitor dries up or opens. Low hum in the speaker might result from a dried-up decoupling capacitor. An increase in resistance of the voltage or isolation resistor results in a low voltage source.

POWER TRANSFORMER PROBLEMS

The power transformer might be located on the electronic chassis or off by itself on another PC board or mounted separately *(Figure 3-15)*. The primary winding of the step-down power transformer is wound with smaller copper wire than the secondary winding. If a silicon diode electrolytic capacitor becomes leaky or shorted, the primary winding might open up in the electronic chassis that has no fuse protection. An overloaded circuit connected to the same voltage source can make the power transformer run very warm. When lightning hits the power line, the transformer can be damaged since there is no fuse protection or the on/off switch is located in the secondary winding. Check the resistance across the AC male plug to determine if the transformer winding is open.

The damaged power transformer might blow the house fuse when the unit is turned on or appears real warm with no sound. The leaky transformer can keep blowing the main line fuse. Sometimes a loud hum and groaning noise from the transformer indicates an overloaded transformer. The electronic chassis might vibrate or buzz with a defective power transformer. Remove the secondary leads of the transformer from the fullwave rectifiers or bridge diodes. Make sure all secondary leads of the transformer are removed from the circuits. Replace the transformer if it has a loud hum and runs quite warm. No doubt the transformer has burned and shorted internal windings. Check the following components that can cause the transformer to run warm and overheat *(Figure 3-16)*.

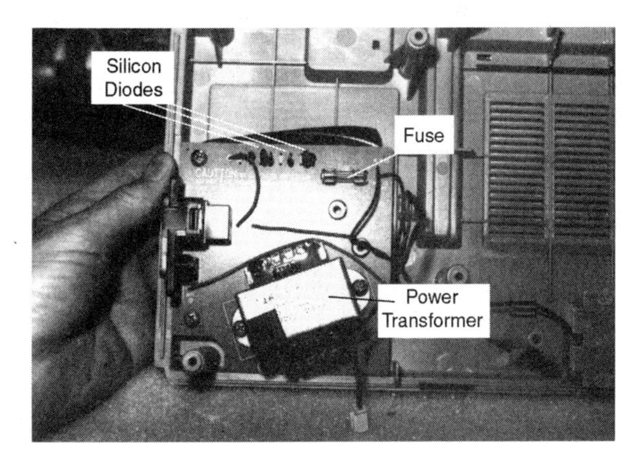

Figure 3-15. The power supply might be mounted upon separate PCB in the portable radio or cassette player.

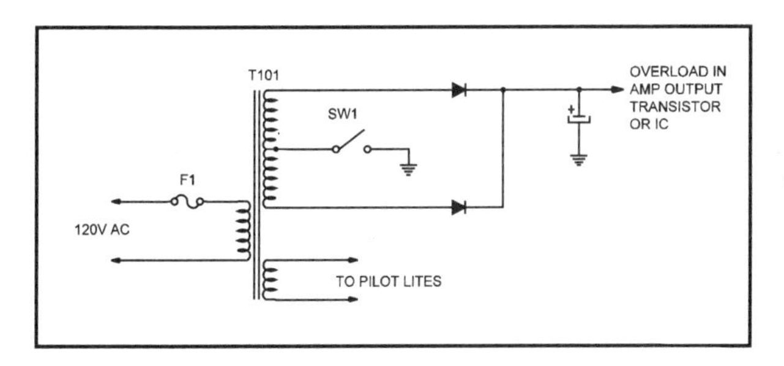

Figure 3-16. Check the following components that might make T101 overheat.

Try to replace the power transformer with the original part number; it fits in the right spot and has the correct AC voltage. For instance, in a Sylvania R53-14/-18 AM/FM/MPX radio, phonograph and tape player the original transformer part number (55-14146-13) supplies an 18V, 13.6V, and 11.8 volt source. The center tap from the secondary winding provides 5.3 AC volts to several pilot lights. The secondary winding feeds 17.3 volts AC to a half wave rectifier with three resistors and three filter capacitors providing different voltage sources *(Figure 3-17)*.

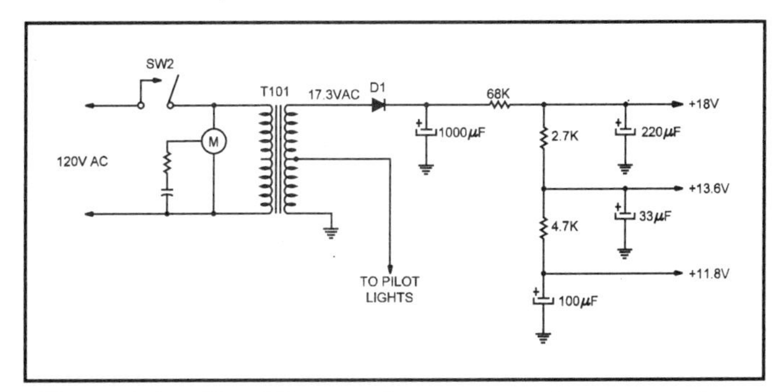

Figure 3-17. The different voltage sources form a step-down power transformer in the Sylvania R53-141-18 radio, phono and tape player.

The 11.8 volt source feeds both tape preamp transistors (Q1 and Q2). A 13.6 volt source feeds both 2nd tape preamp transistors, while the 18 volt source supplies voltage to the 3rd tape preamp player. A separate power transformer and low voltage power supply provides several voltage sources to the volt amp, driver and push-pull output transistors. This is a special type of transformer and should be replaced with the original part number.

NOISY SOUND PROBLEMS

The most pronounced noises found within the audio amp circuits are filter hum, pickup hum, and frying or hissing noise. Hum heard in the speakers with the volume control turned down is caused by defective filter capacitors. A pickup hum might result from a poor microphone or tape head connection or open tape head. A loud rushing noise with no music can be caused by broken tape head wire. Sometimes an increase in resistance of a base resistor of a preamp stage can cause a low hum noise. A constant frying or hissing noise can be caused from a defective transistor or IC. The small ceramic capacitor connected to the collector or base terminals of a preamp or driver transistor can also cause a frying noise.

Shunt another known electrolytic across the suspected filter capacitor to see if the hum disappears in the speakers. Discharge both capacitors. Clip the new electrolytic capacitor across the suspected capacitor with wire clip leads. Do not shunt the capacitor while the unit is operating. Now, notice if the hum is eliminated. Replace all electrolytic capacitors within one can if one is found to be defective; they will all go sooner than later.

Inspect the microphone cables, plugs and jacks for a poor solder connection. Replace all frayed cords and poor shielded connectors with gold plugs. Check all patch cables for poor connections. Pickup hum can result from poor input ground connections of the preamp circuits. Inspect all ground return wires and cables at the input microphone and phono cables.

Suspect a defective transistor or IC for a constant frying or hissing noise. Critical voltage, resistance and transistor tests will not locate the noisy transistor. Determine if the frying noise is in the output or front-end circuits. Turn down the volume control and if the noise disappears, the defective component is in the front-end circuits. Several coats of coolant might spot a frying or hissing transistor or IC.

Shunt the base terminal of each transistor with a 100 uF 50 volt electrolytic capacitor to ground. Discharge the capacitor each time so as not to damage the transistors. Another method is to short out the base terminal to the emitter terminals of each transistor. Make sure you have the correct base and emitter terminals or you can destroy the transistor if the base or emitter is shorted to the collector terminal. If the noise stops each time, the defective component is ahead. When the noise is present and the base terminal is shunted, you have located the noisy circuit. Replace the suspected transistor.

The frying noise might be caused by a defective ceramic or electrolytic capacitor in the base and collector circuits *(Figure 3-18)*. Suspect a 0.1 μF to 47 μF coupling capacitor to cause a frying noise within the preamp or AF circuits. Shunt each coupling and bypass capacitor within the located noisy stage. You can signal trace the noisy component by going from base to collector of each preamp and AF transistor with an external amplifier. Likewise check the output and input terminals of the preamp and AF-IC components for a frying noise. Sometimes replacing the noisy component is the only answer.

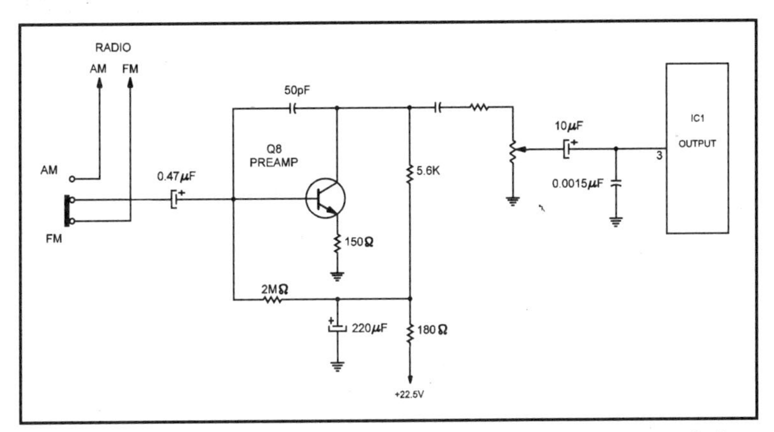

Figure 3-18. Check the following components that can cause a frying noise in the amplifier circuits.

SERVICING THE CASSETTE PLAYER POWER SUPPLY

The small cassette player might have a very simple fullwave or bridge rectifier circuit. Only two silicon diodes may be connected to a center-tapped power transformer. SW3 (AC/ battery SW) switches the batteries out of the circuit when the AC cord is inserted. SW2 is a leaf type switch that turns on the cassette motor and power to IC1. The dc ripple is filtered out with C22, C21, and C20 *(Figure 3-19)*. IC1 is powered from C22 and the erase head voltage is taken from C20 and R20 (470 ohms).

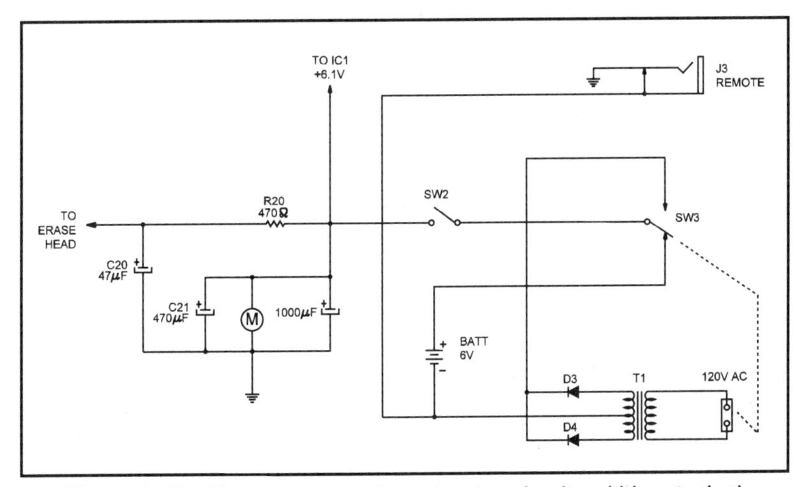

Figure 3-19. The power supply and motor circuits within a typical cassette player.

Suspect the dc power supply when the tape player operates on batteries and not from the AC receptacle. Measure the dc voltage across C22 (1000 μF). Check the dc voltage at either cathode of silicon diodes D3 and D4. If a dc voltage is found here and not at the filter capacitor (C-22), suspect dirty switch contacts at SW3 and SW2. Clean up the switch contacts with a piece of cardboard or fingernail file. Spray cleaning fluid into the switch area; work the switch back and forth.

Check each silicon diode with the diode test of the DMM when AC voltage is measured on the anode terminals. Suspect a poor AC cord or power transformer with no AC voltage at either diode. Remove the AC plug from the wall and take a resistance check of the primary winding, ACross the AC plug prongs. A continuity measurement indicates the cord and primary winding are good. Often the primary winding of the transformer will be open when the silicon diodes (D2 and D3) become leaky or shorted.

Test each battery for correct voltage in a battery tester, when the cassette player will not operate in battery mode. Make sure the AC cord is not plugged into the cassette player. Clean up all battery contacts. Suspect a dirty switch contact of SW2 or SW3 after installing a new set of batteries *(Figure 3-20)*. Double check each battery polarity as one might be turned around, when they were replaced.

Figure 3-20. A dirty leaf switch might cause a dead or intermittent cassette player operation.

TROUBLESHOOTING THE BOOM-BOX POWER CIRCUITS

The high-speed dubbing cassette deck in a boom-box player might have a transistor and IC regulator circuits in the AC power supply. SW6 applies AC voltage to the bridge diodes (D501-D504) from the power transformer (T501). C502 (2200 μF) provides filtering action before the 18 volts are fed to transistor voltage regulators Q501 and IC501. The 12 volt source out of regulator IC501 provides power to the audio AF and output circuits, and to the motor drive circuits. Q501 and D503 provide transistor-diode regulation to R353 and electrolytic filter capacitors C505 and C344. The +13.5 voltage source is fed to the Dolby, tape head and preamp circuits *(Figure 3-21)*.

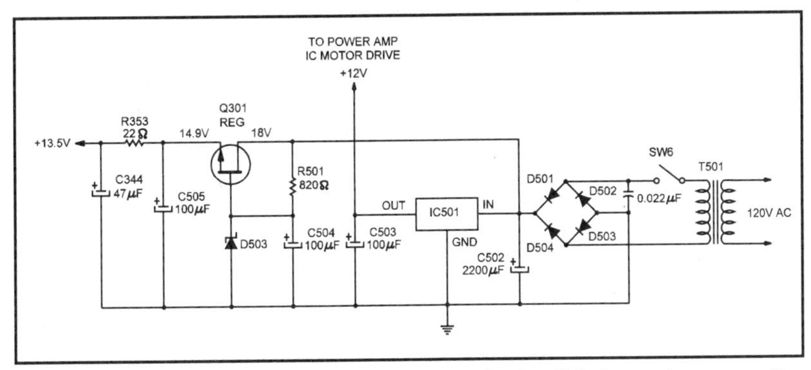

Figure 3-21. A typical regulated power supply circuit in boom-box cassette player.

Notice, T501 is on all the time when the AC plug is plugged into the wall receptacle. No fuse protection is found in this boom-box player. Check the dc voltage (18V) across the largest filter capacitor C502 when the boom-box player is entirely dead. Suspect defective silicon diodes (D501 through D504), SW6 and T501. Measure the AC voltage across the secondary power transformer winding. Check for a bad cord or open primary winding with no secondary AC voltage.

When the tape drive motors will not rotate and there is no sound at either speaker, suspect regulator IC501. Measure for 12 volts at the "out" terminal of IC501. Sometimes the silicon diodes and IC regulators are damaged when the boom-box is hit by lightning or a power line outage condition. An open IC501 might have no or very little output voltage. A leaky IC501 may have a lower than normal output voltage. Remove the 12 volt lead from C503 and out terminal of IC501 to see if the voltage increases, indicating an overloaded circuit connected to the regulated IC501.

Suspect a defective Q501 regulator circuit when the motor rotates and no voltage is applied to the front-end circuits. Rotating the volume control up and down can indicate if voltage is present in the amplifier circuits with a rushing type noise in the speaker. No voltage can be measured at the emitter terminal of Q501 if the transistor is open. Replace the intermittent (open) Q501, if the player is on for a few minutes and shuts down with the motor operating. When Q501 appears leaky or shorted, the output voltage might be lower with an overheated D503. Replace both D503 and Q501 when the regulator transistor becomes leaky. Critical voltage measurements within the low voltage circuits can quickly locate defective diodes and regulators.

REPAIRING THE DELUXE AMP POWER SUPPLY CIRCUITS

The integrated stereo component system might have a main and a sub power transformer with several different voltage sources. The main power transformer might supply +60 volts and -60 volts to a high powered amp output IC of 100 watts on each stereo channel *(Figure 3-22)*. The push on-push off power switch applies the 120V AC power line voltage to the primary winding of T1. F1 provides fuse protection in the primary winding.

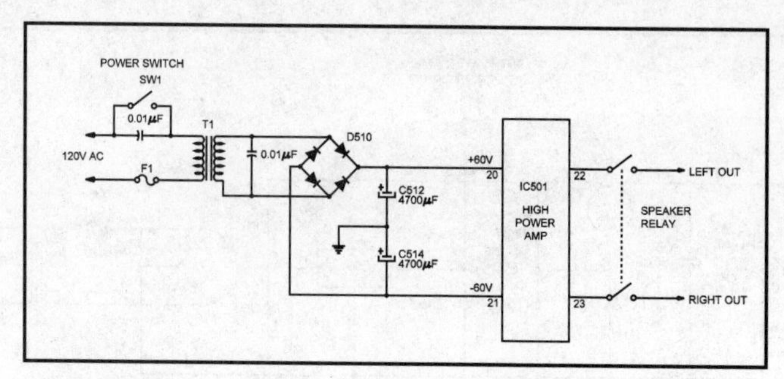

Figure 3-22. A +60 and -60 voltage source feeds the high-powered output amplifier.

A blown fuse (F1) might indicate a shorted bridge rectifier (D510), leaky C512 or C514, and a leaky power IC501. Suspect a defective component within the power supply when no sound is heard in the speakers. Locate the largest electrolytic capacitors (C512 & C514) on the main power board and check for correct voltage across each filter capacitor. No voltage might indicate a blown fuse, shorted or leaky bridge rectifier (D510), and C512 or C514 capacitors. Low or improper voltage might result in a leaky power amp IC or defective filter capacitors. You might find a higher supply voltage (33 to 75 volts) feeds the high-power amp IC or power output transistors.

Check each silicon diode with the diode test of DMM when the fuse keeps blowing *(Figure 3-23)*. Take a resistance test across each filter capacitor (C512 and C514) after discharging each capacitor. A low resistance measurement to common ground might indicate a leaky filter capacitor or IC501. Remove the positive or negative lead from each capacitor and take another test. Suspect a leaky or shorted power amp output IC with normal capacitor tests. Check terminals 20 and 21 to common ground for a low leakage test, indicating a leaky power output IC501.

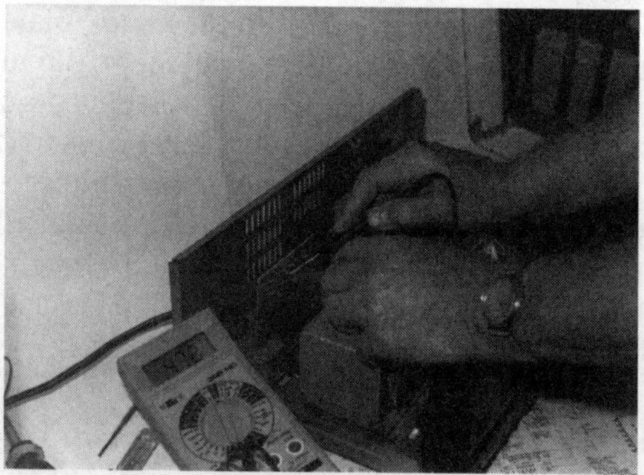

Figure 3-23. Check each silicon diode with the diode test of DMM.

SERVICING HIGH AND LOW TV POWER SOURCES

Most TV chassis today have a low voltage power supply that operates directly from the AC power line without a power transformer. The only transformer that might be found in the TV chassis is a small stepdown power transformer for the standby remote circuits. The 120 VAC power line voltage feeds into a protection fuse, line filter, degaussing coil, on/off switch and bridge rectifier circuits *(Figure 3-24)*.

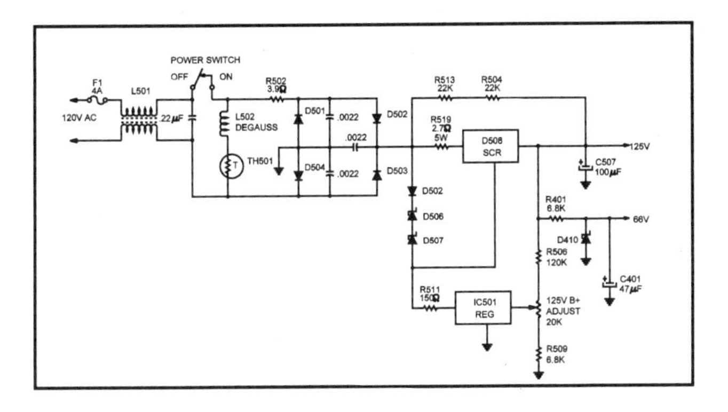

Figure 3-24. A power line-operated power supply found in the Emerson model MS1980R TV.

In an Emerson MS1980R TV power supply, a large filter capacitor C506 (1000 µF) helps filter out the dc ripple current and is fed to an SCR (D508) regulator circuit. The 125 volt source is fed to the horizontal driver and output transistors. The 125 volt source is also fed to the UHF and VHF tuner circuits through a 30 volt regulator (IC502) providing a voltage source to the tuners. The zener diode (D410) 6.6 volt source provides a dc voltage to the horizontal and deflection circuits. A +155 volt source feeds to the stereo sound driver and output transistors.

The primary winding of the standby stepdown power transformer connects to the 120 VAC line after the 4 amp fuse. This power transformer is on all the time. The secondary AC voltage feeds into a bridge rectifier circuit and is filtered by C104 (470 µF) and IC104, a 6 volt regulator IC. The standby voltages are a 13 volt, two 5 volt, 5.3V, and a 3.9 volt source. The 5 volt source provides voltage to IC101, IC102, IC103, OS101 Recocon receiver, AFT balance, screen reset, power relay predriver Q104, LED drive and reset transistor Q113 *(Figure 3-25)*.

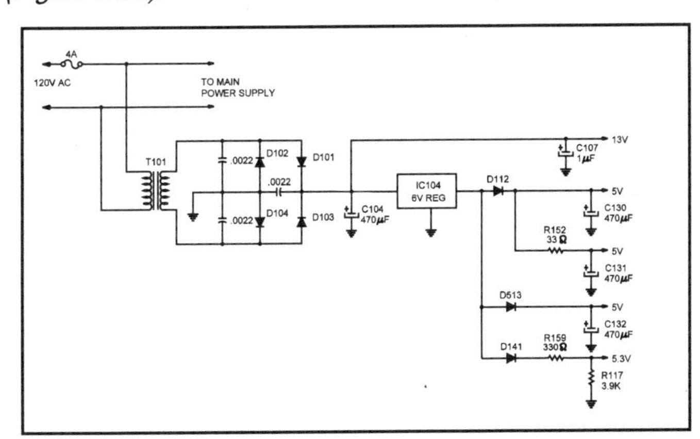

Figure 3-25. The different standby voltage sources provide remote control operation in the TV.

A 13 volt source provides voltage to operate power relay drive transistor (Q103) and relay (Ry 101). The other 5 volt source supplies power to the screen generator IC102. The 5.3 volt source feeds voltage to the F. Syn Micon (IC101) and reset transistor Q102. A 3.9 volt source goes to hold terminal (34) of IC101 Micon.

Before the scan-derived secondary voltage sources of the flyback (FB401) can operate, the low voltage power supply, horizontal, and flyback circuits must perform. The high voltage winding of FB401 provides a 27.5KV to 30KV voltage to the anode button on CRT, a focus voltage from 4.0KV to 8.0KV, and a screen voltage around 411 volts for the picture tube.

Besides these voltages three different scan-derived windings provide 8 different voltage sources *(Figure 3-26)*. The 183 volt source furnishes voltage to the color amps and output transistors tied to the cathode elements of the CRT. A 12.8 volt source feeds the tone control IC371 and IC103. The 8.6 volt source feeds Q201 IF preamp transistor. A 9 volt (A) source provides voltage to the IC201 VIF/SIF/chroma, AGC, 1st transistor buffer Q203 and 2nd transistor buffer (Q604), the sound SIF of IC201 and sound buffer transistor (Q301), the chroma section of IC201 and the brightness (y) buffer transistor (Q602).

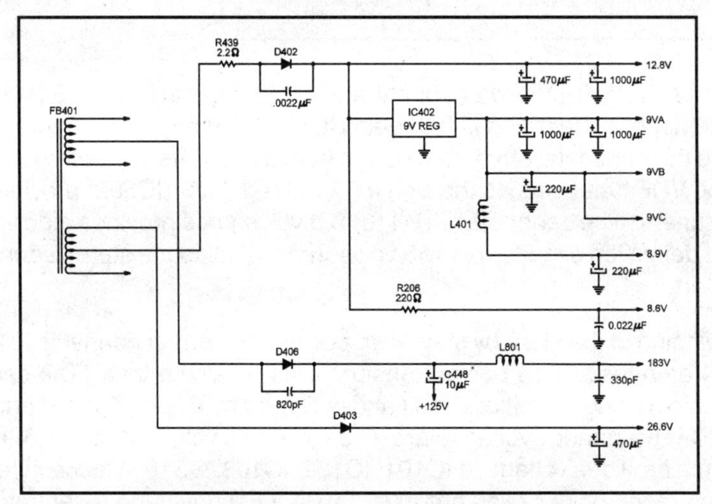

Figure 3-26. Besides the HV, screen and focus voltage, the flyback circuits provide many different voltage sources.

The 9 volt (B) source feeds voltage to the MTS IC951, SAP amp transistor Q951, SAP amp Q952, SAP amp Q953 and 9 volts to noise reduction IC952. A 9 volt (C) source powers the SAP LED indicator, SAP SET LED, switching transistors Q111 and Q110, and also lights up the stereo LED connected to IC951. The 8.9 volt source feeds the kinne bias of IC901 and a peak ACL transistor (Q903).

In this Emerson model the TV sound stages are powered by a high +155 volts and the 9 volt A, B, and C voltage sources. You may find several different voltage sources feed the stereo sound circuits. Most mono TV sound stages are fed from one high voltage source. Always check the low voltage sources feeding the audio circuits when a dead, weak, distorted or intermittent sound symptom is found. Try to secure a wiring diagram when repairing the stereo TV audio circuits. Remember, when the voltage source feeding the sound circuits in the TV chassis are protected by the flyback, both power supply, horizontal and HV circuits must operate.

Most TV sound problems occur in the audio output stages. Locate the output transistor and IC circuits. Take a quick voltage measurement of the collector (body) terminal of a power transistor or voltage supply pin (VCC) of the audio output IC. If low or no voltage, go directly to the TV low voltage power supply sources.

If the TV symptom is that the whole set is dead, the TV chassis must be repaired before any sound stages can be tested. When the TV symptom is no sound, distorted audio or weak sound, suspect the audio output circuits. Rapidly, rotate the volume control up and down to hear a rushing noise. Go directly to the audio output circuits without a rushing noise symptom. Proceed to the audio input stage if a noise or hum can be heard in the speakers.

Check all voltage sources that feed the sound circuits from the main or flyback power supply circuits. When one stage has low or improper voltage, see what voltage source feeds this audio circuit. Look for open transistor regulators and burned zener diodes. The suspected IC regulator might cause no or improper voltage at the sound circuits. Critical voltage measurements within the low voltage source can quickly determine if the power supply circuits are defective. Do not overlook an open resistor in series with a leaky silicon diode within the scan-derived voltage sources.

SERVICING THE CD POWER SOURCES

The portable CD player has a very simple power supply compared to the CD player found in the compact entertainment system. A small portable CD player might have a power line adapter to supply a voltage source. The portable boom-box CD player with AM/FM/MPX radio and cassette player might have transistors, IC's and zener diode regulators.

Take a quick voltage measurement across the large filter capacitor (C201) when the CD-boom box player is weak or dead. Check the bridge rectifier or silicon diodes for possible leakage. A shorted diode might open up the primary winding of the power transformer (T201) *(Figure 3-27)*.

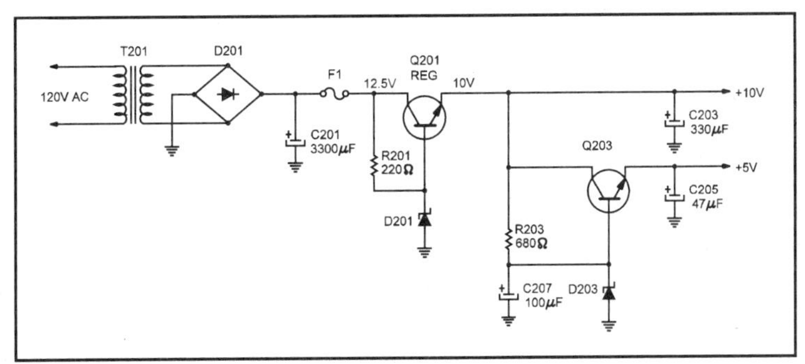

Figure 3-27. A typical boom-box CD player power supply voltage circuit.

Suspect a blown fuse (F1), open Q201 and damaged D201 when no or real low voltage is found at the emitter terminal of Q201. A weak sound problem might be caused by low voltage in the power supply sources. Suspect dried up filter capacitors, leaky transistors and IC's for low voltage sources. Shunt C203 and C205 for a low +10V and +5V source.

Remove one end of a silicon or zener diode when they do not measure up in the circuit. Check each diode with the diode test on a DMM. Test each transistor regulator in the circuit for an open or leakage measurement. Monitor the output voltage of each emitter of Q201 and Q203, when the sound is intermittent. The regulator transistors have a tendency to operate and then break down under load or after they have been operating for several hours.

REPAIRING THE AUTO RADIO-CASSETTE POWER SUPPLY

The typical early auto radio-cassette chassis operated directly from the car battery. A 3 to 5 amp fuse is found in the "A" lead harness with the stereo speakers and ground connections. L901, C905 and C906 provide a choke input network to eliminate hash-noise from entering the battery line *(Figure 3-28)*. The power switch (S1) turns power on to the sound circuits. C906 filters out any hum that appears in the power hookup cable. Suspect an open or dried-up electrolytic capacitor (C906), when hum is heard in the audio output stages.

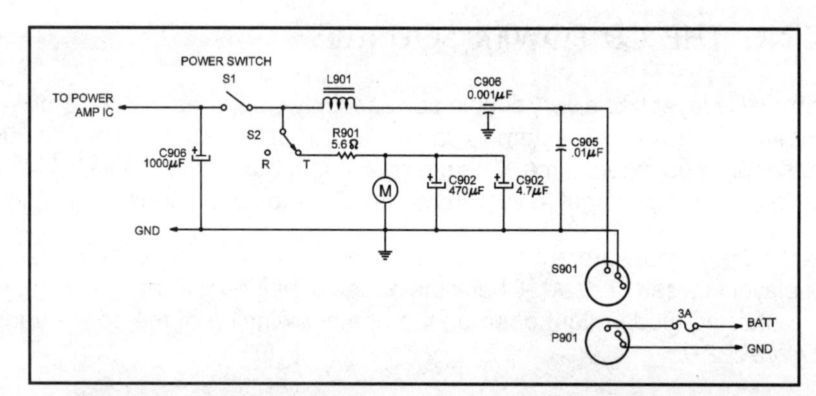

Figure 3-28. Switch S1 provides voltage to the output IC while S2 switches in either the tape player or radio operation in the auto player.

Besides a blown fuse, poor switch contacts and a defective filter capacitor (C906) provide most of the service problems within the power source of the auto radio. Check the 3 amp fuse and poor switch contacts of S1 for a dead auto receiver. Clean up switch contacts of S2 for intermittent radio and tape operation. Suspect R901 for open conditions when the tape motor will not operate. Take a quick voltage test across motor terminals to determine if R901 or S2 is defective. Shunt C906 when a loud hum is heard in the speakers with the volume control turned down.

The audio circuits within the auto CD player are fed from a +5V, 10V or 12 volt source. The D/A converter and line amplifiers might have a positive and negative voltage feeding the audio IC component. Many of the auto CD players contain a DC to DC converter unit connected to the 14 volt battery source. You might find several IC, transistors and zener diode regulator circuits in the compact disc player. Alot of the headphone amp circuits are fed from a +5 volt source.

The DC to DC converter is fed from the 14.4 volt battery source and the transistor or IC oscillators provide a pulsating voltage to transformer (T101). The secondary winding might have a half wave silicon diode or a bridge rectifier circuit *(Figure 3-29).* You might find transistors, zener diodes and IC components as regulators. Several electrolytic capacitors provide filtering action in the dc output circuits.

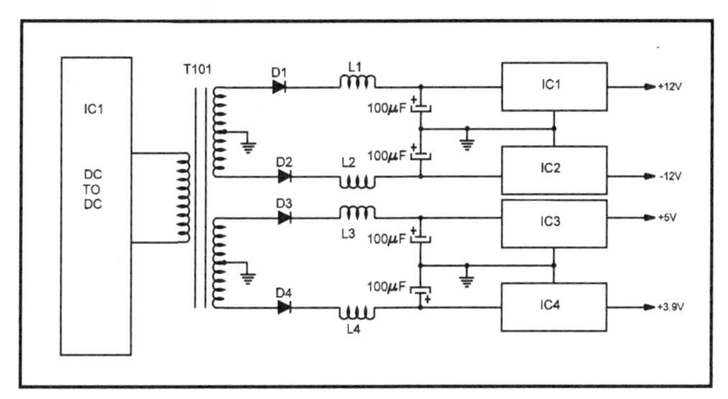

Figure 3-29. The block diagram of an auto CD player-DC-converter power supply.

Check the voltage supply source feeding the audio and headphone amplifiers within the CD player. Suspect a defective transistor or IC regulator with no voltage output. Measure the voltage into the regulator. Check the dc voltage at the silicon diodes in the secondary winding of the power transformer. Accurate voltage measurements and scope waveforms of the IC or transistor operated dc to dc converters solve most auto CD power supply sources.

CHECKING THE MOSFET POWER SOURCES

To provide high output power in a 100 to 200 watt auto stereo amplifier, higher dc voltage must be greater than the battery (14.4) voltage source. In a 170 watt auto stereo amplifier, the positive and negative (34V) source was acquired with an IC oscillator and several dc to dc converter MOSFET transistors *(Figure 3-30).* The 14.4 volts DC battery voltage was fused by a 30 amp fuse and fed into the primary winding of transformer T101. The positive and negative 34 volt source might feed 10 to 14 transistors of only one stereo channel in the high-wattage amplifier.

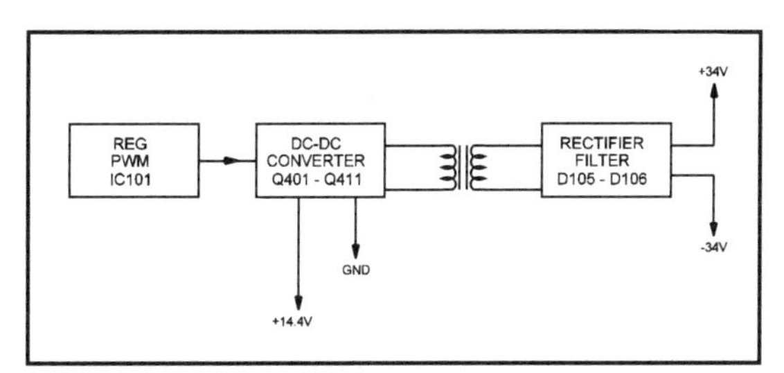

Figure 3-30. A block diagram of the auto DC-power supply circuits.

The PWM signal is fed from U3 to Q401 and Q405, with MOSFET transistors Q409 - Q411 connected in parallel with power transformer T401. Likewise MOSFET transistors Q406 - Q408 are connected in parallel with the other T401 winding. The dc-dc converter transistors, U3 and T401 provide a positive and negative (34) volt source. D405 and D406 rectify the output voltage of the secondary winding of T401 *(Figure 3-31).* A 2200 μF electrolytic capacitor filters the dc voltage source.

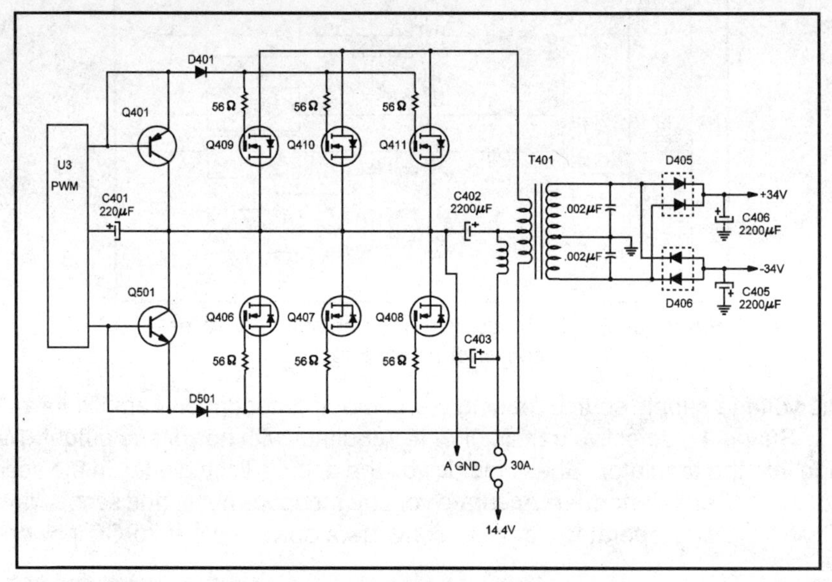

Figure 3-31. The MOSFET DC-DC power source to power high wattage amp circuits.

Suspect a blown fuse when the pilot lamp is out and no sound from the high-wattage amplifier. If the fuse keeps blowing, check for a shorted or leaky MOSFET transistor (Q406-Q411) and a leaky D405 and D406. Do not overlook an overloaded power output transistors in the high-powered amplifier channels. Inspect transformer (T401) and L101 for poor soldered connections. Check and test all MOSFET transistors when a 14.4 volt source is found on the primary winding and no output voltage.

Remove one end of each diode when a leaky measurement is found across the terminals. Inspect the silicon diode terminals D405 or D406 with single 20 amp silicon diodes when the original parts are not available. The six power MOSFET transistors can be replaced with a YTF541 universal type. Shunt electrolytic capacitors C402, C403, C405 and C406 when a low hum is heard in speakers. Replace C402 and C403 with a 35 volt working voltage and C405, C406 at 50 volts.

TROUBLESHOOTING THE TUBE RECTIFIER CIRCUITS

In early audio amplifiers, a fullwave rectifier tube such as a 5R4GY, 5Y3GT, 5U4GB, 5Z4, 6X4, 5AR4, and GZ34 provided a high B+ voltage to the amplifier tubes. The twin diode tube was replaced with the selenium rectifier and then silicon diodes. Since vacuum tubes require higher voltages to operate, the power transformer provides a much higher AC voltage to the rectifier tube *(Figure 3-32)*. In some tube chassis you might find a voltage doubler circuit to boost the dc voltage.

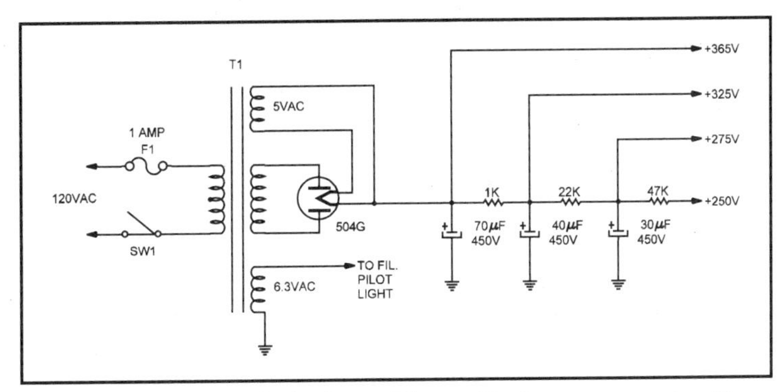

Figure 3-32. A typical tube or PA amplifier's power supply with higher voltage sources.

Besides providing a high AC voltage to the rectifier tube, a separate winding for the tube filaments (6.3V AC) and another winding for the pilot lights and bias circuits might be found in the P.A. amplifier. The power supply circuits were protected with a 1 or 2 amp fuse and in larger amplifiers with a 5 amp fuse. A 5 volt AC winding provided heater voltage for the rectifier (5U4) tube. Most tube rectifiers have a choke or capacitor input filter network; others contained a resistor-capacitor filter circuit. The filter capacitor working voltage averaged 450 volts, since the output voltage to the amp were quite high compared to today's amplifiers.

Most problems found in the tube rectifier circuits were defective or weak tubes, dried-up filter capacitors and blown fuses. A quick voltage test across the filter capacitor will indicate if the power supply is normal. Suspect a leaky rectifier tube or filter capacitor with a blown fuse. A shorted or leaky output tube can blow the line fuse. Check the tube heaters or filament for open conditions with the ohmmeter.

A loud or low hum in the speakers indicates an open or dried-up filter capacitor. Shunt each capacitor until the defective one is located. Choose an electrolytic capacitor with the same working voltage and capacity or one with a higher rating. Replace the entire container of electrolytic capacitors when only one is found to be defective. After replacing or shunting the electrolytic capacitor and a low hum is still heard, suspect a burned choke winding or voltage dropping resistors. Usually the windings are burned, providing poor voltage regulation. A leaky electrolytic capacitor can damage the power transformer.

Remove the rectifier tube and replace the open line fuse. If the transformer runs hot or smokes, suspect shorted windings in the power transformer. Remove all transformer secondary wires from the power supply circuits. Now if the fuse blows or the transformer appears warm or hot in a few minutes, replace it.

LAST BUT NOT LEAST

Shorted or leaky silicon diodes in the power supply can cause the fuse to keep blowing after replacement. A poorly soldered connection at the base or collector terminal of the voltage regulator transistor can cause no sound or relay-click in the speaker of a receiver. The audio might cut out, become intermittent, and the speaker relay clicks off when one of the diodes in the bridge rectifier becomes defective in the receiver. No audio in the receiver might be caused by a shorted or leaky decoupling capacitor lowering the dc voltage. A defective regulator transistor or IC might cause audio distortion after several hours of receiver operation.

The radio receiver volume might go all the way up or down with one or two defective zener diodes. A defective voltage regulator transistor might cause audio distortion after the receiver has warmed up. Weak or no audio, and/or relay click, might result from a defective voltage regulator transistor and diode in the receiver power supply circuits of a deluxe receiver. A protection relay might not turn on the speakers with a leaky zener diode in the B+ line of a receiver.

No sound within the TV chassis might be caused by shorted filter capacitors in the low voltage power supply. Check for burned or open isolation resistors in the TV power supply for no audio symptom and a dead chassis. No or weak audio might result from defective zener diodes, transistor or IC regulator circuits. The no sound symptom might result from open or dried up 1 µF to 10 µF decoupling capacitors in the TV supply sources. Suspect open resistors with shorted diodes in the flyback voltage sources feeding the sound circuits for a no audio symptom. No audio might result from burned resistors and poorly soldered joints on capacitors in the power supply.

Check for dried-up or open electrolytic capacitors in the voltage sources for weak audio in the TV chassis. The weak sound symptom in the TV chassis might be caused by a dried-up 4.7 µF 50 volt electrolytic capacitor. A garbled sound in the TV speaker might result from a defective regulator transistor. Check for an improper voltage source when distorted sound is found in the TV chassis. Low hum in the audio can be caused by defective electrolytic filter capacitors in the decoupling circuits. Garbled audio might be heard in the TV speaker from a defective regulator transistor. Check the large filter capacitors with no audio and a loud hum in the speakers. Suspect a defective zener diode in the low voltage sources with weak audio and poor sound adjustment.

The low hum symptom with no audio might be caused by an open low-ohm resistor and regulator transistor. Check for a defective zener diode by listening for a buzz or hum in the speaker. Intermittent sound with no audio might result with poorly soldered connection of capacitors in a voltage source. Do not overlook a defective scan-derived voltage source when the sound quits with vertical foldover in the TV. Remember, the dc voltage source must equal that found on the schematic and feeding the sound circuits, for normal sound conditions.

Chapter 4

Troubleshooting Preamp And AF Circuits

The sound amplifier stage provides a boost of audio signal at the input of an audio amplifier circuit. Often, the preamp stage is the first electronic amplifier (analog) connected to a tape head in the cassette player, phonograph, or an audio amplifier *(Figure 4-1).* In the early tube audio circuits, the preamp tube was the first audio circuit connected to the microphone in the PA system. The preamp circuit usually consists of a voltage amplifier.

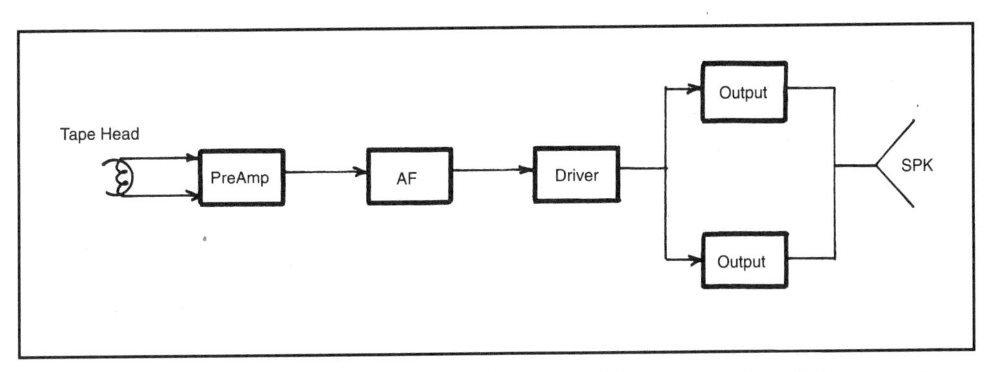

Figure 4-1. A block diagram showing the location of the preamp stage in the cassette player.

You might find two transistors or one IC that provides amplification of the weak cassette tape or microphone audio signal. In the AM-FM-MPX radio-cassette circuits, the radio signal is switched directly into the audio circuits while the cassette sound is picked up off of the moving tape, amplified by the preamp circuits, before being switched into the audio circuits. Just about every consumer electronic product has a transistor or IC preamp audio circuit.

The preamp stage might amplify the weak signal of a phono cartridge, before it is switched into the AF audio circuits. A magnetic phono cartridge signal is very low and a separate audio stage must amplify the weak signal of the moving coils before it is coupled to the preamp or AF audio circuits. The preamp circuits amplify the very weak audio signals to the AF or audio driver circuits of the audio amplifier.

THE SECOND BEST

The audio frequency (AF) amplifier might be the first stage that amplifies the radio, phonograph or cassette preamp circuits. The AF amplifier operates within the 20 Hz to 20 kHz frequency range. This audio frequency amplifier can be the second audio stage within the cassette player. You might find more than one or two AF transistor or IC circuits within the large stereo amplifiers. Often, the AF circuits amplify the audio signal and are coupled to the driver transistor within a push-pull output amp circuit.

The audio signal from the radio or cassette circuits is coupled into AF transistor (Q210) by a 4.7 μF electrolytic C210, R212, R213, and R210 provide forward bias for transistor Q210. The different voltage upon the base and emitter terminals to ground should have a forward bias of 0.6 volts on an NPN transistor. A 0.6 volt DMM measurement from base to emitter is the forward bias voltage *(Figure 4-2)*. When a correct forward bias voltage is found, nine times out of ten, the transistor is normal. An in-circuit forward bias voltage test can quickly determine if the transistor is open, leaky or normal.

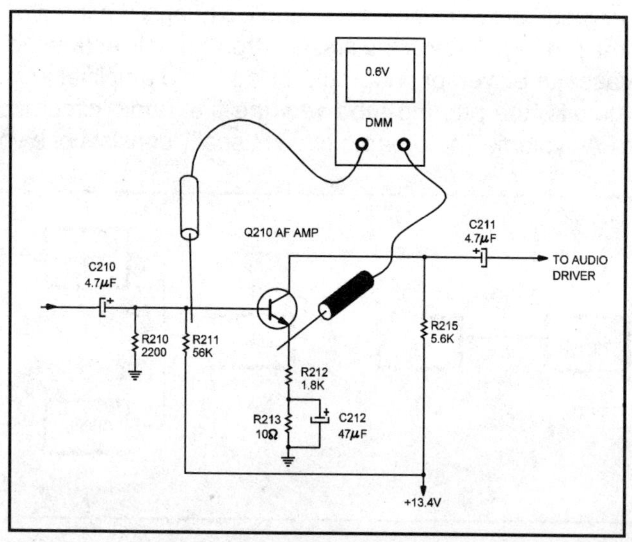

Figure 4-2. A critical forward bias voltage measurement (0.6V) between emitter and base terminal of a silicon transistor.

The audio input signal is applied to the base terminal of Q210 by capacitor C210, amplified and capacity coupled to the audio driver stage with C211 (4.7 μF) capacitor. R215 (5.6k) serves as a collector load resistor. The supply voltage (13.4v) from the low voltage power source is connected to R215 and R211.

DIRECTLY-COUPLED FOR LIFE

Two preamp or AF transistors directly-coupled to one another can provide more audio gain with less electronic components in the audio circuits. A direct-coupled amplifier is where the output of the first transistor is wired directly to the input of the following stage. The collector terminal is wired-direct to the base terminal of the second audio transistor. This type of amplifier circuit has a wide frequency response and can handle either ac or dc signals. There are no capacitors or resistors between the collector of the first audio amp and the second transistor *(Figure 4-3)*.

Figure 4-3. Locate the power output transistors upon the heat sink and the preamp transistor circuits are nearby.

Although *Figure 4-4* is an AF audio circuit, an identical preamp audio circuit with R136 is eliminated from the circuit. Notice that the collector terminal of Q102 is wired directly to the base terminal of Q103. The preamp or AF input transistor supply voltage (1.85v) is very low to prevent pickup noise. A critical voltage measurement from base of Q103 to ground and from emitter of Q103 to ground should equal the forward bias voltage. This voltage difference should be quite close to 0.6 or 0.7 volts for a normal NPN transistor.

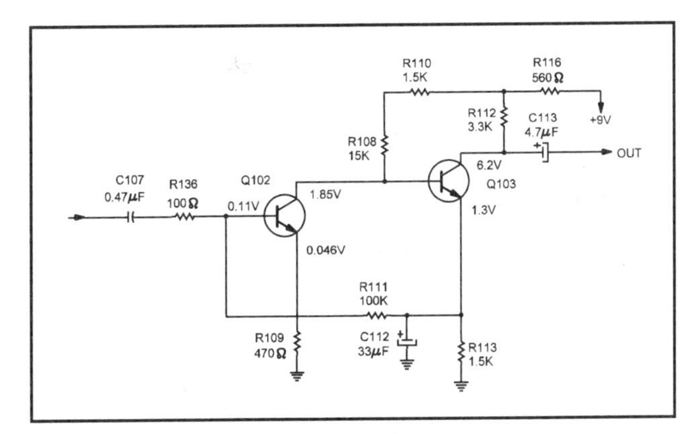

Figure 4-4. The directly-coupled Q102 from collector to bass terminal of Q103.

A preamp directly-coupled amplifier circuit is usually found in the tape, cassette, or phono input circuits. The AF directly-coupled stage is found in the radio, cassette and other amplifier circuits. You might find a volume and tone control circuit after the AF direct-coupled amplifier. Several different AF circuits might follow the directly-coupled circuit within the higher-powered amplifiers.

The defective preamp or AF transistor might become leaky, open or intermittent. If the first or second amp transistor becomes leaky or open in a directly-coupled amp circuit, the voltage will also change on the other transistor. When Q102 becomes leaky, the base and emitter voltage goes to zero volts with only .003 volts on the base terminal of Q104 *(Figure 4-5)*. This voltage should measure 1.7 volts with a normal transistor. You will find the voltages are also lower upon the emitter and collector terminals of Q104.

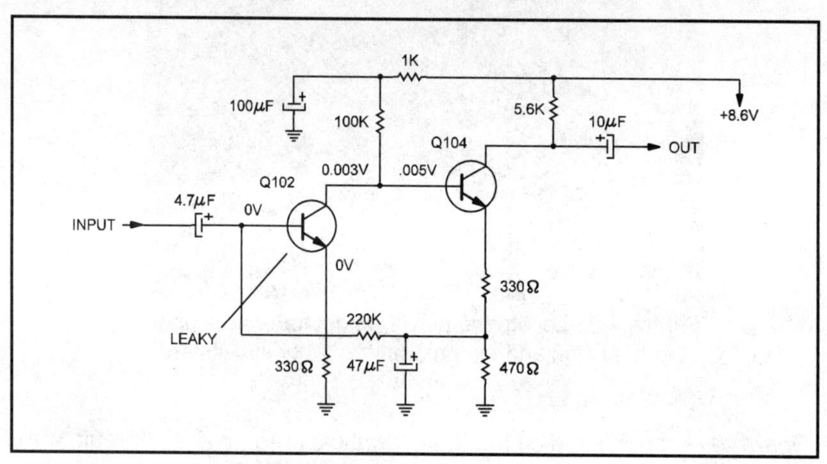

Figure 4-5. Notice the low voltages on all preamp transistor terminals when Q102 becomes leaky.

When the first direct-coupled transistor goes open, the measured voltage on the emitter and base terminals will decrease to about one half of the original normal voltage. Since the transistor opens up, the collector voltage will almost increase in voltage equal to the supply voltage for the directly-coupled transistors; this also increases the dc voltage to the second directly-coupled transistor. Sometimes when the meter probe touches the base or emitter terminal, the intermittent transistor might return to normal operation. At other times, the suspected transistor might test good out of the circuit and open up under load.

DUAL-IC PREAMP CIRCUITS

Instead of two preamp or AF transistors directly coupled, you might find one dual IC performing audio amplification to both stereo circuits. The IC preamp circuit might contain more than one transistor in each stereo channel within the IC component. You will find the IC preamp and AF circuits within the latest auto receiver, cassette players, receivers and amplifiers.

The auto receiver with a cassette player might have a single IC preamp stage with switched input tape heads. The stereo tape heads are switched into the input terminals of IC101 with S5-1 and S5-2 *(Figure 4-6)*. C102 and C202 couple the tape audio to the input terminals 1 and 8 of IC101. The amplifier output audio signal is found on terminal pins 3 and 6. C203 and C204 connect the preamp audio to the volume and tone controls. Pin 4 of IC101 is the supply voltage terminal.

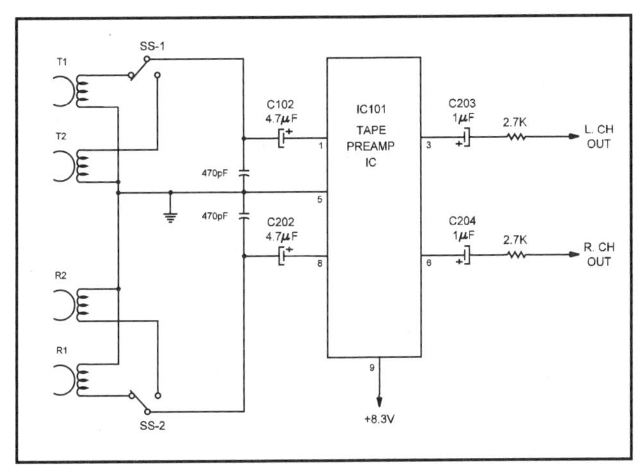

Figure 4-6. A typical preamp stereo IC circuit within the auto stereo receiver.

The defective preamp or AF IC might become leaky, shorted or open. Extreme distortion with weak tape audio is heard with a leaky or shorted IC. An open front-end IC might have no or weak audio reception. When the IC becomes leaky or shorted the supply voltage will decrease in voltage with a slight change in voltage on the other IC terminals. The voltages on all IC terminals might match those on the schematic when the suspected IC opens up. Critical voltage measurements on each IC terminal and an input-output signal tracing method can locate the defective IC component.

MICROPHONE INPUT AUDIO CIRCUITS

The microphone input circuits might be found within the cassette player, phono amplifier, sing-a-long amp, compact disc, boom-box players, and PA systems. The stereo microphone input jack in a high-powered amp may be switched into the AF audio circuits after the preamp stage. Usually the mike jack is connected ahead of the volume control in the same amplifier that amplifies the tape and receiver circuits.

The built-in microphone of the cassette player is switched out of the circuit when the auxiliary or external microphone is used *(Figure 4-7)*. Here, the built-in auxiliary jack (J2) and external microphone jack (J1) are switched into the AF or preamp circuit of the cassette player. SW1-A places either the tape head or built-in microphone into the front-end circuits of a low-powered amplifier.

The built-in microphone is usually a condenser or electret type with a low dc voltage applied to the microphone. The built-in mike signal is coupled with C101, through R101 to the AUX switching terminals, and to the shorting jack terminal of J1 and switched by SW1-A. SW1-A switches in either the tape head or microphone audio signal. Then the audio signal is amplified by two preamp or AF transistor stages before the volume control.

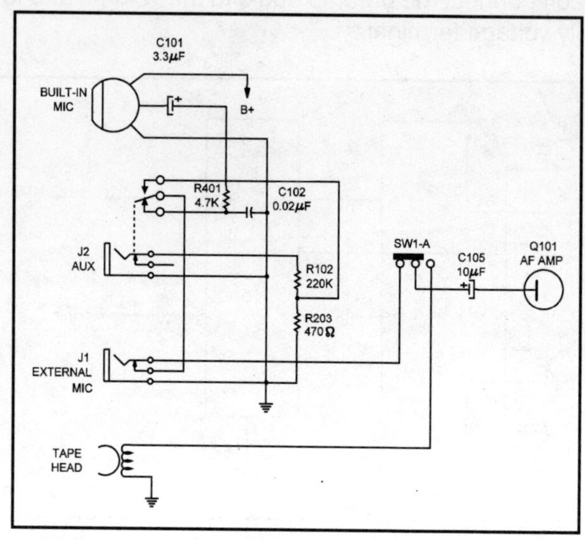

Figure 4-7. The built-in microphone audio passes through the external mic (J1) before being switched to the base of Q101.

In the early monaural cassette player, the built-in microphone, auxiliary J1, and external mic J2 are switched in the same manner to SW1-D and coupled to C103. C103 couples the audio signal to pin 14 of IC101. The entire tape and microphone signal is amplified with a single IC chip *(Figure 4-8)*. Pin 1 is the ground terminal while pin 9 is the voltage supply terminal.

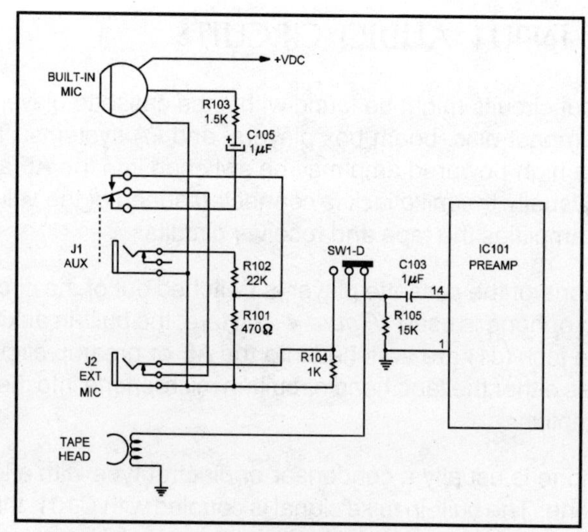

Figure 4-8. IC101 amplifies the tape head and microphone audio signal.

TALK TO ME

The dynamic microphone has a small coil attached to a moving diaphragm that moves freely over a magnet, somewhat like the construction of a speaker. The coil moves in the magnetic field generating the ac output voltage *(Figure 4-9)*. This output voltage is very low. The dynamic microphone is a popular microphone with great frequency response and used in large PA systems, churches and broadcast stations. The dynamic mike has a low impedance and must be matched to the input sound circuits. The typical dynamic mike might have a frequency response of 60 Hz to 15 kHz and a 500 to 600 ohm impedance.

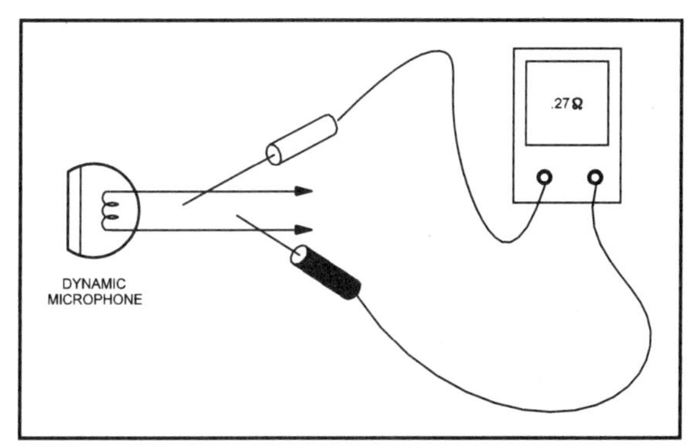

Figure 4-9. Taking a continuity test of the dynamic mike and cord with the ohmmeter.

The crystal mike employs a piezoelectric effect somewhat like the crystal phono cartridge. When sound pressure is applied to the diaphragm, this places mechanical stress upon the crystal that generates a low voltage. A crystal microphone has a high impedance output and can operate directly into the audio circuit. The crystal element is made from Rochelle salt and can be damaged when dropped on a hard floor. The output level is fairly high with a fair frequency response.

A condenser microphone has two small plates that are separated. The front plate is a flexible diaphragm while the other is rigid. The movement between the two plates changes the capacity between them. The condenser mike has an excellent frequency response with low distortion. A dc voltage must be applied to the microphone elements.

The electret microphone is constructed like the condenser mike with an external voltage applied. The electret microphone might operate from 2 to 10v DC *(Figure 4-10)*. When sound waves strike the electret mike, an output voltage is generated. The electret microphone must have an outside dc voltage applied before it will operate. The electret condenser microphone is unidirectional with a frequency response from 50 Hz to 14 kHz. The condenser and electret mikes are found in cassette players, sing-along amps, and camcorders.

The different microphones can be repaired by replacing the internal elements. Simply replace the low-cost mike, instead of repairing it. The high-priced dynamic microphone can be sent to the factory for element replacement. Most problems related to the microphones are damaged cords, broken or worn plugs, and damaged internal elements when dropped on a hard surface. The ceramic and electret microphone can be damaged by children poking sharp objects into the built-in microphones. Replace the defective built-in mike with original part number.

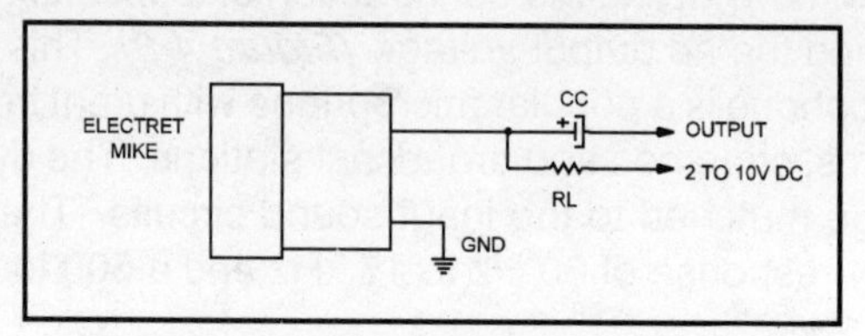

Figure 4-10. A typical electret microphone circuit with external voltage applied.

RECORD CHANGER MAGNETIC PREAMP CIRCUITS

The typical early phono players and changers contained a crystal cartridge with sapphire and diamond stylus or needles. A crystal cartridge might develop a higher output voltage than the magnetic cartridge. Very little voltage is developed with the magnetic cartridge which must have an additional preamp circuit to boost the audio signal from the record. A variable-reluctance phono pickup is called a magnetic cartridge.

The magnetic cartridge has a high frequency response compared to the crystal cartridge. The stylus is attached to a piece of magnetic material that moves between the two coils, producing ac voltage. Since the variable-reluctance cartridge output voltage is so low, a preamp stage is needed to boost the signal to a typical AF circuit *(Figure 4-11)*.

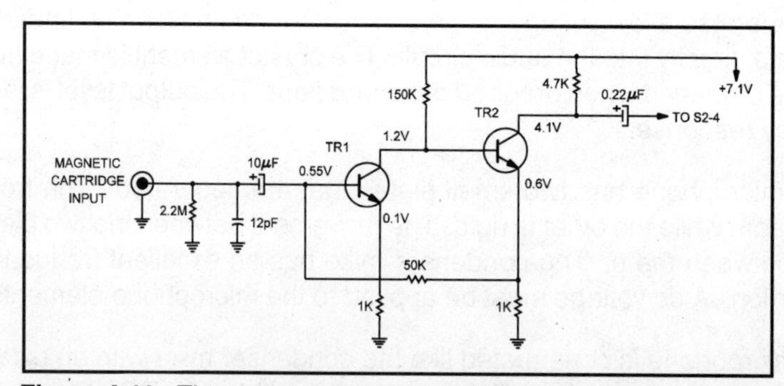

Figure 4-11. The magnetic phono cartridge signal is amplified by two preamp transistors.

Check the resistance of each coil of the stereo phono cartridge to determine if the cartridge or preamp stages are defective. The resistance should be under 200 ohms and both coils should measure the same resistance or within a few ohms. Notice the dc voltage on TR1 is very low to prevent additional pickup noise. Perform a transistor and voltage in-circuit test to locate the defective component.

DIRTY TALKING FUNCTION SWITCHES

You will find some type of function switch in just about every electronic product that provides more than one function. The deluxe receiver might have a rotary function switch to provide AM-FM radio, phono, cassette, and recording features. A boom-box AM/FM/MPX receiver and cassette player might switch to the different functions with a sliding type switch. Likewise the auto receiver might have a function switch that removes the AM-FM radio reception when a cassette is placed into the compartment. The small portable radio might have an AM-FM switch that switches in the various reception bands.

Since most switching contacts contain a silver-type switching connection, the contacts tarnish and become dirty. A dirty AM and FM switch might result in no FM with normal AM reception or vice versa. The dirty or poor contacts within the sliding bar switch of many contacts might produce a loud howling noise and erratic operation. The cassette player may not record with a dirty record/play switch. A bad switch might cause a dead-no sound symptom.

You might locate a broken plastic shaft that engages the R/P switch when the cassette player plays but no record. Garbled and distorted audio might result from a dirty radio-cassette switch. A record/play switch which is not fully engaged might cause the cassette-recorder to erraticly record. A loud rushing or hissing noise on one stereo channel might be caused by a defective function switch. A no left stereo channel might be caused by a defective channel selector switch.

Clean up the function switch contacts with a silicone based cleaning fluid. Place the plastic tube down inside the switching area *(Figure 4-12)*. Work the function switch back and forth to help clean up the contacts. Check for poorly soldered contacts if the switch is still erratic. Resolder all switching contacts on the PCB. Replace the defective function switch with worn or broken contacts.

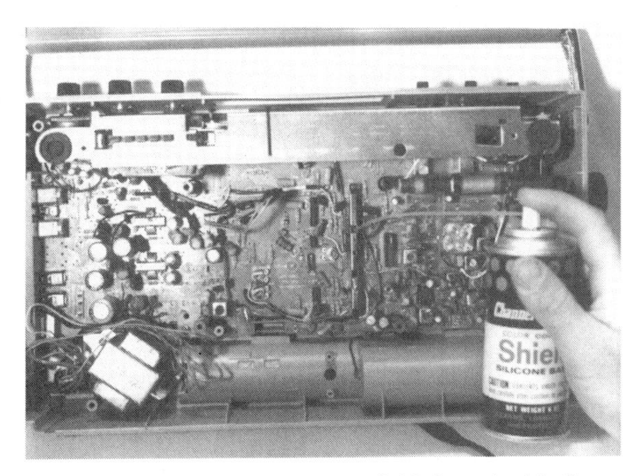

Figure 4-12. Spray cleaning fluid down inside the boom-box function switch to clean up the dirty contracts.

LOUD AND CLEAR PROBLEMS

After several years of usage the volume control might become worn and cause noise in the speaker, when rotated. Sometimes the audio will cut in and out with a defective volume control. The right channel of the amp might be dead with a poorly soldered connection on the volume control. The left channel might be dead with an open volume control. One channel might be intermittent, and again, have no volume change with a defective volume control. The radio section might be dead with an internal short within the volume control. No volume change will be noted with a broken ground connection or open volume control. Replace the volume control when the volume cuts up and down.

Clean up the noisy volume control by spraying cleaning fluid down into the open slot where the terminals come out of the control. Rotate the control up and down to clean up the carbon control. Resolder all terminal connections with an intermittent or erratic volume control.

Replace the volume control with worn contacts or open condition. While special type volume controls must be replaced with the exact part number, the most common controls can be replaced with universal controls. Choose the correct resistance and correct taper. Replace the audio volume control with a known audio taper.

BALANCING ACT

When one channel becomes weaker than the other channel, the balance control must be rotated towards the normal channel to provide equal volume out of the stereo speakers. The left channel might be louder than the right when the balance control is at zero, indicating a weak right channel. The stereo channels will not balance up when one audio channel is a lot weaker than the other.

The equal adjustment of the audio signal in a stereo amplifier might have a separate balance control or individual volume control to balance the audio in the speakers. The balance control is usually connected ahead of the volume and tone controls. The stereo signal from the left and right AF stages are applied to the outside terminals while the center (wiper) terminal is grounded *(Figure 4-13)*.

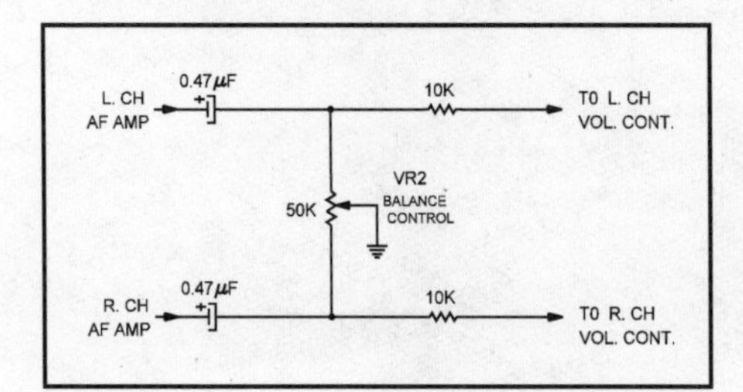

Figure 4-13. The balance control is connected between the left and right stereo channels.

When the amplifier stereo channels will not balance, suspect a weak audio channel. Connect a sine wave signal to the stereo auxiliary, phone, or function switch input. Clip both input terminals to both L and R channels. Check the waveforms at the high end of the volume control with the scope *(Figure 4-14)*. Both audio channels should be equal on the dual-gated oscilloscopes. Another method is to check the audio sine wave at each end of the balance control with the balance control at zero or balanced point on the dial. Although, the balance control can balance a weak channel, the very weak channel should be signal traced and repaired.

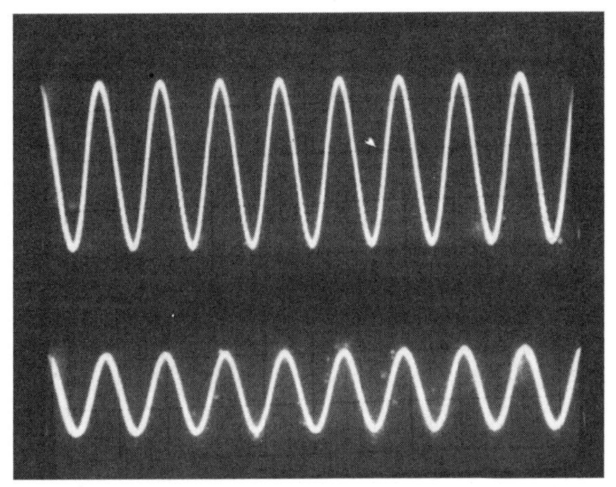

Figure 4-14. The bottom waveform indicates the right channel is weak compared to the left stereo channel.

A defective balance control might not balance the stereo channels with an open or defective control. Check both terminals of the control for broken wire connections. Inspect all wiring connections. Make sure the center terminal is grounded. Check for correct resistance of the control. Replace the control, if open or erratic, with a linear audio taper.

TREBLE AND BASS CIRCUITS

The treble and bass controls might be located in the same circuits, after the volume control. Some larger audio amplifiers might have a tone amp circuit with the volume control in the base circuits and the bass and treble controls in the collector output circuits. Notice the balance control is located across the volume control center terminals. The controlled-volume is amplified by Q102 and Q202, while the output signal is taken from the bass control *(Figure 4-15)*.

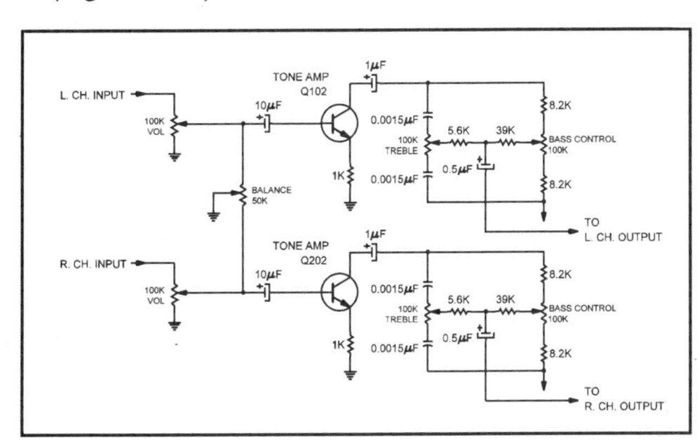

Figure 4-15. Q102 and Q202 are separate tone amps with the bass and treble controls in the collector circuits.

The defective bass or treble controls might become noisy or erratic. Clean up each ganged-control with cleaning fluid. If the control is worn or broken replace with the original part number.

DEFECTIVE TAPE HEAD CIRCUITS

The dirty tape head in the cassette player might cause weak or distorted music. A heavily-packed tape head with oxide dust might be dead or produce a weak and distorted signal. Broken wires on the tape head might result in a loud rushing or howling noise in the speakers *(Figure 4-16).* A loud rushing noise with no music might result from a broken ground wire in the tape head. A weak right channel and normal left channel might be caused by a packed-oxide tape head. A worn tape head can cause a missing of the high notes of the recorded music.

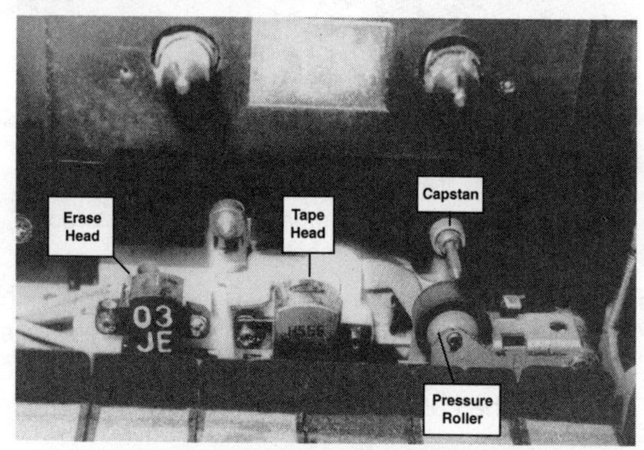

Figure 4-16. Check for broken tape head wires with a loud rushing sound.

Replace the defective tape head when it will not record in the left or right channel. A weak and rushing noise recording might result from a defective tape head. Sometimes the dead recording will come to life when pressure is applied from an insulated tool on the tape head. Moving the tape head might remove the distorted recording on the right channel. A jumbled recording might be caused by a broken ground wire broken on the erase head. A packed oxide on the tape head might result in a weak record on the right channel and no record on the left channel, and on it goes.

Most weak and distorted cassette music can be cured by simply cleaning up the tape head. Clean up the front of the tape head with alcohol and cleaning stick. Remove stubborn oxide with an insulated plastic tool. An old toothbrush dipped in alcohol can clean up the packed-oxide tape head.

Check the tape head for broken wires or poorly soldered connections. The suspected tape head can be checked by clipping the external amp across the tape head terminals. Measure the continuity of the tape head. The typical cassette heads have a resistance from 200 to 830 ohms. A defective stereo channel might have a different resistance measurement than the other tape head winding *(Figure 4-17)*. Both stereo tape head measurements should be within a few ohms of one another. For instance, if one head measured 235 ohms and the other 230 ohms, the stereo head is normal. If one side measured 315 ohms and the other 380 ohms, replace the defective tape head.

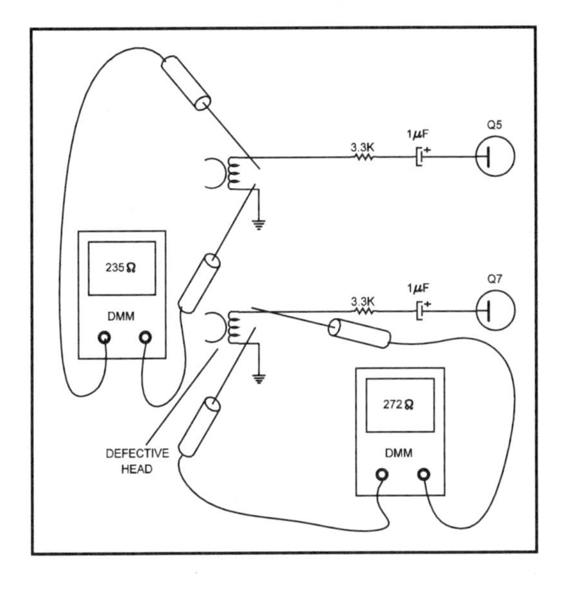

Figure 4-17. The normal stereo tape heads resistance should be within a few ohms of each other.

A SNAKE IN THE FRONT-END CIRCUITS

Hum and noise in the front-end circuits, including the preamp or AF, might be caused by a leaky transistor. A defective or dried-up decoupling capacitor within the voltage source of the preamp or AF circuits can cause no sound with only a hum noise. A poor ground in the front-end circuit can cause a low hum symptom. Motorboating sound in the speaker can also be caused by an open decoupling electrolytic capacitor, defective transistor or IC. The noisy sound condition that goes dead might result from an intermittent AF transistor.

A loud rushing or hissing noise might be caused by broken tape wires on the tape head. An open tape head might cause a rushing sound in the speakers. A no volume with a rushing sound results in a defective tape head. The loud howling noise can be caused by a broken tape head connection. Suspect a noisy preamp transistor when the noise disappears by turning the volume control down. Check each electrolytic coupling capacitor in the front-end circuits for a noisy channel. Double check ceramic and electrolytic capacitors in the emitter circuit for a frying noise in the sound. The intermittent noisy channel can be caused by poor board or soldered connections on the volume control.

Signal trace the noisy audio to the preamp or AF circuits with the external audio amplifier. After locating the noisy stage, determine if the transistor or IC is making the noisy condition. Apply several coats of coolant to the semiconductor devices. Shunt electrolytic and bypass capacitors with a good known value. Replace the suspected transistor or IC preamp when all other tests fail.

ALL IN ONE

The stereo preamp or AF circuits might be included in one IC component. Within the auto-cassette receiver the preamp tape head circuits are found in IC201. When a cassette is inserted into the auto radio, the tape switch applies a dc voltage to the preamp IC circuit. The left tape head is coupled to pin 1 of IC201 with C203 (1 μF). Likewise, the right tape head audio is coupled to pin 8 with C204 (1 μF).

C209 and C210 (4.7 μF) couple the output signal to a fixed diode to the volume and tone controls *(Figure 4-18)*. D203 and D204 prevent the radio AM and FM signal from entering the preamp circuits. Diodes D206 and D205 apply the radio input signal, while D201 and D202 connect the FM stereo audio signal to the audio amplifier circuits. Instead of a large function switch, the fixed diodes provide AM, FM and tape music to be switched to the audio circuits without any moving parts.

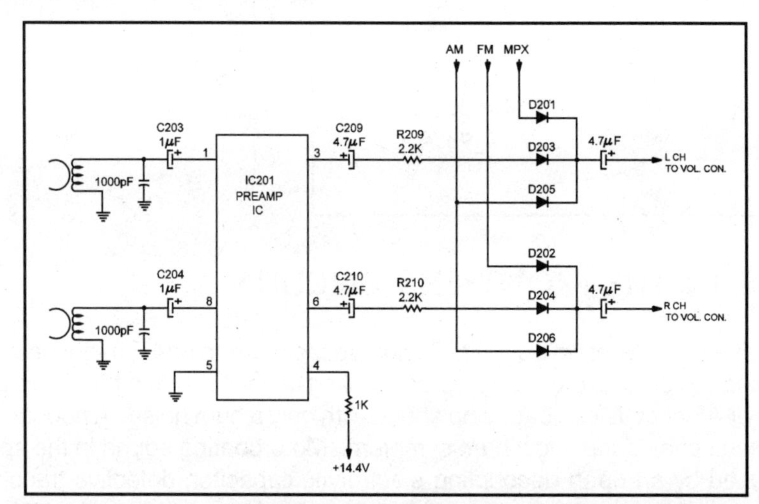

Figure 4-18. IC101 amplifies the stereo tape heads and is switched in the output with fixed diodes in the auto receiver.

WEAK PREAMP CIRCUITS

Check and shunt electrolytic capacitors within the preamp circuits for weak audio symptoms. The leaky AF transistor can cause weak reception. A weak audio stage can be caused by a leaky preamp or AF transistor or IC. An open preamp transistor might result in no audio signal in that channel. A very weak right channel with a normal left channel can be caused by a defective preamp IC. The weak and intermittent sound symptom might be caused by a defective preamp transistor. Shunt electrolytic capacitors in the emitter circuits for weak sound in AF circuits. Do not overlook a dirty tape head in the cassette player for a weak audio channel. A weak preamp circuit might result from an improper voltage supply source.

Insert a 1 kHz test cassette to test the preamp and AF circuits of the radio-cassette player. Signal trace the preamp circuits with the scope or external audio amp. Clip the ground wire of the audio amp to common chassis ground. Start at the tape head and work back to the volume control to determine what stage produces the weak audio signal. Proceed through the circuit by placing the amp probe on each side of the coupling capacitors.

You can quickly signal trace the preamp circuits by taking critical waveforms through the preamp circuits. Start at the volume control and work towards the tape heads in the preamp or AF audio circuits. The test cassette should show a dual waveform on the scope if the preamp stages are normal *(Figure 4-19)*. Check the weak waveform against the normal channel at the volume control, coupling capacitors, collector and base terminals to locate the weak component.

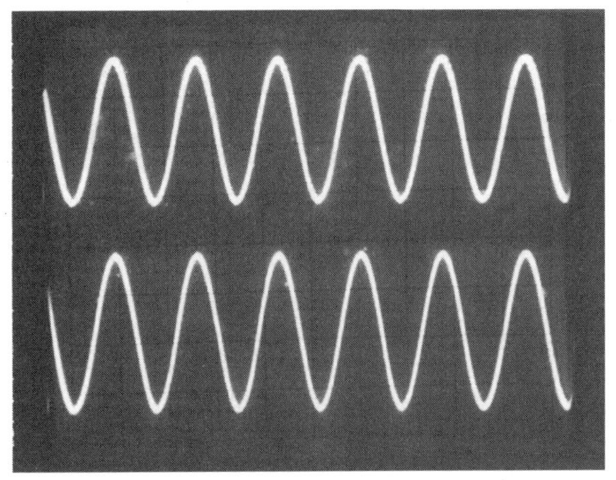

Figure 4-19. Both stereo preamp waveforms are fairly equal in amplitude indicating a normal preamp circuit.

After locating the weak stage, check each transistor with in-circuit transistor tests. Measure the bias resistors and check the resistance with the schematic or color code of the resistor. Sometimes a resistor might overheat and change the color bands on the resistor. Remove one end of the resistor for a correct measurement. An increase in resistance of a bias or base resistor can cause a weak audio signal with some distortion.

DEAD-NO AUDIO IN PREAMP OR AF CIRCUITS

The dead chassis with no audio is much easier to service than a weak or intermittent one. No audio in the preamp or AF circuits can be signal-traced with an external amplifier or scope. A quick method to check the front-end circuits is with a clicking test. Turn the volume control wide open and listen for a hum sound. Touch the tape head ungrounded wire, tape head coupling capacitor, or base of the first preamp transistor with a small screwdriver blade. A clicking noise or a loud hum should be heard in the speaker. No click, no hum, no operation.

A dead preamp circuit can be located with an injected audio signal at the input auxiliary jack, tape head or phono input jack. Clip a 1 kHz sine or square wave from the function generator to the input and common ground; tie both stereo input channels together. Scope the input and output of the preamp or AF circuits to determine where the waveform stops *(Figure 4-20).*

Figure 4-20. When the sine or square wave form stops, you have located the defective stage.

Likewise, insert a 1 kHz or 3 kHz test tape in the cassette player and signal trace with the external audio amp. Go from tape head to electrolytic coupling capacitor, base of first preamp transistor or input IC terminal, and trace the missing signal through the preamp circuits. If no signal is found at the volume control, the dead audio component must be from tape head through the preamp and AF audio circuits.

The no audio symptom might be caused by an open preamp or AF transistor. A leaky preamp transistor can cause no audio sound. An open or poor electrolytic coupling capacitor can cause a dead audio problem. A dirty or worn input switching terminal can cause a no sound symptom. A burned or open emitter resistor can cause a no audio condition. The broken ground connection of the tape head in the cassette player might produce a dead chassis.

The dead front-end circuits can be caused by an open preamp IC with a normal voltage source. Improper or no voltage applied to the preamp or AF circuits can produce a dead chassis. The dead preamp stage can be caused by an open voltage regulator transistor in the low voltage power supply source. When both stereo channels are dead, suspect a leaky preamp IC or shorted decoupling capacitor in the voltage source feeding the preamp circuits. A leaky AF transistor might result in no audio, with hum in the speaker. A dead receiver with low hum noise can result from a leaky decoupling capacitor in the voltage source of the preamp circuits.

DISTORTED PREAMP CIRCUITS

Although audio distortion is found in the output circuits, leaky components in the preamp can cause distortion. A leaky preamp transistor or electrolytic coupling capacitor can cause distorted audio. The weak and distorted audio symptom might result from a leaky preamp IC. Suspect a leaky AF transistor for a weak and distorted audio problem. Check for a bad or dirty radio-cassette switch with distorted and garbled music. Slight distortion in the left channel might be caused with an open AF transistor and when tested in-circuit, the transistor was normal.

A change in resistance or burned bias resistors of the preamp or AF circuits can cause a distorted symptom. The right channel in a tape player can be distorted with a defective tape head. A dirty tape head can cause distortion in the recorded sound. Replace the preamp or AF transistor for a microphonic-distorted noise. Apply several coats of coolant on each transistor to locate the one that produces the microphonic sound. Most distortion problems in the audio amplifier are found in the audio output circuits.

REPAIRING THE PREAMP RECORDING CIRCUITS

The preamp recording circuits within the preamp stages are the same in the play mode of the portable, boom-box or auto receiver cassette player. These preamp circuits might be two directly coupled transistors or one IC component. You will find two identical preamp circuits within the stereo channels. In early portable cassette players all of the audio transistors operated in the play and record circuits, while in other players the preamp circuits provided recording features.

The first preamp IC might include both play and recording operations. You might find several different switches engaged in the input and output circuits of the R/P functions. When SW1-1 is switched to record mode, the built-in mike and external microphone jack is switched into the input circuit of IC1101 *(Figure 4-21)*. The microphone sound is amplified by the IC and comes out of pin 13 and coupled to another switch that returns the recorder audio to the left channel R/P tape head. The recorded message is applied by the tape head to the recording cassette. The right channel recording circuit is identical to the left channel of IC1101.

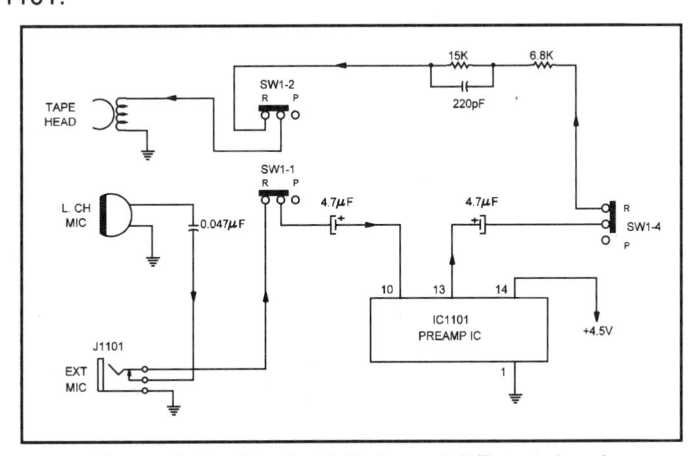

Figure 4-21. A typical left channel R/P tape head.

No recording in the left channel might be caused by a defective tape head or packed tape gap. When a cassette player operates in play mode but not in record, this indicates a broken plastic shaft engaging the R/P switch assembly. Intermittent recording can result from a poor tape head connection or dirty switching contacts. Check for a defective electrolytic coupling capacitor where it ties into the PC board with intermittent recording.

A recording with only a rushing sound might indicate a defective tape head. The record intermittent and warbling sound can be caused by a dirty record switch assembly. A defective function switch might produce intermittent recording and play modes. When one recorded channel is weak and the other will not record, check for a packed oxide-tape head. The loud noise in one channel, when recorded, can be repaired with a cleanup of the record/play switch.

The jumbled recording can be caused with a defective erase head circuit or a ground wire off of the erase head. A noisy recording can be caused by a defective built-in microphone. Poor erasing of the erase circuit can cause a messed up recording. Scope the erase head signal from the erase head bias oscillator and check for poor shielding of the erase head circuits. No erasing of previous recordings might be caused by a poor ground connection in the erase bias circuits.

Make sure the playback functions are normal before attempting to service the recording circuits. Clean up the tape head, erase head and R/P switches. Scope the waveform from the recording oscillator at the left and right tape heads. Check for a dc voltage switched into the erase head on low-priced cassette players. If no voltage is found, check for a dirty recording switch or open voltage dropping resistor. Repair the recording bias-oscillator stage when no scope waveform is located at either recording tape head. When the playback mode is normal, the preamp and AF R/P circuits are good. Suspect defective switching and erase head problems with normal playback circuits.

SERVICING TWO HEADS IN ONE

The dual-tape cassette recorder usually has two different tape head compartments where one cassette is used only for playback and the other for record and playback modes. The record/playback tape heads can record from the built-in microphones or from the other set of playback tape head sections *(Figure 4-22)*. The record/play compartment is to the left and play only to the right side of stereo cassette player.

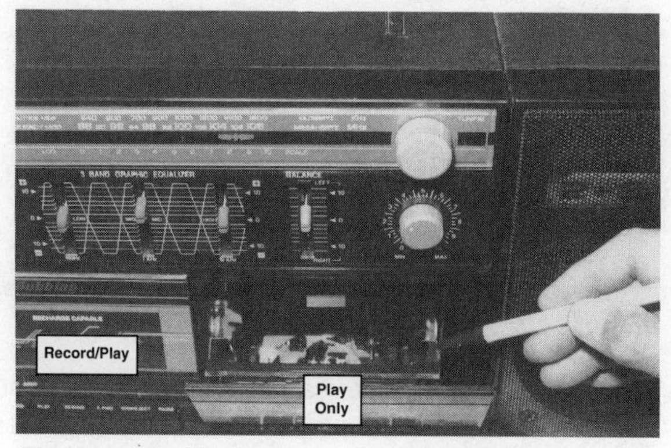

Figure 4-22. The cassette compartment on the right plays only with no recording features.

You will find one erase head in the record/play cassette and one play head in the play only mode cassette player. One tape motor might rotate both cassette departments with a large motor belt. Some portable dual-boom-box players might have a separate motor to rotate the tape in each tape deck. The stereo play tape heads are connected directly into the preamp IC circuits while the stereo record/play heads are switched in play/record modes.

A block diagram of the dual-deck cassette player is shown in *Figure 4-23*. Tape head-1 contains playback stereo tape heads and tape head-2 has both R/P tape heads. The left and right built-in microphones are coupled to IC102 amplifier and to a buffer amp IC before the stereo signal is applied to the Dolby B/C NR IC111. The recording output signal from IC111 is applied to the record amp Q201. The recording signal is amplified by Q201 and switched to the R/P heads of tape(2).

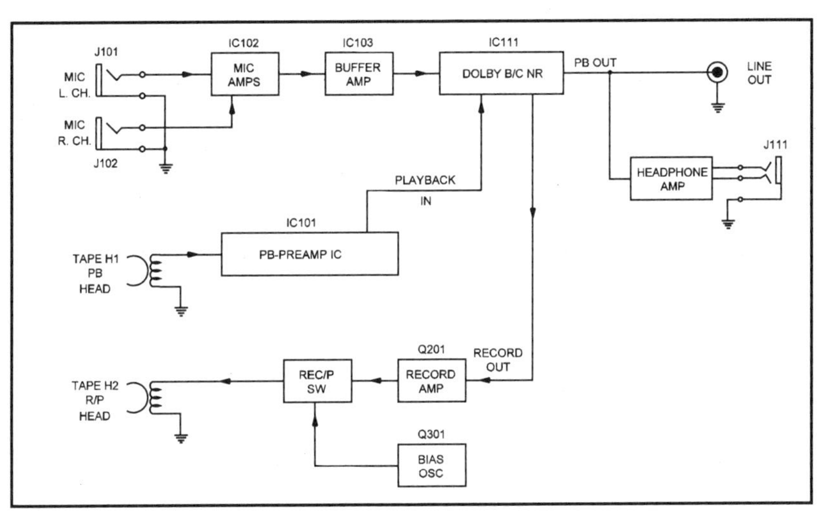

Figure 4-23. A block diagram of a two-deck cassette player left channel record/play circuits.

Notice the stereo play tape heads are connected directly to the microphone amps while the R/P heads (2) are switched in and out of the input circuits. The different tape deck circuits are switched into the amplifier circuit with an R/P switch, supplying a dc voltage source from the regulated power supply.

Service the dual-cassette player like any cassette player. If the playback only tape deck (1) is normal and no R/P in deck-2, you know the playback circuits of preamp IC101 and IC111 are normal. Clean up the R/P switching contacts and tape heads (2). When tape head (2) does not play back or record, suspect the tape head (2) R/P switch assembly and record amp. Inject a 1 kHz or 3 kHz signal at the input of Q201 (record amp) and scope the record amp R/P switches, and tape head. Check for a bias oscillator signal at the record tape heads (2) with the oscilloscope.

Suspect one of the built-in microphones when one channel is dead in the record mode. Switch microphones to determine if the mikes are normal. To check the microphone amps (IC102), inject a 1 kHz audio signal at one of the microphone jacks and check the output of

IC102, IC103, and IC111 with the external amplifier. The entire record and playback circuits can be checked by injecting an audio signal and using the scope or external amp as an indicator. When the audio signal stops, take critical voltage and resistance measurements.

TROUBLESHOOTING THE AUTO AF AND PREAMP CIRCUITS

The early auto radio AM and FM stereo circuits are switched into the transistor preamp or AF audio circuits. When the cassette player is included within the auto receiver, the R/P stereo heads might be switched into the preamp transistor or IC circuits. The early IC preamp circuits might have a single IC for each stereo circuit or one dual IC for both left and right tape head circuits *(Figure 4-24)*.

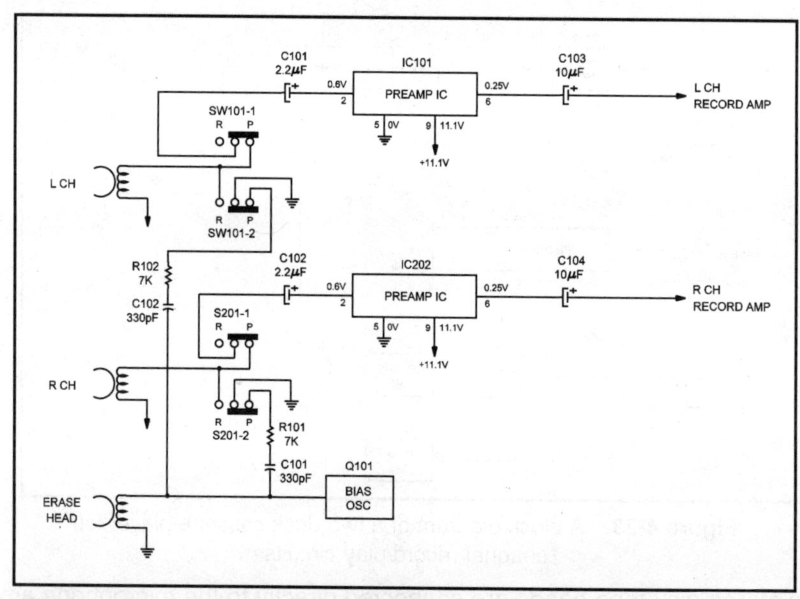

Figure 4-24. In the early auto receivers the cassette player has separate preamp IC's.

After tape head and switching cleanup, take critical voltage measurements on each terminal of IC101 and IC202. Determine if the left or right channel is defective. Suspect a leaky IC if the voltage is low at pin 9 of either the left or right IC. Remove pin 9 from the PCB with solder wick and iron. If the supply voltage returns to normal, replace the leaky preamp IC. Suspect a voltage regulator transistor when the voltage is still low at pin 9.

Insert a musical cassette and signal trace the playback signal from tape head to the input terminal with the external audio amp. For instance, if the right channel is weak and distorted, start at the tape head, check both sides of C102, and then on pin 2 of IC202. If the playback (PB) signal is normal check the audio at output pin 6. Then proceed to both sides of C104 (10 μF) coupling capacitor. Most problems within the auto cassette preamp circuits are dirty tape heads and R/P switching contacts. Intermittent or no music can be caused by loose or torn wires on the tape heads.

TUBE AF CIRCUITS

In the early audio circuits you might find a triode as the preamp or AF amplifier tube. The preamp stages in a musical instrument PA system can be a dual-triode vacuum tube. The various triode tubes used were the 12AX7, 12AY7, 12AT7, and 7025. The volume control might be found between the two different triode elements. The bass and treble controls might be ahead of the volume control *(Figure 4-25)*.

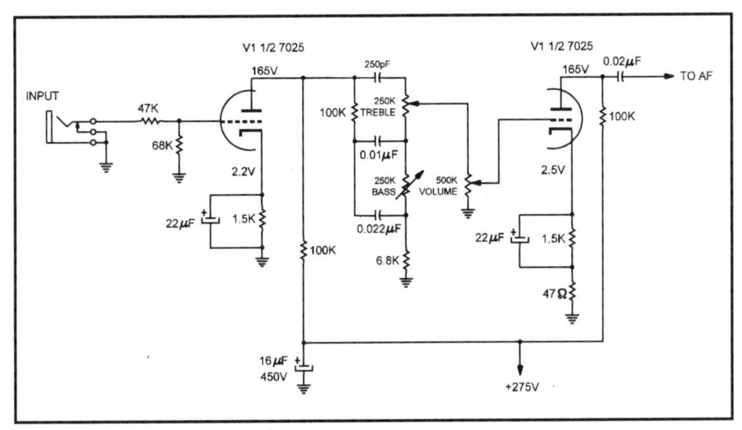

Figure 4-25. The early tube preamp circuits might have a dual-purpose tube with the different controls between triode sections.

Check the preamp tube in a tube tester or sub a good tube in the preamp socket with a dead, weak or microphonic sound. Take critical voltage measurements of the tube elements. Notice that the voltages found on the plate socket terminals might range from 150 to 350 volts. Make sure the voltmeter is able to measure these high dc voltages. Clip the black meter lead to chassis ground and the red probe to take voltage measurements.

A leaky or gassy tube, or an increase in grid or plate load resistors, might cause a weak and distorted sound in the speaker. Check for a burned cathode resistor with a leaky or shorted preamp tube. The preamp tube might become weak, the cathode bypass electrolytic might become open or dried-up, and cause a weak sound symptom. Excessive hum can be caused by a defective decoupling or filter capacitor. The leaky coupling capacitor might cause a weak-distorted sound. A low pickup hum might be caused by an increase in resistance of a grid resistor or poor grounds. Spray cleaning fluid down into the different controls when a scratching or intermittent volume is noted.

Chapter 5

Repairing Power Output Circuits

In the early transistor radios the output circuits consisted of a driver transistor coupled by an interstage transformer to the output transistors in push-pull operation *(Figure 5-1)*. The early car radio might have only a driver transistor and one higher-powered output transistor. The early portable tape player contained a PNP driver transistor and push-pull PNP output transistors, while the early TV audio stages consisted of two transistors or an IC circuit driving two output transistors.

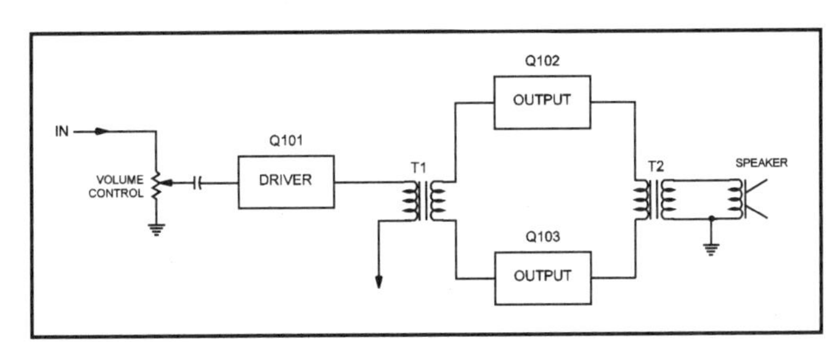

Figure 5-1. A block diagram of the early transistor radio and cassette player PNP output circuits.

The power amplifier delivers audio to a load, such as the speaker. The power amplifier might be a tube, transistor or IC component. Today, the power amplifier delivers audio power to drive several speakers. In stereo circuits the power amplifier is found in both output stages of the left and right channels. The vacuum tube power amplifier is still found in a tube amplifier chassis. A high-powered amplifier might contain many directly-coupled transistors to provide up to 1000 watts of power or more.

THE BLOCK DIAGRAMS

Although the output circuits of the table radio are easily serviced without a schematic, the high-powered (100 watts & up) car amplifier block diagram might come in handy. A schematic diagram is a must when servicing high-powered receivers or amplifiers. You can quickly locate the defective circuit on the block diagram of the various IC's and transistors found in the high-powered amplifier. Now locate the suspected stage from the block diagram to the components found on the chassis.

With the block diagram, you can see how the preamp circuits tie into the audio circuits, where the tone controls and volume contacts are located, and what section of the power supply connects into the various circuits. You can see how the left and right stereo channels are tied together with dc to dc converter stages *(Figure 5-2)*.

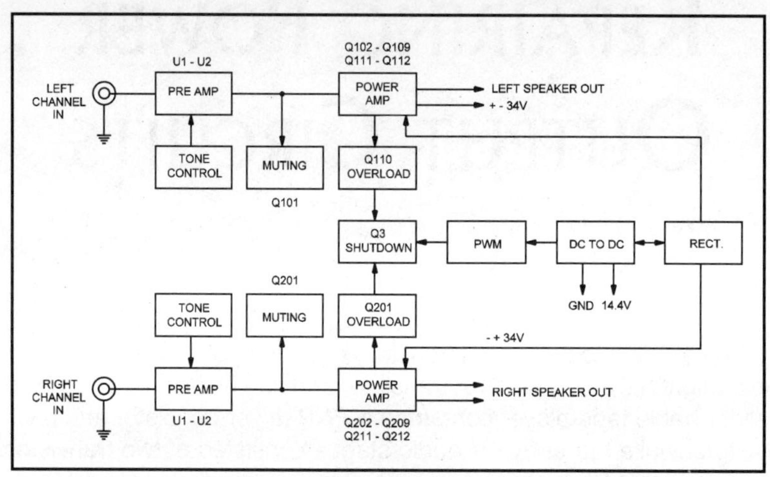

Figure 5-2. The block diagram of an auto receiver 170 watt high-powered amplifier.

In the high-powered auto amplifier the preamp stages might be two IC's (U1 and U2). Notice the tone control is tied into the preamp circuits. A muting transistor Q101 is found ahead of the output power amplifiers. Q104, Q105, Q106 and Q107 are directly-coupled driver transistors. The power output transistors consist of Q108 through Q112. There are eleven transistors used in each stereo channel to produce a 170 watt stereo amplifier.

Q110 is the overload transistor in the left channel while Q201 prevents overload in the right channel. Q3 provides shutdown protection. A 14.4 volt input from the car battery provides the dc to dc converter to step up the +34V and a -34 volts to the power output transistors.

THE TRAIL DRIVER

The driver transistor provides audio power to the output circuits. A driver transformer couples the driver transistor to the driver output stage. Sometimes the driver transformer is called an interstage transformer. The driver circuit in the early transistor radio or portable cassette player consisted of a driver transistor, transformer-coupled to a push-pull PNP transistor output circuit *(Figure 5-3)*.

Notice the PNP driver transistor has a higher negative voltage on collector and a -0.9 volts on the base terminal. Although the emitter has a -0.8 volts and the base 0.9 volts, the forward bias voltage is only -0.1 volt difference. The forward bias upon a PNP transistor is around 0.3 volts. The collector terminal of driver transistor (TR3) is a -4.4 volts supplied through the primary winding of T1. The 200 ohm voltage dropping resistor and decoupling capacitor (470 µF) provide a decoupling filter network from a 6 volt battery source.

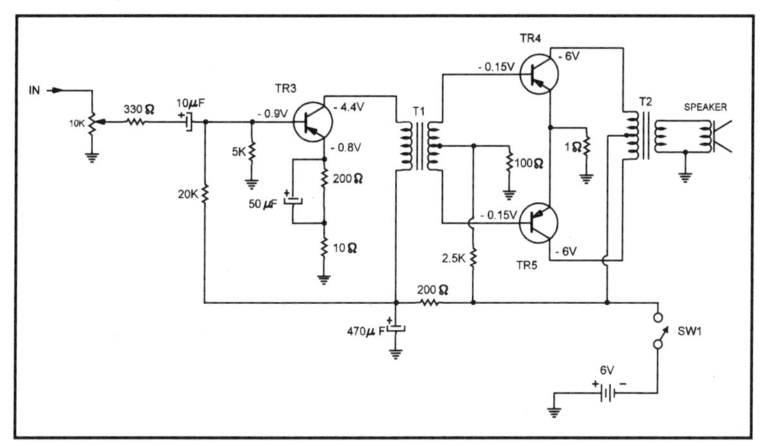

Figure 5-3. The early PNP transistor transformer-coupled driver transistor to two push-pull output transistors.

The driver transistor might be directly-coupled or provide capacity-coupling to the output transistors. In some higher-powered directly-coupled circuits a silicon diode might be found from the driver collector terminal to the base terminal of the two output transistors. The AF transistor audio signal is fed from the volume control to the base terminal and the output of the collector terminal is tied to a coupling capacitor C216 (1 µF). The negative terminal of C216 is tied to the base of driver transistor Q206 *(Figure 5-4)*. Q206 driver transistor collector terminal is connected directly to the base of output transistor Q208.

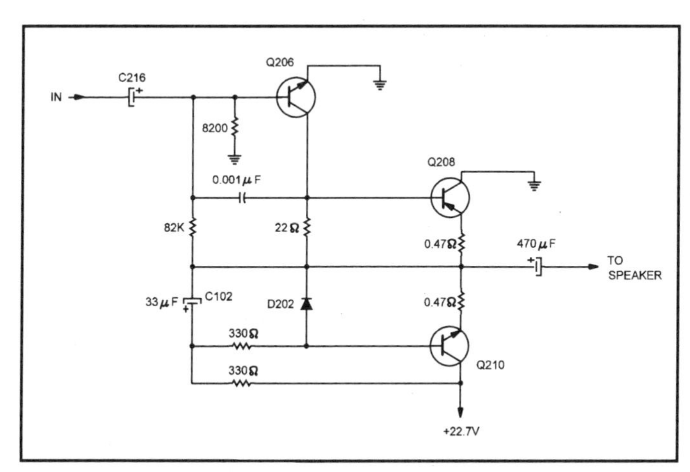

Figure 5-4. The directly-coupled power output stages found in the cassette auto receiver or tape deck.

A no left channel symptom might be caused by an open driver transistor and an increase of voltage on the base terminals of Q208 and Q210. The leaky driver transistor in the auto radio might create only a hum in the speaker. The open driver transistor might result in a dead right channel. The driver transistor was found open and a leaky output transistor was located in a Motorola auto receiver with a dead left channel. A very weak and distorted audio was found in a Craig auto receiver with a leaky driver transistor. The weak and distorted music might result from an open driver transistor.

When a leaky driver transistor is located, always check the directly-coupled output transistor for possible damage. Check the emitter bias resistor for a burned, or a change in resistance with a shorted, driver transistor. Suspect the driver transistor with a noisy right or left stereo channel. Simply replace the driver transistor. The leaky driver transistor might cause the voltage dropping resistor to run warm or change value. A weak and distorted or distorted with no sound can result from a leaky or shorted driver transistor in the high-powered amplifier.

The dead right channel and a normal left channel can be caused by burned bias resistors, leaky driver transistor, and leaky output IC in a J.C. Penney 683-3845 amp. The speaker relay might click on and then hum with poor soldered driver transistor terminals; solder all three driver transistor terminals. Check for a leaky driver transistor when only the power switch operates in a Kenwood receiver. Intermittent audio can result from an intermittent driver transistor under load; the driver transistor might test normal in-circuit, replace it anyway.

SINGLE-ENDED OUTPUT CIRCUITS

Only one power output transistor was found in the early car receiver or table radios. The single audio output transistor might be coupled with an interstage transformer from the driver transistor. The single output transistor in the auto radio might be driven by two directly coupled AF amp transistors. A tapped output transformer had only one winding with the tap tied to the 10 ohm PM speaker *(Figure 5-5)*. The rear metal chassis serves as a heat sink for the output transistor.

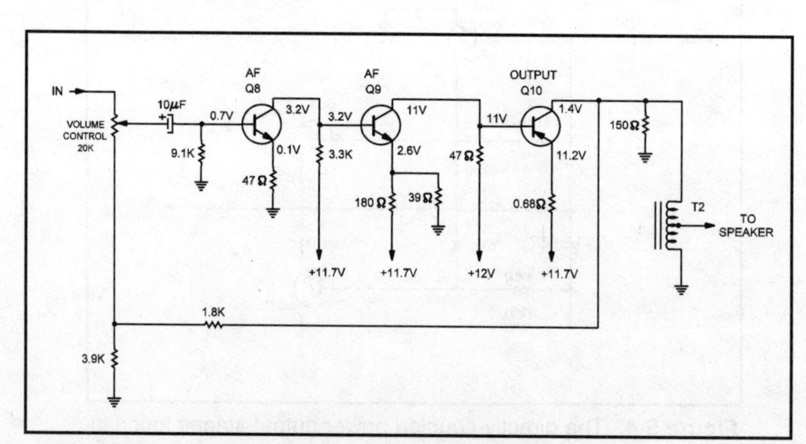

Figure 5-5. Only one power output transistor is located in the ouput circuits of the car radio.

The AF amplifiers and output transistor work in a directly-coupled circuit. All three audio transistors are tied together. When Q9 becomes leaky or shorted the power output (Q10) might also be destroyed. A weak and distorted audio symptom might be caused by an open AF transistor. Replace the emitter bias resistor (0.68 ohms) when Q10 becomes leaky or shorted. If the car radio is left on too long, the T2 winding might be damaged. Replace T2 when the outside cover shows a dark brown and burned area. Check Q10 for a dead and distorted hum in the speaker. Replace output Q10 if there is a loud popping noise in the speaker.

YESTERDAY'S TRANSISTOR OUTPUT CIRCUITS

Besides the single-ended audio circuit, the early low-powered transistor audio circuits contained a driver and two output transistors in push-pull operation. This type of circuit was found in the table radio, auto receiver, tape deck, and phono amplifier. Driver transistor Q103 was coupled to the output transistors Q104 and Q105 with an interstage transformer. Notice that the driver transistor is an NPN and the outputs PNP types *(Figure 5-6)*. Q104 collector terminal is at ground potential.

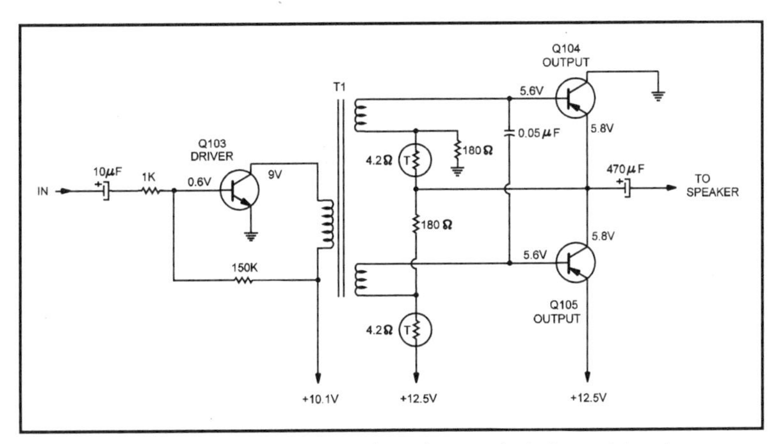

Figure 5-6. The driver transistor is coupled via an interstage transformer in the early low wattage amp circuits.

An open driver transistor might provide a weak or dead audio symptom. You might find Q104 leaky and Q105 open with distorted reception. Replace both output transistors when one is found leaky or shorted. Check the bias resistors and thermistors for a change in resistance with a shorted output transistor. Suspect the 470 μF electrolytic speaker coupling capacitor when intermittent reception occurs.

TODAYS TRANSISTOR OUTPUT CIRCUITS

The transformerless audio output circuits have returned in the low-priced 13 and 19 inch TV chassis (*Figure 5-7*). In the RCA CTC145 chassis, an NPN audio amp is coupled through two diodes to the base terminal of Q1202. The output base terminal of Q1203 is directly coupled to the collector terminal of AF amp transistor (Q1201). The collector terminal of Q1203 is grounded, while a higher voltage, 18.2 volts, is fed to the other audio output Q1202. The 32 ohm speaker is coupled to the emitter terminals by C1207 (100 μF) capacitor. The audio voltage source is fed from a scan-derived flyback secondary circuit.

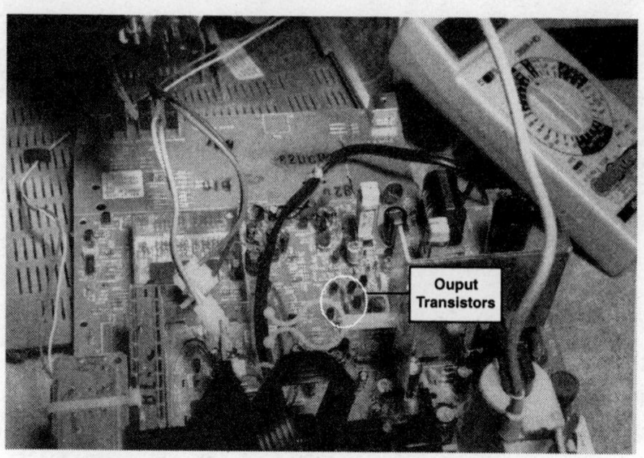

Figure 5-7. The transformerless audio output transistor circuits are found in today's 13 inch TV chassis.

A weak and distorted audio symptom might be caused by a leaky AF transistor Q1201. When either Q1202 or Q1203 become leaky or open, replace both output transistors. Low distortion with weak sound might be caused by CR1201 or CR1202. Check bias resistors R1209 and R1210 (2.2 ohm) when a leaky output transistor exists *(Figure 5-8)*. Intermittent and no sound can result from a defective C1207 (100 μF). Replace both Q1202 and Q1203 and bias resistors R1209 and R1210 with a no audio or dead speaker. Check CR4120 and C4135 (470 μF) for no or improper 18.5 volt source in the flyback power supply circuit.

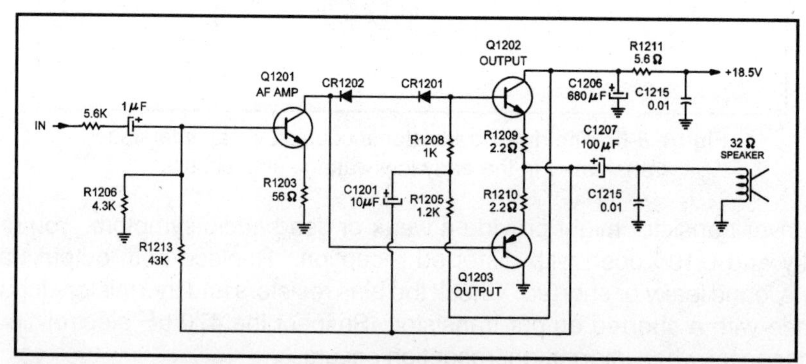

Figure 5-8. The AF amp and output transistors in recent RCA CTC145 TV chassis.

Suspect open R1211 (5.6 ohm) with no voltage on the collector terminal of output transistor Q1202. Replace a leaky Q1202. Shunt electrolytic capacitor C1206 (680 μF) when motor boating is heard in the speaker. Check both Q1202 and Q1203 when a frying noise is heard in the speaker. Replace both transistors since they cost only a few cents.

POWER IC OUTPUT CIRCUITS

The early IC circuits were found in the preamp and AF circuits of the cassette player, auto radio, and home receiver. Then along came the power IC mounted on a large heat sink. Next appeared the IC with both stereo output channels in one IC component. Today, you see large IC's with the entire audio stages in one component of a high powered receiver *(Figure 5-9)*.

Figure 5-9. You might find a complete audio circuit within one large IC of a Technics radio-receiver AM-FM-MPX chassis.

The power output IC might have a single IC part in each stereo channel of the portable radio-cassette player, tape deck, console radio-phonograph and auto receiver. A preamp IC and recording amp IC might precede the single output IC in a deluxe cassette car stereo recorder. The single output IC might be driven by transistor AF and driver stages in audio circuits of the table and auto radios. The main output IC has capacity coupled input and output to the PM speaker and headphone circuits *(Figure 5-10)*.

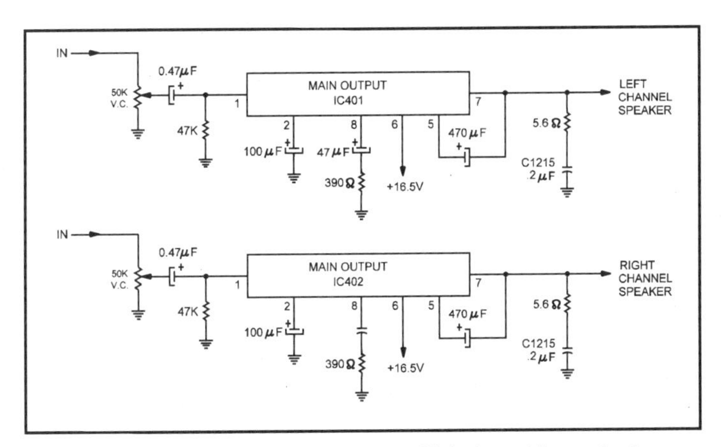

Figure 5-10. A single power output IC is found in each stereo channel of the early portable radio-cassette tape deck and auto receivers.

DUAL CHIP-DUAL SOUND

The dual-IC output might be included in the home AM/FM/MPX, tape and phonograph player with a couple of speakers. A boom-box, table-top receiver and cassette player, compact stereo amp and cassette player/recorder deck might contain a dual-IC in the power output circuits. The dual-IC component might have only the output IC's with driver IC or transistor AF driven circuits. The compact amp, radio and cassette recorder might have a smaller dual-IC for both stereo channels. This dual-IC might contain all of the audio circuits, from volume control to the speaker terminals *(Figure 5-11)*.

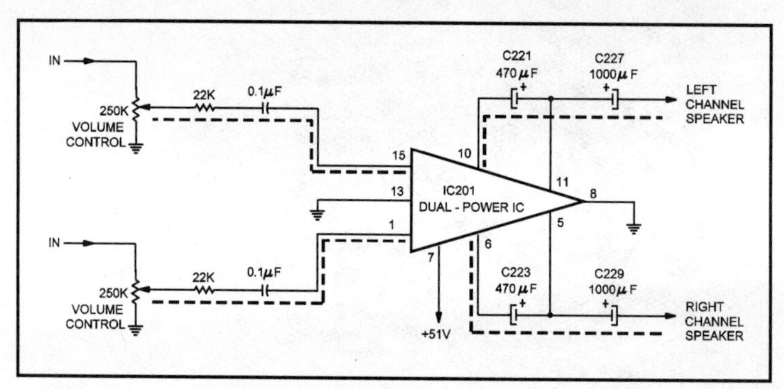

Figure 5-11. The black lines represent the audio path from volume control to each stereo speaker through dual-IC201.

The left and right input signal is controlled by a volume control and coupled to the power IC circuit by a 0.1 μF coupling capacitor to terminals 1 and 15. The audio signal is amplified by IC201 and capacity coupled to the left and right channel speakers through C227 and C229 (1000 μF). Notice the large dc supply voltage on pin 7 of IC201. Terminal pin 8 is the ground terminal.

Replace the dual-output IC when both left and right channels are dead. You might find the left channel is dead with a normal right channel caused by a leaky power IC. Replace dual-IC output (LA4126) when both channels are dead in the Panasonic RX5100 amplifier. You might find a leaky driver transistor, burned bias resistors and a leaky output IC with a dead stereo channel. Replace the dual-IC with distortion in both channels. Check for burned or open low-ohm resistors with a leaky output IC.

Suspect the dual-output IC with a noisy-crackling sound in both output speakers. Replace the power output IC with a frying noise in both or one audio channel. Check for open bypass electrolytic capacitors off the output IC terminals for a low hum noise. A constant motorboating sound in the speakers can result from poor ground connection of the power IC or the IC. The loud-crackling noise can result form a defective output IC. Suspect a defective output IC with a loud-popping noise in the speakers. Replace the open protection IC for no sound in a Hitachi receiver. Replace the dual-IC when a hum is heard in the left channel and no sound in right channel.

HEAT SHIELD PROBLEMS

The large heat sink can cause many audio problems within the amplifier section. When the large power IC is grounded directly to the heat sink and the metal sink has a poor ground, a dead, noisy, and intermittent sound can be heard in the speakers *(Figure 5-12)*. Intermittent sound might result from a defective output IC terminal and heat sink not being properly grounded.

Figure 5-12. The large heat sink helps dissipate the heat from the large power IC in a stereo receiver.

Check the heat sink for overheating of the shorted power IC instead of touching the IC component. Be sure and check for open jumper wire connections that tie from power IC to PC wiring. Tighten the power-IC mounting screws or those on the PCB where the IC is mounted, for intermittent low hum noises. Check for poor IC screw connections when moving or touching the cabinet, when the speakers make a cracking noise. Sometimes it's best to run a separate ground wire from power IC screw and shield to chassis ground.

SPEAKER RELAY PROBLEMS

The defective relay might have dirty or bad points that can cause no audio or intermittent sound. One channel might cut out with weak volume with a bad relay. The volume was intermittent with a bad relay in a Fisher CA272 amplifier. The relay might not click on or off with a dried-up 4700 µF filter capacitor and open 4.7 ohm resistor in the power supply.

The relay might click on and off with a hum in the sound caused by poor soldering connections of driver transistor. Replace the relay for intermittent sound in the speakers. Replace relay or volume control when the sound is intermittent in a Fisher amplifier. Check the relay for one channel intermittent and the other channel low in volume.

An unusual problem existed in a Pioneer receiver when the audio was intermittent and both the power amps were replaced with a defective relay.

The speaker relay cuts out when the power IC's were normal, the audio was intermittent, and the speaker relay clicks off and on with an open D519 bridge rectifier in a J.C. Penney receiver amplifier. Always replace the defective relay instead of trying to repair it. The relay might not energize with a shorted output power IC.

NO SOUND-DEAD AMP

Check for an open or blown fuse in the receiver or large amplifier. If the fuse keeps blowing, check for leaky silicon diodes, filter capacitor, and output transistor or IC components. Next, measure the voltage across the filter capacitor to determine if the low voltage supply is normal. Check the supply voltage source at the audio output transistors or power IC's to determine if the voltage source is dead or improper voltages *(Figure 5-13)*. A defective power switch can cause a dead amp or receiver.

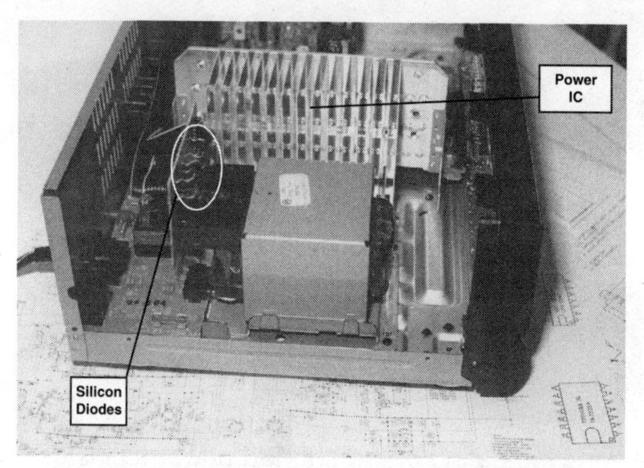

Figure 5-13. Check the supply voltage at silicon diodes, large filter capacitor, or upon the large power IC to determine if voltage source is normal.

A dead left audio channel might be caused by a leaky driver and power output transistor. The left channel might be dead with a burned bias diode and leaky output transistor. Both channels were dead in a Pioneer amplifier with a leaky dual-output IC. In a J.C. Penney 3226 amplifier the fuse kept opening with leaky right channel output transistors, Q209, Q210, Q211 and Q212. The audio channels can be dead with poor soldered connections on the volume control or a shorted volume control.

A defective channel switch can produce a dead amplifier. The Soundesign amplifier circuits were dead with an open winding in the power transformer. Both channels might be dead with a leaky 2200 μF electrolytic capacitor in the power source. A shorted decoupling capacitor can cause a dead left channel with a good right channel. The dead right channel might result from a hole blown out of the output IC. The blown speaker fuse produced a dead channel with high voltage on the speaker terminal, and caused by a shorted output transistor or IC. A dead left channel might result from a power amp poor board-ground terminal.

Most dead receivers or amplifier circuits are caused by leaky power output transistor or IC components. Replace the driver transistor when the output transistors are found open or shorted. Check for leaky bias diodes and burned bias resistors in the transistor output circuits. Sometimes all components should be replaced in the output circuits when leaky output transistors and components show signs of overheating *(Figure. 5-14)*. Do not overlook a dead channel with open (470 μF or 1000 μF) coupling capacitor to the speakers.

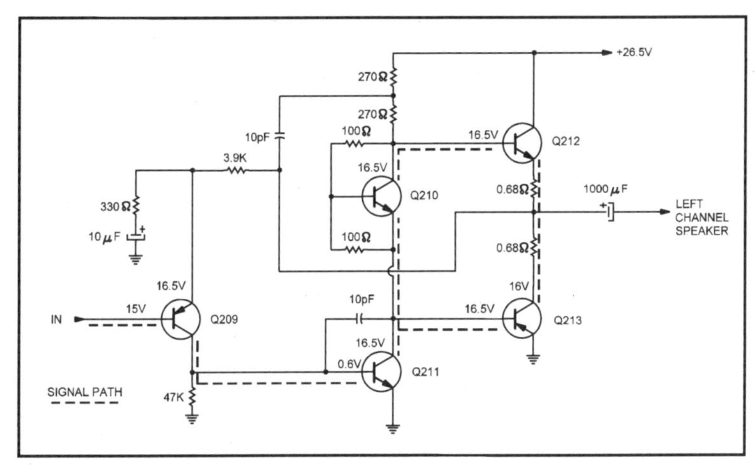

Figure 5-14. Check the following parts when replacing defective audio output transistors.

WEAK SOUND

Check all transistors, IC's, and electrolytic coupling capacitors for weak sound in the amplifier. Inject a 1 kHz signal at the volume control to determine if the front-end preamp or AF stages are weak or the AF, driver and output transistors or power IC's are defective. Measure the voltage at the power output transistors and IC components to determine if proper voltage is found at the output circuits. Signal trace the weak front-end circuits with an external amp or scope. Compare the good channel with the weak one in stereo audio circuits.

Suspect a leaky or open driver transistor with weak sound. A weak volume with distortion can occur with a leaky driver and output transistor and low voltage on the emitter terminals. Replace the leaky bias diode with a distorted left channel and low volume. Check the electrolytic coupling capacitors (1 µF to 4.7 µF) when one channel is weaker than the other. Replace a 10 µF coupling capacitor for a weak right channel.

The output transistors, driver transistor, and bias resistor were replaced with a weak left channel. Suspect an open bypass capacitor on the emitter terminal of the driver transistor for a weak audio channel.

In a Silver Marshall amplifier the driver transistor was replaced with a universal replacement (SK3122) with a weak left channel. Replace both high powered output transistors when the right channel is weak and distorted. Suspect a defective volume control with no volume in one channel. Check for low ohm resistors or an open voltage regulator source that feeds the AF or driver circuits *(Figure 5-15)*.

DISTORTED AUDIO

Go directly to the audio output circuits for extreme distortion in the speakers. Check the output transistors for open or leaky conditions. The left output transistor might be open and the right channel leaky causing distortion with very weak audio. Remove one end of each bias resistor or diode and check for correct resistance or leakage. A leaky driver transistor might destroy both output transistors and bias resistors.

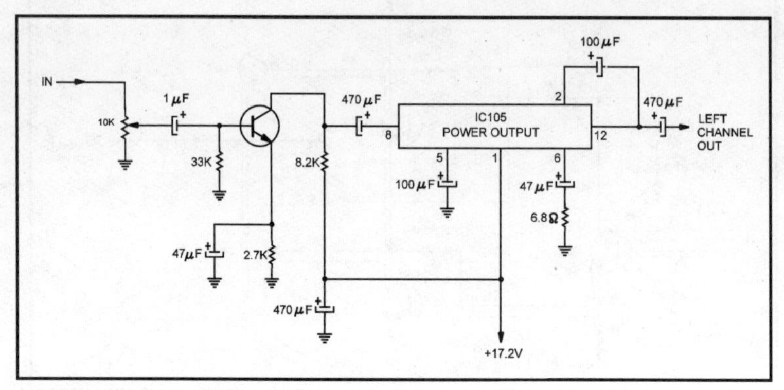

Figure 5-15. Check the following parts for a weak audio channel.

Check for a leaky coupling capacitor that can produce distortion in the audio circuits. The right or left channel can be distorted with a leaky AF or driver transistor. Replace both output transistors and bias resistors when either channel is distorted and has a loud hum.

Suspect a leaky dual-power IC with distortion in both stereo channels. After 5 minutes the right channel became distorted with a defective power output IC. Spray coolant on the body of a suspected IC for distortion and noisy reception to make it act up.

A slight hum was heard in a J.C. Penney amplifier circuit when the output IC voltage source dropped to 36.1 volts from 40.1 volts, caused by a defective 6800 μF filter capacitor. A loud popping noise with distorted sound can result from a defective output IC. Check the output IC for overheating when distorted audio is heard in the speaker.

ERRATIC OR INTERMITTENT SOUND

Erratic or intermittent sound might be caused by a defective driver or AF transistor. A defective speaker relay can cause intermittent sound in the speakers *(Figure 5-16)*. A bad power switch might cause a noisy sound when turned on and off. The right side of a Sony large boom-box player had erratic sound with a loud rushing noise and was caused by a 47 μF 50 volt coupling capacitor to the volume control.

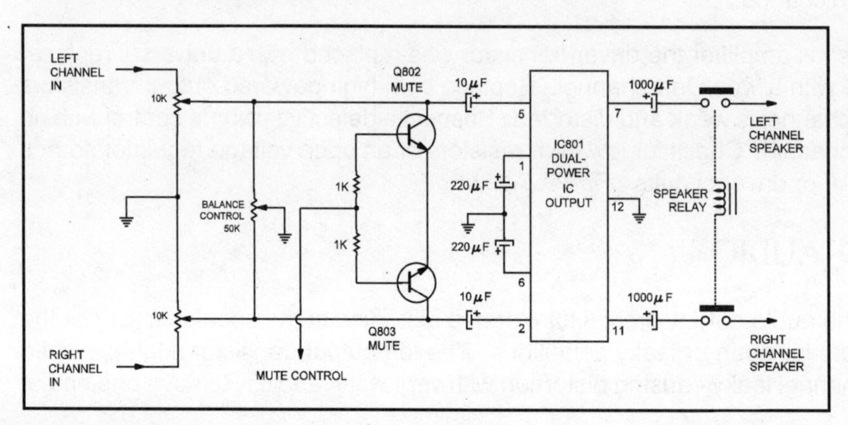

Figure 5-16. Check the following parts for intermittent reception in the power output circuits.

Sometimes just moving components around with an insulated tool can cause a capacitor or resistor to act up. Pushing up and down on the PCB can uncover a poor soldered connection or a tie-wire circuit connection. By snugging up the mounting screws of the main chassis PCB solved the intermittent noise in a Westinghouse output amplifier. A bad soldered connection on the volume control can cause an intermittent and noisy left audio output channel.

The right channel was intermittent within a Soundesign amp caused by a poor speaker ground connection. The volume increased all the way up and then cut out and was caused by a 10 volt zener diode.

When one channel is weak and the other cuts in and out, suspect a dual-power output IC. Intermittent noise in the left channel and then the audio jumps in and out can be caused by a large power output IC. Check the large power IC when the sound pops in and out with distorted audio. Resolder all output IC terminals with intermittent sound symptom in both channels.

LEVEL METER PROBLEMS

The VU (volume units) meter indicator monitors the volume level in the audio channel. Very few electronic problems are found within the VU sound meters of a large amplifier or cassette deck. The stereo meters might be connected after the AF or driver transistors within the transistor output stage or after the power IC in the output circuits *(Figure 5-17)*. Both VU stereo meters have the same type of audio circuits. When one or both audio meters do not register, signal trace the audio signal right up to the VU meters with an external amp or scope tests.

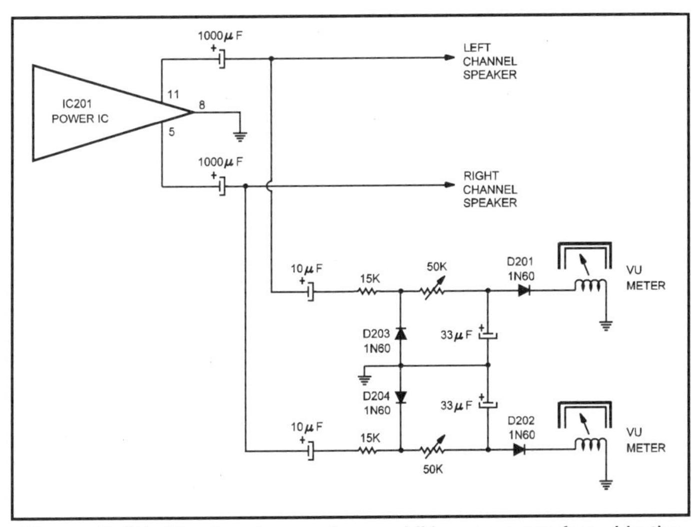

Figure 5-17. The level recording on VU meters are found in the output circuit of audio amplifier.

The LED amp signal strength or VU indicators might be operated from one large or two separate IC's. In the large AM/FM stereo receivers a separate array of LED's are operated from one large IC. The input audio signal from the left channel appears at pin 11 and pin 10 is the B+ supply source (Vcc). Each stereo channel has its own IC amp and set of LED indicators. The various LED's are tied to terminals 1 through pins 7 *(Figure 5-18)*.

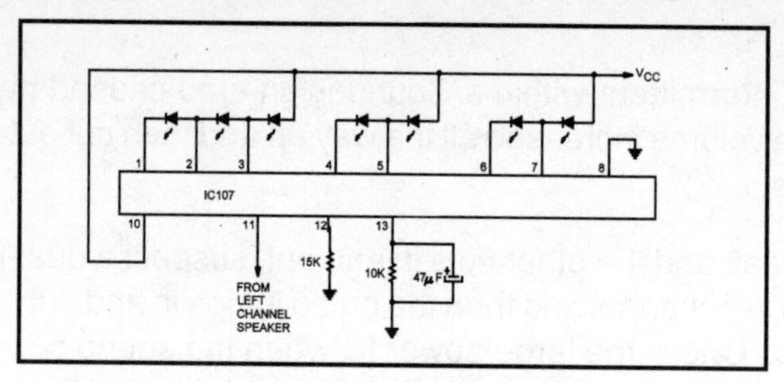

Figure 5-18. LEDs are used to indicate the signal strength or recording level.

Some LED sound indicators might have an array of LED's for each channel while on small recorders one LED array is switched into the play and record circuits. In record mode, the LED's might be switched at the output of preamp IC and in the playback mode at the output of power amp IC. When found in large stereo receivers, the LED's are used as VU indicators.

The audio signal can be signal-traced right up to the LED IC or VU meter with scope or external audio amp. Very seldom do LED's cause any problems. Check the suspected LED with the diode tests of the DMM. Most VU problems are related to a defective IC or improper applied voltage source. When the VU meter becomes erratic or intermittent, suspect a bad wire connection or meter. Sometimes the meter hand might stick in one spot if the cardboard indicator warps. Remove the front cover and re-glue the meter scale. If the meter hands go backwards to the incoming sound, reverse the meter leads.

SERVICING OUTPUT CIRCUITS IN THE EQUALIZER/BOOSTER

The stereo equalizer is a circuit that compensates attenuation to achieve equalization. The stereo equalizer/booster provides stereo frequency equalization with a boost of power. Besides a stereo frequency equalizer stage with separate frequencies (60 Hz, 150 Hz, 400 Hz, 1 kHz, 24 kHz, 6 kHz, and 15 kHz), the output circuits contain separate power output IC's to boost the audio to several speakers *(Figure 5-19)*. The stereo frequency equalizer-booster circuit might be found in the auto CD, cassette or receiver output circuits.

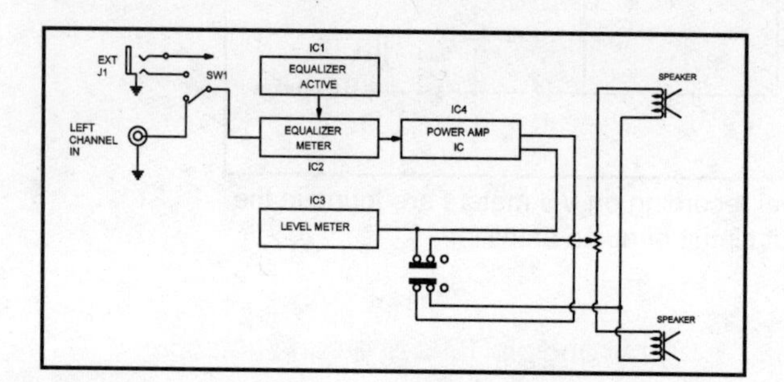

Figure 5-19. A block diagram of an equalizer/booster output circuit.

The stereo frequency equalizer IC2 is coupled through C201 to input terminal 3 of IC4. The 20 watt power IC output is fed to a fader control and to the auto speakers. The output audio signal is also coupled to the level meter driver amp IC3 of the left stereo channel. IC3 provides drive signal to the five different level indicator LED's. The audio level LED's are tied to terminals 1 through 5 of IC3 *(Figure 5-20).*

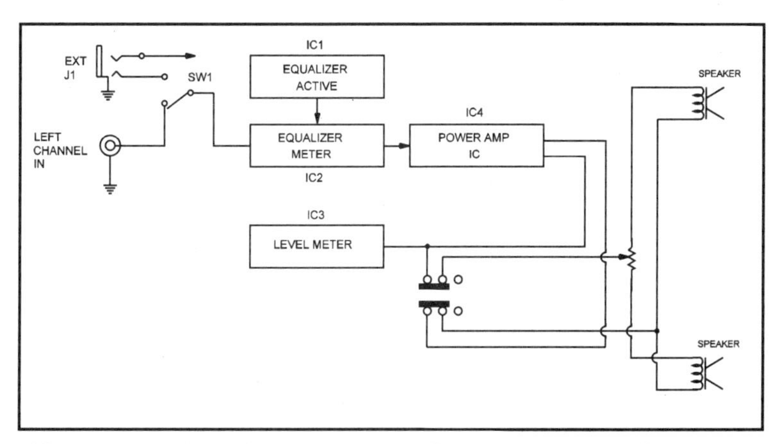

Figure 5-20. The left channel output IC4 and level meter IC3 circuits.

Signal trace the audio signal from the input of IC2 through IC4 and the output speakers with an audio external amp or scope. When the audio signal stops suspect that stage or circuit. Check for a blown fuse with no pilot light or dead amp. Measure the dc voltage input at pin 6 of IC4 and pin 9 of IC3. If no sound, suspect defective IC2 or IC4, with the pilot lamp on. Check all electrolytic coupling capacitors for open conditions.

Check IC4 applied voltage at pin 6. Suspect IC4 for distortion or insufficient sound. Check all electrolytic capacitors attached to IC4 terminal pins for open or leakage conditions. A defective fader control and speaker can cause audio distortion.

Shunt C126 (470 μF) for a motorboating sound in the audio. Check C127 (0.1 μF) for oscillation within IC4 circuits. Suspect IC3 when all of the LED's do not light up. Check each LED with the diode test of the DMM when one of them will not light up. Measure the 14.3 volts on the anode terminal of each LED.

SERVICING BOOM-BOX PLAYER OUTPUT CIRCUITS

The early boom-box radio and cassette players might have transistor output circuits while others can have a separate output IC or dual-IC. The dual-output IC is found in the radio, cassette-CD player amp circuits. The dual-IC output is usually switched by earphone jacks into a tweeter and woofer speaker in each audio channel. The headphone jack is connected to each speaker output terminal through an audio dropping resistor (220 ohms).

The left volume control controls the input audio to the left input (pin 5) of IC201 *(Figure 5-21).* A right volume control (20 kohms) provides audio signal to the right input terminal 8 of IC201. The left amplifier audio signal is found at pin 2 and right signal at pin 10. Notice the large coupling capacitors of 1000 μf that couple the audio to the speaker and earphone circuits. Pin 1 of IC201 is the voltage supply pin and pin 12 is at ground potential.

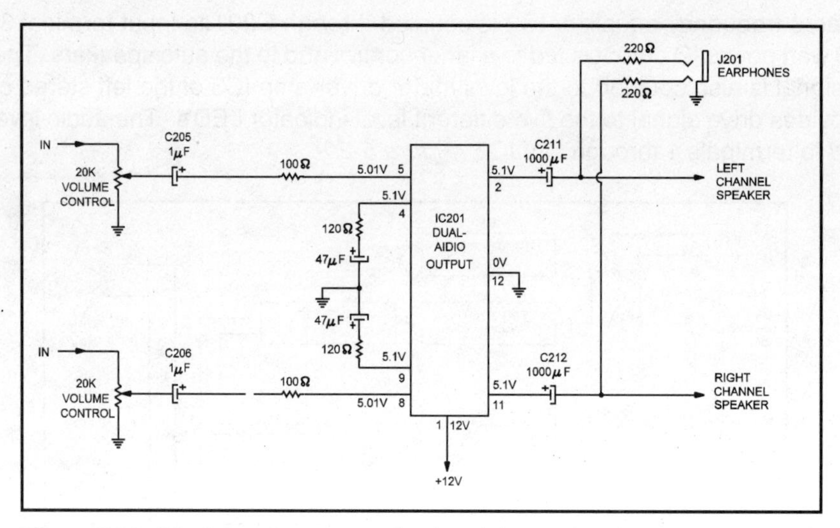

Figure 5-21. The boom-box output circuits might consist of one dual-output IC.

The dead left channel might be caused by an open C205 or C211. A defective IC201 or open capacitors tied to the IC output pins can cause a dead circuit. Check the volume control for an open control or broken connection. Both channels might be dead from an improper voltage supply source.

If the headphones are operating and no sound to the left speaker, suspect a dirty earphone switching terminal or speaker. When only the tweeter speaker is operating and no woofer sounds, sub another speaker across the woofer terminals. Sometimes too much volume applied to the speakers can blow open a voice coil. Suspect a dropped cone in the woofer speaker with a mushy sound.

Check for improper voltage at pin 1 and defective IC201 when both channels are distorted. A distorted right channel can be caused by a leaky C206, C212, or IC201. Remember the power output IC might be defective when either channel or both are distorted. Leaky output C212 (1000 μF) can cause distortion and a dc voltage upon the speaker terminals. C211 and C212 have been noted to cause an intermittent channel. Check the earphone jack for poor connections when one channel earphone is dead, erratic or intermittent.

TROUBLESHOOTING THE CASSETTE PLAYER OUTPUT CIRCUITS

The early cassette player might have a driver transistor transformer coupled to the output transistors. The radio-cassette player might have transistor or IC output circuits. The AM/FM/MPX receiver with a cassette player can have a separate IC or a dual-IC in the output stereo circuits. A monaural cassette player might have a transistor preamp and a power IC output.

The monaural output IC input signal is coupled from the volume control by C10 to input terminal 9 *(Figure 5-22)*. The output audio is switched into the earphone and speaker by headphone jack (J1) from coupling capacitor C14 (470 μF). Both the preamp transistor and power output IC1 provide play and record audio signal.

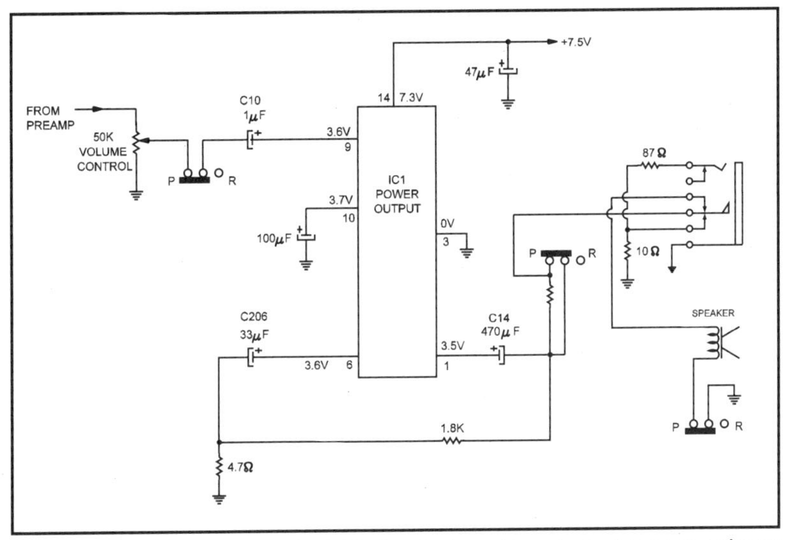

Figure 5-22. The power output IC circuits found in a typical cassette player.

Signal trace the audio signal from the tape head to the output IC with an external amplifier or scope. When the audio stops, check the coupling capacitors, transistors, and power IC's for open conditions. Check all voltages upon the suspected IC or transistors. Take critical resistance measurements upon each IC pin terminal to locate a leaky IC. Suspect a leaky IC when low or improper supply voltage is found upon the output IC.

A weak sound with a loud hum can be caused by an open tape head. The weak channel can be caused by weak batteries or improper voltage source. Replace the power IC for a slightly weak right channel in a J.C. Penney cassette player. The weak left channel can result from a poor grounded tone control. Check for a leaky low value coupling capacitor for low audio in the play mode. A broken wire in the leaf switch might result in a weak or dead audio symptom.

Check and clean up the tape head for extreme distortion. An open preamp transistor under load might cause a slight distorted channel. Test the output transistor in-circuit for leakage and burned bias resistors with a distorted speaker. A bad radio-cassette switch can result in distorted and garbled audio. Suspect a power dual-IC with distortion in both channels.

Suspect coupling capacitors, transistor and IC for intermittent audio. Clean up the bias variable resistors for intermittent audio in one channel. Check the function switch for intermittent record or playback modes. First clean up the function switch. Replace if broken or has worn contacts. Check the volume control when you cannot turn down the volume and the sound becomes intermittent. Check all bypass electrolytic capacitors off of the power output IC for an intermittent left channel. Replace the output IC for intermittent right or left channel. Check for a bad muting switch for intermittent audio in the cassette player.

Check for open tape head or lead connection for a rushing noise in the cassette player. A low rushing or frying noise can result form a defective output IC. The noisy right channel can be located by pushing up and down on output IC with insulated tool. Check for a noisy right channel with an open electrolytic capacitor across the emitter resistor.

Inspect poor soldered connection or tie wires on the PCB. A bad function switch can produce a hissing noise in the speakers. A noisy and intermittent left channel can be caused by a defective output IC. Replace dual-output IC when one or both channels are noisy. Replace the power switch for an arcing noise when first turned on.

Within a large Sony portable radio and cassette player, the right channel had a rushing or frying noise like that found in a transistor or IC. The large output IC was replaced (TA7215P) and the results were the same. The cassette output circuits were signal traced with the external audio amp. The noise was found upon pins 15, 16, 17, and 18 and not on the volume control *(Figure 5-23)*. The 47 µF 50 volt electrolytic capacitor was found to be open. Replacing the electrolytic coupling capacitor solved the rushing noise.

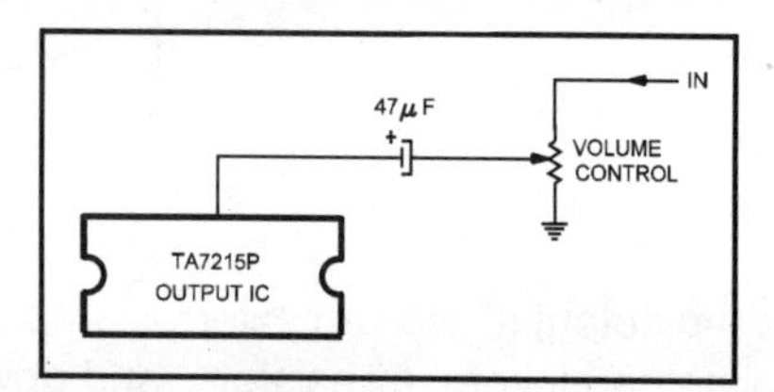

Figure 5-23. The open 47 µF electrolytic coupling capacitor caused a rushing noise in a Sony portable radio and cassette player.

A ground wire that had come off of the erase head caused a jumbled recording in a Sanyo cassette player. Check for a defective function switch in record or playback for intermittent recording. The intermittent recording can be caused by a defective microphone coupling capacitor. A noisy recording might result from a defective microphone. The weak recording on the left channel was caused by a packed tape head. The same sound circuits in the small cassette player might combine the record and play modes. Check both play and record modes to locate outside recording symptoms.

REPAIRING THE CD PLAYER OUTPUT CIRCUITS

The analog sound circuits within the CD player starts at the D/A converter (digital-analog). The left and right stereo circuits are developed at the output of the D/A converter. Often, one or two stages of audio in each channel amplifies the audio signal to the line output jacks and an earphone jack. Most CD players have a muting transistor in each output channel.

The early or low priced portable CD players might have a transistor amp in each stereo channel. One dual-IC might be located in today's CD line and output circuits. The boombox stereo radio, cassette and CD player might have the radio, cassette and CD signals from the D/A converter switched into a dual IC preamp and dual-output IC to each speaker. A separate headphone IC amp might be connected to the line output jacks *(Figure 5-24)*.

Signal trace the audio signal from pins 10 and 9 through the preamp IC2 to the line output jacks for weak, intermittent or distorted audio. Often, distortion is caused by leaky coupling capacitors, transistors, IC's and defective mute transistors. The weak channel might be caused by an open electrolytic coupling capacitor, transistor and IC amplifier. Improper voltage source to the IC amplifiers can result in a weak and distorted audio symptom. The normal stereo channel can be used to check out the defective or loss of signal in the other stereo channel.

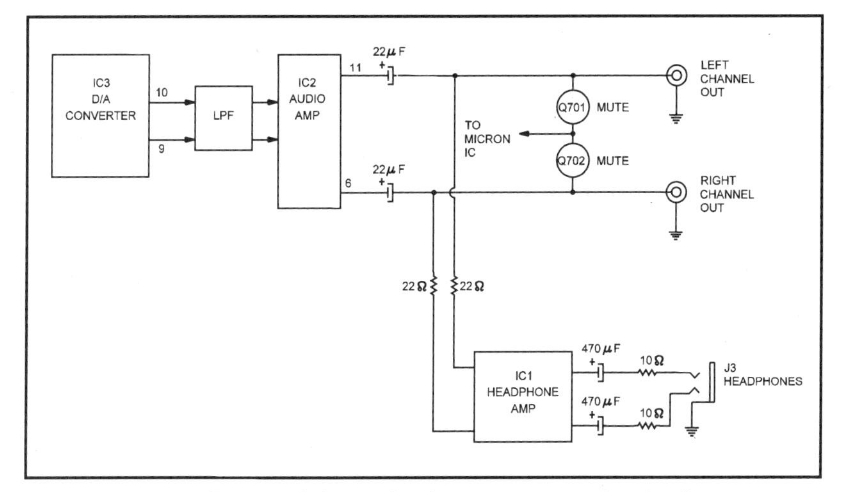

Figure 5-24. The headphone circuits are connected to the line output circuits in a portable CD player.

Check the low pass filter (LPF101) for no audio in the left channel. Replace the D/A converter IC with no audio output. Check the Mechanism Micro IC when no muting during program mode. Replace the audio preamp IC for no audio at the line output or headphone IC. A dead-no power output can be caused by defective filter capacitors in the power supply. Check the 14 volt regulator zener diode when the protection relay won't turn on the speakers. Suspect a defective voltage regulator with no audio output and LCD display.

Check the left preamp IC with distorted left channel. Suspect a defective RAM IC with distorted sound. Replace the audio output IC with distortion after the CD player has warmed up. Check all IC preamp terminals for poor connections with audio distortion. Replace the D/A converter IC for distorted audio.

Suspect the preamp IC with noise in either stereo channel. Replace the headphone amp IC with noise in both earphones. A background noise can be caused by a defective D/A converter IC. Check those small electrolytic coupling capacitors for a noisy channel. Replace the RAM IC when there is a ticking noise in the audio. Remove and replace the preamp IC for a popping noise in either channel.

SERVICING TV AUDIO CIRCUITS

The defective sound IF stages might produce hum and distorted sound. Check the discriminator coil adjustment for garbled and distorted audio. Simply touch up the coil adjustment until the sound returns to normal. Poor soldered connections upon the coil can cause intermittent and buzzing audio. Resolder pins 15 and 16 off of IC101. Distorted audio can result from a coil or terminal connection of T101.

The complete audio circuits might be included in one large IC201, except the sound IF circuits. The audio signal is taken from pin 23 of the SIF (IC101). The volume is controlled by an MPU (IC001). C209 couples the audio from IC101 to pin 2 of audio output IC201 *(Figure 5-25)*. The amplified audio signal at pin 8 is coupled through a 220 μF electrolytic to output transformer T203. Some speaker circuits might be switched out of the circuit when the headphones are plugged in.

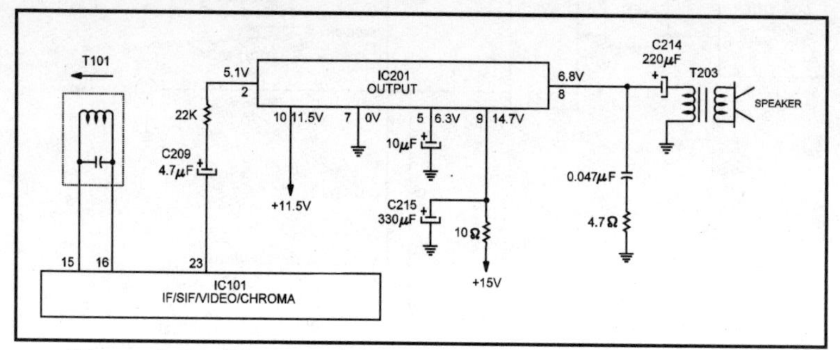

Figure 5-25. The sound output circuits are found in one C201 component.

Check IC201 for weak and distorted audio. Suspect IC201 for intermittent and noisy sound. No audio might be caused by leaky IC201 or open 220 μF capacitor. Check all electrolytic capacitors on IC201 pin terminals for weak or garbled audio. Inspect the heat sink screws for weak and intermittent sound. Inspect for poor IC201 grounds for repeated failure of the output IC. Check IC201 for a hum and buzzing noise. Suspect IC201 for hum when the set is turned off.

Suspect C209 (4.7 μF) and C214 (220 μF) for intermittent and weak sound. The improper voltage source might cause weak or no sound. The dead audio speaker might be caused by open C209 and C214.

SERVICING THE HIGH-WATTAGE AUDIO AMP CIRCUITS

A 200 watt or higher audio amplifier circuit might have ten or more power amp transistors with several preamp circuits. The typical 170 watt auto amplifier has nine power amps, one muting, and two transistor for overload protection *(Figure 5-26)*. Of course, the high-powered stereo amplifier has twice this amount of transistors. Both left and right stereo channels are identical. You can use the normal channel to compare audio signal and voltages with the defective channel.

After removing the top cover, inspect the chassis for burned or charred parts and wiring. Signal trace the audio from stage to stage with the external audio amp or with a scope and function generator. Connect high-powered loaded resistors across the left and right speaker output terminals. Do not fire-up the high-powered amplifier without speakers or a load attached to the amplifier. Keep the gain or volume control as low as possible. Inject a 1 kHz audio signal at the input terminals. Check and compare the signal in each stage until the signal is lost in the defective channel.

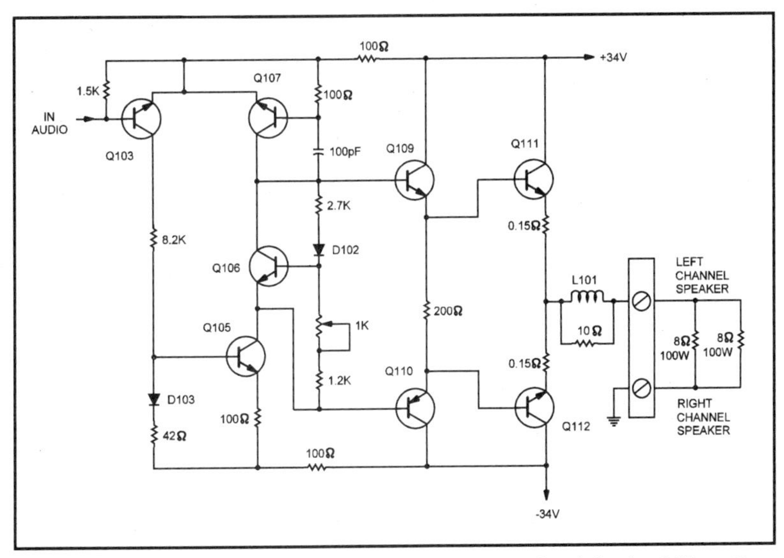

Figure 5-26. The final directly-coupled output circuit in the 200 watt high-powered audio amplifier.

Some technicians prefer to check all transistors first, before taking signal or voltage tests. Sometimes transistor tests might not be accurate when directly-coupled transistors or fixed diodes are found in the base and collector circuits. Take critical voltage measurements upon each transistor and compare them to the normal channel. A schematic is a must item where ten or more transistors are found in the audio output circuits. When a dc voltage is found upon the speaker terminals, the last two or more power transistors might be leaky or open.

Check the bias resistors for correct resistance when the power output transistor is leaky, shorted or open. Test both directly-coupled transistors when one is found leaky or open. A leaky driver transistor can damage one of the directly-coupled power transistors to open and the next transistor to become leaky. Replace all three transistors.

A dead left channel can be caused by a leaky coupling capacitor, driver and leaky output transistors. The weak and distorted output circuits can result form a missing 40 volts of either positive or negative voltage.

The woofer speakers can be damaged when the output transistors become open or leaky, applying a dc voltage upon the voice coils; this might occur when no coupling electrolytic capacitors are found between speaker and amplifier. The right channel might be distorted with a leaky driver transistor. The weak and distorted left channel might be caused by the power output transistors.

Suspect output power transistors for a popping and cracking sound in the speakers. Replace both power output transistors when the fuse keeps blowing. Test each transistor for a short or leakage. Check the filter capacitors and voltage regulator circuits when a slight hum is heard in one channel. Suspect leaky or shorted power transistors when coupling or bypass electrolytics have a blown top. Check the output transistors when the speakers begin to hum after operating for 5 or 10 minutes. Check for burned bias resistors in both output transistor circuits. All audio problems related to any power amplifiers can occur in the high-powered amp circuits.

SERVICING TUBE OUTPUT CIRCUITS

The dual-triode tubes such as a 12AX7, 12AT7, 12AU7, 6BK11, 6U10, 6SC7, and 7025 were used in a driver stage. A push-pull power output tubes as the 6L6G, 6V6GT, 7027A, 7591, 6AQ5, 5881 and EL34 were found driving power to the speakers. Notice the high voltages found upon the screen grid and plate elements of each output tube. A choke coil and resistance filter network with small electrolytic capacitors are found in the B+ circuits.

The defective tube should be replaced when the amplifier becomes weak, noisy or distorted. Check the tubes in a tube tester, if available. Simply replace the tube with another known good tube is the best method. Sub another tube when one becomes noisy or microphonic. The output power tubes should be replaced with a matched pair. Pickup hum in the input circuit can result from resistor increases or open coupling capacitors *(Figure 5-27)*.

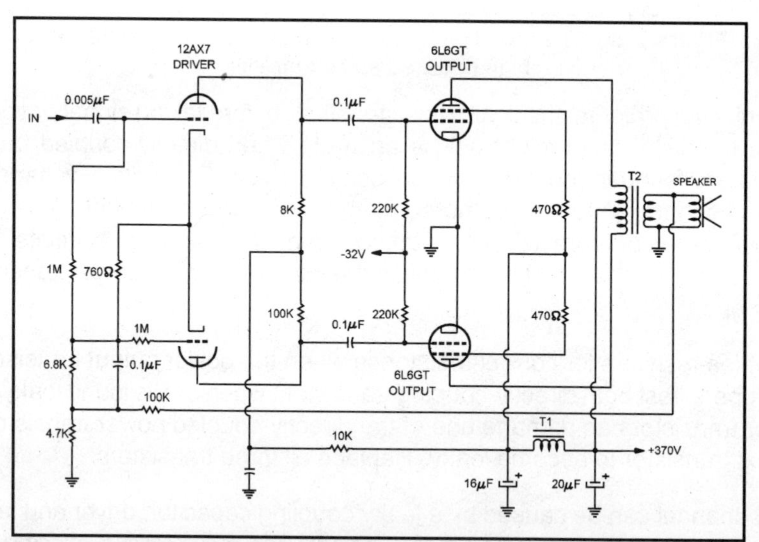

Figure 5-27. A typical tube output amplifier circuit employing a 12AX7 and two 6L6GT output tubes.

A gassy or leaky tube can cause extreme distortion. The leaky or shorted tube can damage transformers and cause resistors to overheat and burn or change value. A quick resistance measurement from grid to common ground upon the output tube can indicate if the grid resistors have changed value. Likewise a check of the screen grid to common B+ can quickly find a change in the screen grid circuits.

Check the resistance from the output plate elements to the center tap of the output transformer can locate a defective winding. Extreme hum and distortion in the output circuits can be caused by output tubes, damaged choke winding, and filter capacitors. An improper negative bias voltage from the power supply can cause distortion in the audio output circuits.

Chapter 6

TROUBLESHOOTING STEREO AUDIO CIRCUITS

The stereo amplifier is a two-channel audio system for music reproduction from input to several connected speakers or line output jacks. The left and right audio channels are involved in a stereophonic reproduction system. The early stereo channels were found in the AM/FM/MPX receivers and phonographs. Just about every audio product today has a stereo system, including the deluxe TV chassis.

TYPICAL STEREO BLOCK DIAGRAM

The early audio stereo system found in the small radio receiver or cassette player might consist of an AF, driver and push-pull output transistors within the separate stereo channels. Next came the AF or driver transistor with a dual-IC output stage. Today, the small stereo amplifier might consist of only one large IC component for the entire audio system.

Several transistors are found in each left and right channels of high-powered amplifiers. *(Figure 6-1)*. The output circuits often consist of the AF, driver and output transistors in a directly-coupled push-pull arrangement. The AF transistors can be capacity-coupled to the next stage while the driver transistors are directly-coupled to the power output transistors. A directly-coupled transistor has a collector terminal of the preceding stage tied directly to the base terminal of the next audio transistor. Often, the AF transistor is a PNP while the directly-coupled transistor is an NPN type. You might find both driver transistors in each channel as an NPN transistor. PNP and NPN power transistors are found in the early output circuits.

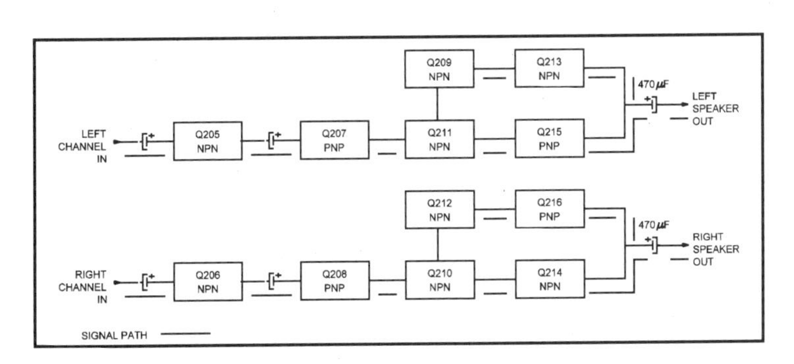

Figure 6-1. A block diagram of a stereo signal path.

The medium-powered stereo circuits found in the boom-box, CD player and TV, might consist of a buffer or AF transistor tied to a volume control IC. This IC controls the volume and amplifies the weak audio signal. The volume control IC varies the volume in both channels with only one volume control *(Figure 6-2)*. A mute transistor tied to the volume control IC provides muting of the audio channels. An electrolytic capacitor couples the controlled-signal to a dual-stereo power output IC. The stereo speakers are coupled to the sound output IC with large electrolytic capacitors (470-1000 µF).

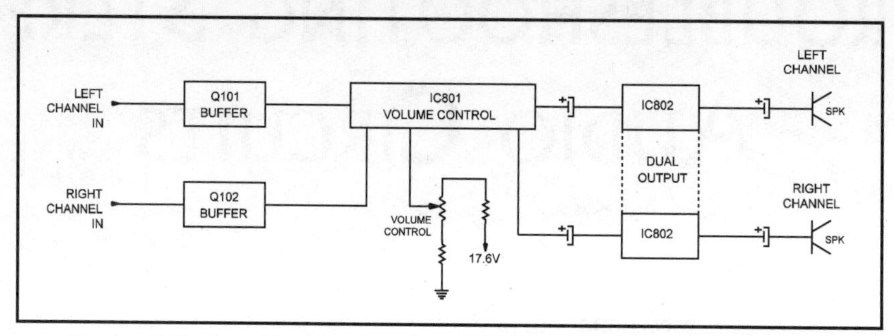

Figure 6-2. A typical TV audio transistor and IC stereo block diagram.

The deluxe stereo channels found in the large screen TV might have several stages of stereo channels. Within the RCA CTC157H chassis, the stereo system starts at the stereo demodular (IC1701) and is coupled-directly to separate Summer transistors (Q1703 and Q1704). The NPN Summer transistor output is coupled to the Matrix IC (U1702). An audio switch (U1402) is coupled through a resistor to the right and left output terminals of the Matrix IC *(Figure 6-3)*.

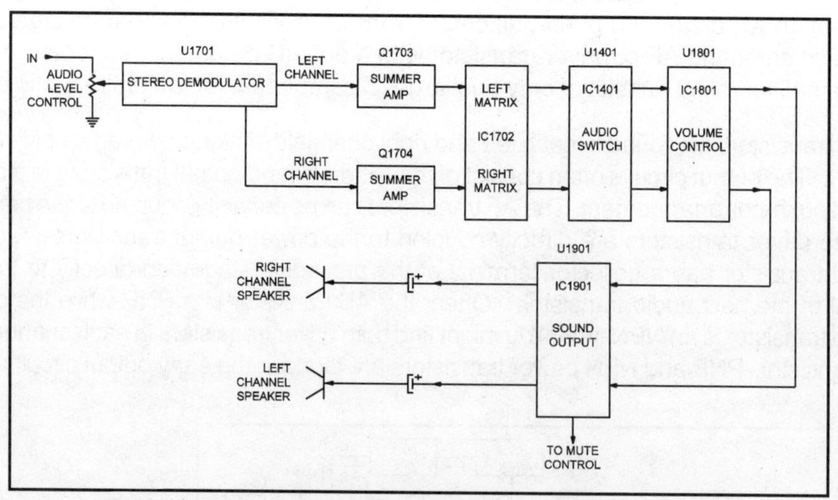

Figure 6-3. The block diagram of a stereo audio circuit in an RCA CTC157H chassis.

The audio IC switch (U1402) provides switching of the TV audio, auxiliary and mute functions. A control logic circuit controls the muting of both channels in the IC switch from a mute signal of the microprocessor. The volume control IC (U1801) controls the volume applied to the sound output IC (U1900). A signal from the AIU controller controls the volume in U1801.

The stereo volume control IC (U1801) couples the audio with a 1 μF electrolytic capacitors to the sound output IC (U1900). Two 220 μF electrolytics connect the 32 ohm speakers to the sound output IC. In the CTC157C and R chassis, two external Hi-Fi output jacks are provided that can be connected to a high-powered stereo amplifier system. Separate audio AF transistors are found after the volume control IC to the Hi-Fi output jacks.

THE PREAMP STEREO AUDIO CIRCUITS

Like most preamp, buffer or AF audio circuits, a separate transistor stage is found in each stereo circuit. The base of the preamp transistor might be coupled with a 1 μF to 10 μF electrolytic coupling capacitor. Most of the buffer transistors are NPN types with the audio taken from the emitter circuits instead of the collector terminals *(Figure 6-4)*. A 22 μF capacitor couples the audio from Q851 and Q852 to a stereo or mono selector switch. Two separate 47 μF electrolytics connect the buffer signal to pins 6 and 8 of the volume control IC.

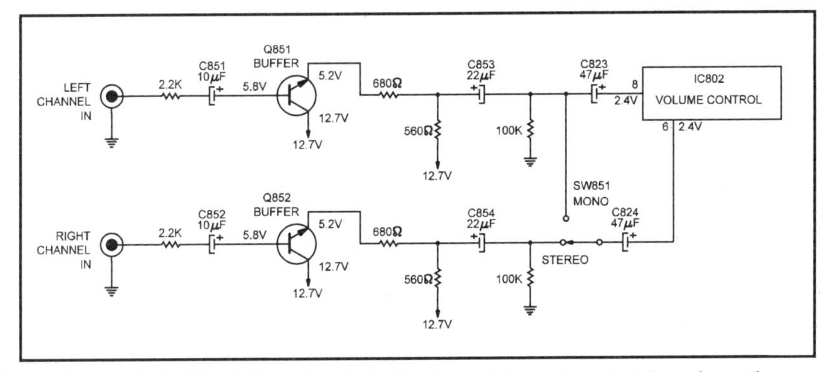

Figure 6-4. The left and right buffer transistor output is taken from the emitter instead of the collector terminal.

The preamp or buffer signal can be signal traced with the audio signal generator or scope. Compare the audio signal at the base and emitter terminals of Q851 and Q852. The amplitude of the audio signal should be quite close with a normal stereo channel. Check Q851 for open conditions with a normal signal upon the base terminal and very little audio at the emitter terminal.

A distorted right channel can be caused by a leaky buffer transistor or electrolytic coupling capacitor. Suspect poor soldered terminals of buffer transistors for intermittent audio channel. Monitor the output signal with the scope or external amp for intermittent reception. Check C851, C853, C824, and Q851 for a weak or lower audio signal applied to pin 8 of volume control IC802.

In a JC Penny AM/FM/MPX stereo receiver, a continuously low level noise was heard in the left channel. Sometimes the noise was very loud. The noise could still be heard with the volume control turned down, indicating trouble occurring after the volume control. Start at the volume control and signal trace the noise from volume control to the left speaker terminals. The noise was heard at the base terminal of preamp transistor Q402 and not at the volume control *(Figure 6-5)*. Replacing C404 (1 μF) electrolytic coupling capacitor cured the low noise sound in the left channel.

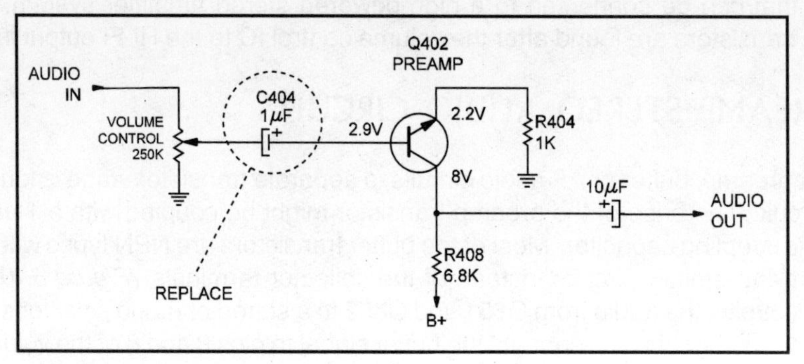

Figure 6-5. C404 was noisy in a J.C. Penney preamp audio circuit.

IN THE DRIVERS SEAT

The driver circuits in a transistor power amplifier stage might consist of a pre-driver and a driver transistor. In the IC stereo driver circuits, the driver circuits might be included in a dual-driver IC or combined in one large power output IC. The pre-driver transistor can be directly-coupled to the driver output stage. The regular driver transistor is connected directly to the output transistors.

When Q202 becomes leaky, the voltage will change on Q203 and Q204, since they are directly-coupled together. A change of voltage upon Q203 can change the base voltage upon the power output transistor *(Figure 6-6)*. Signal trace the audio signal to the base of each transistor to determine where the audio stops, becomes weak or intermittent. Take an in-circuit voltage test upon each transistor and record them upon the schematic. Now check the forward bias of each transistor from emitter to the base terminal. The normal silicon transistor should have a 0.6 volt measurement between the two terminals.

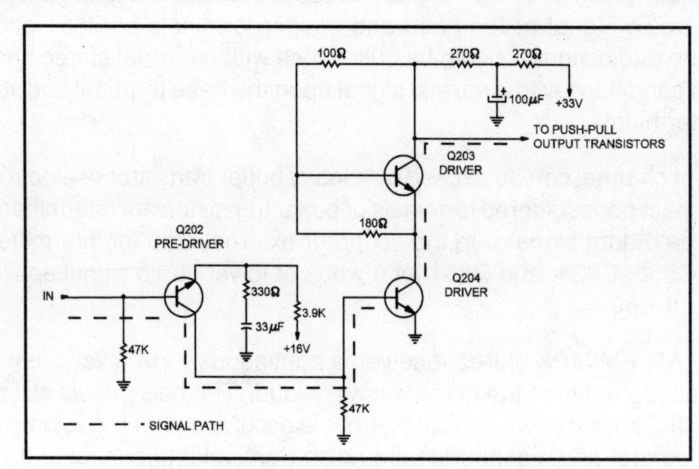

Figure 6-6. The pre-driver signal is directly-coupled to two output transistors with audio signal path.

Take critical resistance measurements from each base and emitter terminals of driver transistors to common ground and compare to the normal stereo channel. Remove one end of a suspected resistor from the circuit and take another measurement *(Figure 6-7)*. Make a in-circuit test of each suspected transistor with transistor tester or diode-test of DMM. Sometimes a leaky transistor can be located with accurate voltage and resistance measurements to common ground.

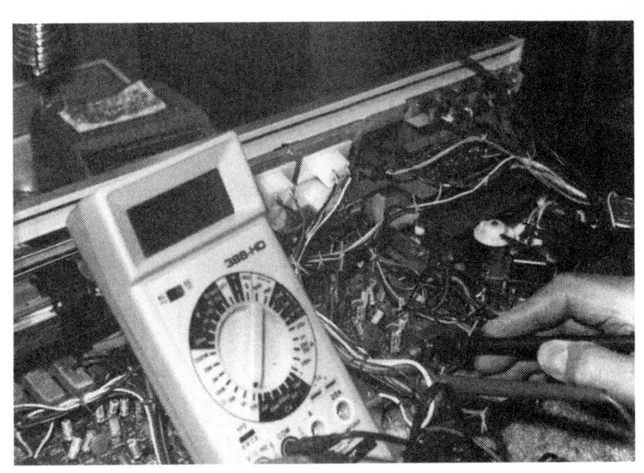

Figure 6-7. Test the suspected resistor within and out of the audio circuits.

Check the audio signal in and out of a driver IC circuit with the scope or external audio amplifier. Compare the input and output signal with the normal channel. If the signal is normal at the input terminal and weak, distorted or intermittent at the output terminal, suspect a defective preamp IC or connected components. Take critical voltage measurements upon each IC terminal and compare with the normal channel. Suspect a defective IC or improper voltage source when both channels are weak or distorted in a dual-preamp IC.

TRANSISTOR OUTPUT STEREO CIRCUITS

The transistor output stereo circuits might consist of two transistors directly-coupled to the driver transistors. The collector terminal of the NPN output transistor might be supplied from a higher voltage source (20 to 70 volts). A PNP matched output transistor collector terminal is at ground or a negative potential. The two emitter terminals are wired together through very low ohm emitter resistors. A large electrolytic capacitor couples the high-powered audio to a set of speakers. Both left and right stereo output circuits are identical *(Figure 6-8)*.

When one of the audio output transistors appears open or leaky, the connected transistor can be damaged. Often, when one transistor becomes leaky the other transistor might be open. Sometimes a directly-coupled driver transistor becomes leaky or shorted can damage both output transistors. Check the bias resistors for overheated and burned areas with a leaky or shorted output transistor. You might find the emitter bias resistor cracked into. Replace the damaged emitter resistor with a 2 or 5 watt replacement. When one output transistor is found to be leaky or open, replace both output transistors. If one is found to be open or leaky, you may end up replacing the driver and both output transistors.

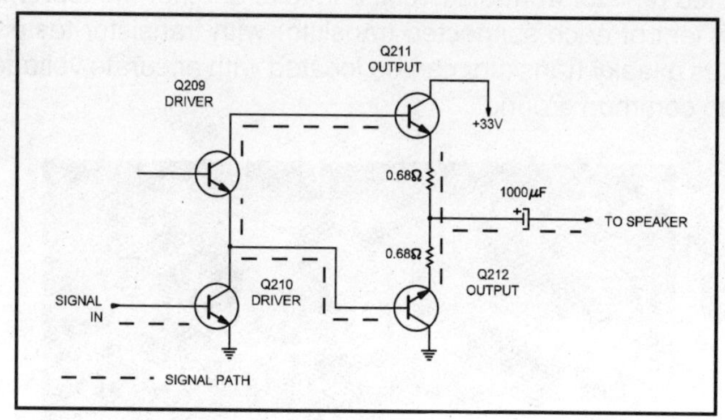

Figure 6-8. A typical driver circuit directly-coupled to two output transistors audio signal path.

TYPICAL DUAL-OUTPUT STEREO IC CIRCUITS

The dual-IC output stereo circuit within the RCA CTC157 TV chassis amplifies both audio channels. The left channel audio from the volume control IC (U1800) is coupled through C1901 to pin 8 of sound output IC (U1900). Likewise, the right channel audio is fed into pin 4 by a 1 µF electrolytic capacitor C1900. The amplified power output signal is taken from pins 1 and 13 and fed to a 32 ohm PM speaker in each audio channel *(Figure 6-9)*.

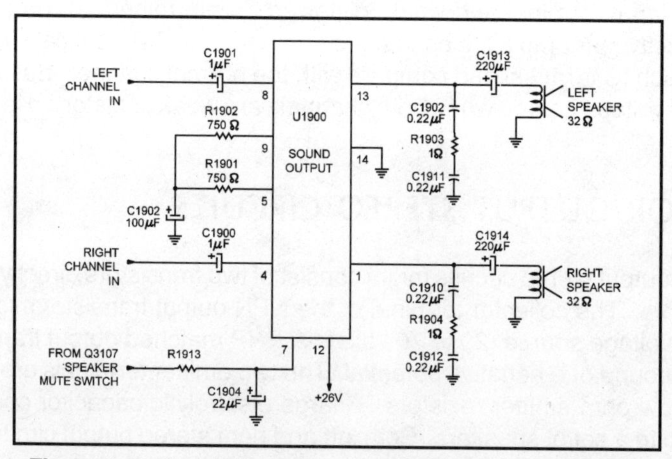

Figure 6-9. The dual-power output IC (U1900) found in the RCA CTC157 TV chassis.

The +26 voltage supply source (Vcc) is connected to pin 12 of U1900. Pin terminal 14 is at ground potential. Both speakers can be muted with a signal from the mute switch (Q3107) at pin 7. Some stereo TV sound circuits are also muted at the volume control IC.

Go directly to the audio output transistors or IC for a distorted speaker. Suspect the dual-power output IC for improper voltage source when both speakers are distorted. Scope the audio signal in and out of the output IC. If a scope is not handy, check the audio at pin 4 and 8 of IC1900. Now check the output signal from the power IC to the PM speakers.

A defective power output IC can produce a weak, distorted, and intermittent audio. Replace the leaky dual-output IC, when one channel is distorted and the other channel is normal. The output IC is often leaky with a low voltage power source. Remove pin terminal 12 from the PCB with solder wick and iron. Flick the pin terminal with the small screwdriver blade to make sure the terminal is free and no longer connected to the PC wiring.

Take a resistance check from pin 12 to the common ground. The power IC is leaky if the resistance is below 100 ohms. Now measure the +26 voltage supply at the pc wiring. If the voltage returns to normal or a few volts higher, IC1900 is leaky or shorted. Before removing and replacing the suspected leaky IC, check all components tied to the pin terminals of IC1900.

WEAK STEREO RIGHT CHANNEL

The weak stereo channel can be caused by a defective AF, driver or output transistors. The open AF, driver or buffer audio transistor can cause a weak audio signal. A defective driver or audio output IC can cause a weak right or left channel. Suspect an open electrolytic or bypass capacitor for a weak audio symptom. Check for a leaky or open bypass capacitor for weak sound on pins connected to the power output IC. Shunt the speaker coupling electrolytic capacitor between speaker and power IC for a weak stereo channel. Check for a defective electrolytic capacitor connected to the volume control for weak sound.

The unusual weak sound problem might be caused by a leaky or open diode off of the volume control pin 37 of the analog interface IC (AIU). The AIU IC (U3300) controls the OSD (on-screen-display), brightness, tint, color, contrast, and audio circuits in the RCA CTC157 chassis *(Figure 6-10)*.

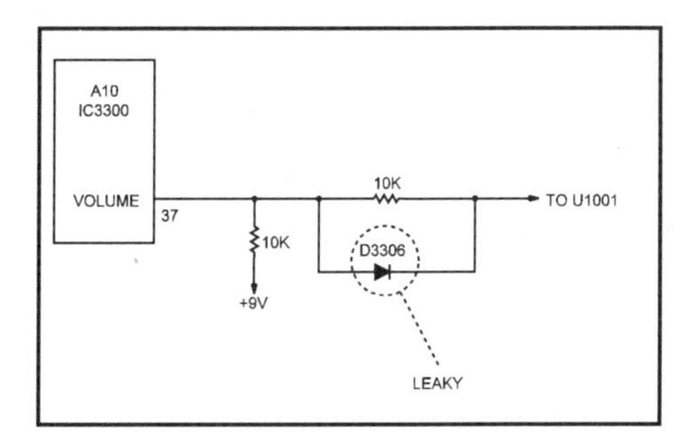

Figure 6-10. The unusual sound symptom was caused by a leaky diode (D3306) in the AIU IC3300 volume control circuit.

Check for weak sound with an open electrolytic coupling capacitor within the MPX decoder IC. Do not overlook improper voltage source feeding the sound circuits. An open voltage regulator transistor or low voltage from the power supply can cause a weak sound problem. Shunt each electrolytic in the voltage source feeding the sound circuits for weak reception.

DISTORTED LEFT STEREO CHANNEL

The weak volume and distorted symptom can be caused by defective electrolytic in the power output IC circuits. Distorted audio can result from a leaky driver or power output transistor. The left channel might be distorted with a leaky dual-power IC. Besides leaky capacitors, suspect a change in resistance for distortion within the driver transistor circuits. Remove one end of the suspected resistor from the circuit and test for an increase in resistance; this is especially true of resistors with large ohm values. After several hours of operation, the audio became distorted with a defective power switch in a Magnavox stereo receiver.

Check for poor soldered connections upon AF or driver transistors for distortion in either channel. The audio was distorted in an RCA CTC156 chassis with a leaky U1900 power output IC. A leaky SIF IC can cause distorted audio. Touch up the discriminator, quadrature or detector coil for weak and garbled sound. Suspect poor IF transformer connections for distorted audio. Check for cracked SMD resistors in the sound circuits for distortion. Suspect a video/Chroma/SIF/IC for distorted audio. Improper voltage or low negative voltage to the power output transistors can cause distortion.

INTERMITTENT STEREO SOUND

Intermittent electrolytic-coupling capacitors or poor capacitor connections can cause intermittent audio. Check for poor board connections on sound ICs or PCB. Place a little pressure on the body of the IC with an insulated tool and notice if the sound cuts up and down. Suspect a defective audio transistor after the sound quits in one hour or so of operation. Spray transistor and IC components with several coats of coolant to make the sound pop on and off.

Intermittent sound can result from a defective preamp or power output IC. Monitor the audio signal in and out of the suspected channel. Do not overlook a defective level control for intermittent audio.

Distorted audio can be caused by broken resistors or an increase in resistance of bias resistors. Check for a defective AF or driver transistor when the sound blares out at high volume. Bad soldered connections on small resistors in the audio voltage sources can cause intermittent sound. Check for poor soldered griplets, feed-through wires or bars for intermittent audio. The defective mute transistors can also cause intermittent sound in the speakers.

In the RCA CTC167 chassis, the intermittent sound would cut in and out of both channels. Both stereo channels were monitored at the Summer transistors (Q1703 and Q1704), to no avail. The sound became intermittent in the left channel upon pin 1 of the left matrix IC (U1702). Voltage measurements upon Q1702 turned up nothing. When taking critical resistance measurements in the matrix circuits, R1753 (120K) had increase in value *(Figure 6-11)*. These same stereo circuits are found in the latest RCA TV stereo chassis.

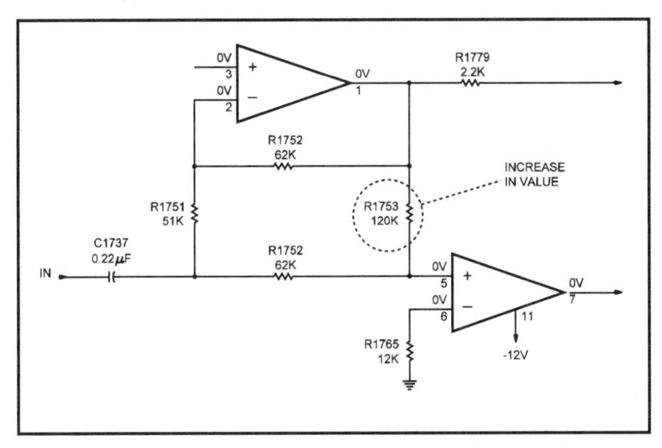

Figure 6-11. R1753 increased in value causing intermittent sound in the RCA CTC167 matrix stereo system.

BOTH STEREO CHANNELS WEAK OR DISTORTED

When both stereo channels are defective, check components or power sources that are common to both stereo channels. For instance, a high voltage source might feed +157 volts to one of the sound output transistors within each left or right stereo channel. A low negative or positive voltage source to the output transistors can cause a weak or distorted channel. A defective voltage regulator transistor, low ohm isolation resistor or open decoupling electrolytic can cause weak and distorted conditions.

A leaky dual-driver IC or power output IC can cause a weak and distorted audio symptom. Likewise, a defective dual-preamp IC can also cause sound problems in both audio circuits. The defective matrix IC can cause sound problems in both audio channels. A defective dual-audio switch IC or dual-volume control IC, in the latest TV audio circuits, can cause distorted and weak sound in both channels. Replace the dual-output IC when the left channel is noisy and fuzzy after 5 minutes of operation.

DEFECTIVE BASS AND TREBLE CIRCUITS

A dirty or worn volume, bass or treble control can cause intermittent and distorted sound within the audio circuits. The open bass or treble control has no action when rotated. The poor terminal or broken connections might cause a no treble or bass sound. Check for a shorted wiping blade inside the control for a no audio symptom.

The audio can be intermittent with a low hum in the sound with an open volume control. The sound might be noisy when the stereo receiver is just turned on and caused by a defective volume or treble control. The left channel might be dead with a shorted left channel volume control. Suspect a defective coupling capacitor for a weak or no audio symptom, when wired to the volume or treble control.

STEREO RECORDING CIRCUIT PROBLEMS

Most problems with stereo playback or recording circuits are dirty or worn function switch contacts. Spray cleaning fluid down inside the switching area. Rotate the function switch back and forth to help clean the silver contacts. Replace the original function switch if there are worn or broken contacts.

A no record symptom in the left channel can be caused by a defective tape head. Suspect a worn tape head when the cassette player operates in the play mode and only a rushing noise in record mode. The distorted right channel might result from a dirty or packed oxide upon the tape head. An open right channel tape head might result in no playback or recording.

The intermittent recording was caused by a defective bias control (10K ohms) in a Silver Marshall cassette player. Check the recording amp transistor or IC for improper voltage in record mode.

The poor or no recording can result from a defective bias oscillator circuit. Poor erasing of the previous recording can produce a jumbled or cross-talk recording. The bias oscillator might operate from 60 to 100 kHz frequency range to erase the previous recorded music and provide linear recordings. In the early bias erase circuits, a dc voltage was switched into the head circuit to erase the previous recording in the low-priced cassette players *(Figure 6-12)*.

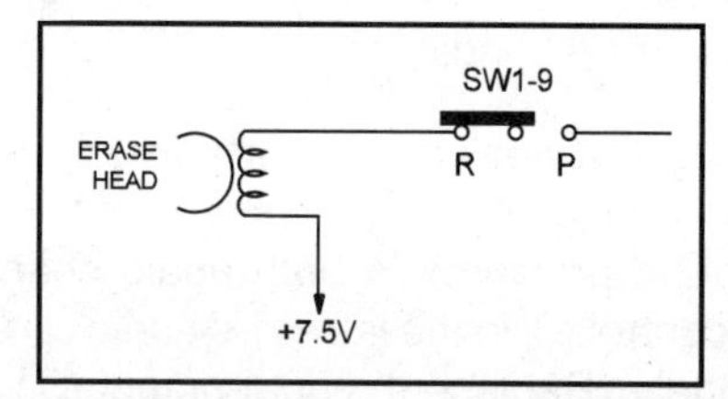

Figure 6-12. Low dc voltage is switched into the inexpensive cassette player to erase the previous recording.

You might find one bias oscillator transistor circuit exciting both stereo erase and recording channels in the cassette player. The erase head is excited from the oscillator coil or transformer as the erase head is switched into record mode. Both stereo tape heads are excited by the bias oscillator through isolation resistors and coupling capacitors. The bias oscillator 11.4 supply voltage source is switched only in the record mode *(Figure 6-13)*.

No recording is noted when Q501 becomes leaky or open. Improper supply voltage can cause poor recordings. Take critical voltage measurements upon Q501 and compare to the schematic. Check for broken or poor oscillator coil connections for intermittent recordings. Scope the left and right stereo tape head terminals for a waveform in record mode. Suspect a defective bias oscillator circuit with no waveform on the erase or stereo tape heads.

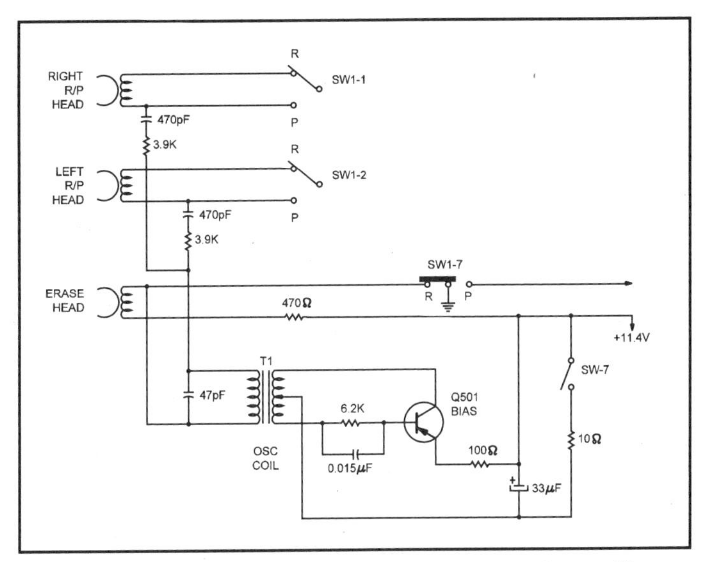

Figure 6-13. A dc voltage is switched into the bias oscillator circuit to excite the stereo and erase tape heads.

STEREO SPEAKER PROBLEMS

Speaker damage can be caused by too much volume applied to the speaker, weather conditions, and a dc voltage applied to the voice coil. Excessive power applied to the speaker can blow out the voice coil and cause it to drag or hang from the cone area. Wet weather can cause the cone to warp in the auto speaker. The speaker cone can come loose upon the framework and produce a blatting noise *(Figure 6-14)*. The solid-state amplifier can be damaged with an open speaker connection.

Figure 6-14. The voice coil and cone can be damaged with too much output power applied to the speaker.

Always check the voltage at the speaker terminals, when the voice coil is frozen upon the magnet pole. A leaky output transistor or IC can destroy the voice coil by placing a dc voltage at the speaker terminals. The defective output component can place a dc voltage upon the speaker terminals if a electrolytic capacitor is not connected between amp and speaker. The voice coil heats up with applied voltage and drags or remains frozen against the magnet pole. A balanced output circuit should have zero voltage where the speaker is directly-connected to the output transistors or ICs.

Check the suspected speaker with an ohmmeter test across the voice coil terminals *(Figure 6-15)*. Next inspect the cone for breaks or holes. Make sure the spider or cone is not loose upon the metal frame. Sometimes the cone or spider can repaired by regluing it into position. Place both thumbs upon the opposite sides of the cone and move the cone up and down. If the cone rubs or does not move, replace the speaker. Small holes punched into the speaker cone can be repaired with speaker or contact cement.

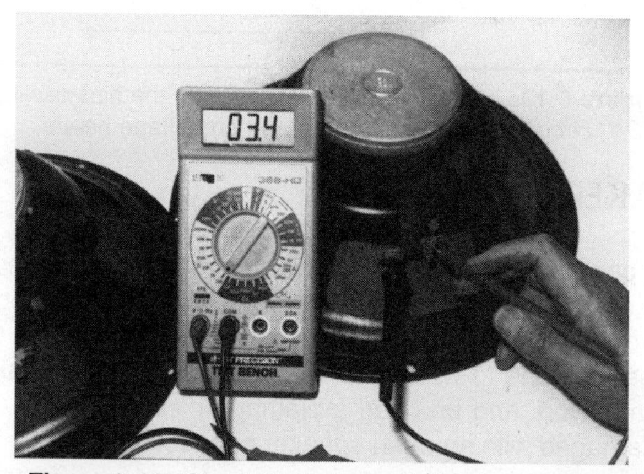

Figure 6-15. Check the voice coil with the low ohm range of DMM.

SERVICING AM/FM/MPX STEREO RECEIVER CIRCUITS

The left audio signal of an AM/FM/MPX receiver or cassette player can be signal-traced from the MPX IC2, through the radio-cassette switch S-4-1, Q13 and Q15, volume control, dual-power output IC3 to the left channel speaker. Signal trace the audio signal with the scope or external audio amplifier *(Figure 6-16)*.

Determine if either the radio or cassette player are functioning. If the FM MPX receiver is not working and the cassette player is normal, the audio problem lies in the FM MPX IC2 or front-end circuits. When the radio circuits are normal and the cassette player sound is weak or distorted, check the tape head and preamp circuits ahead of S4-1 and S4-2.

A weak left channel is caused by a defective component in the left audio circuits. Check the audio signal at the left channel volume control and compare the signal at the right channel volume control. The greatest thing about the audio stereo circuits is that you can compare signal strengths against the defective and normal channel from the input to the output circuits. You can use either the test cassette in the cassette player or audio signal as the source in signal tracing.

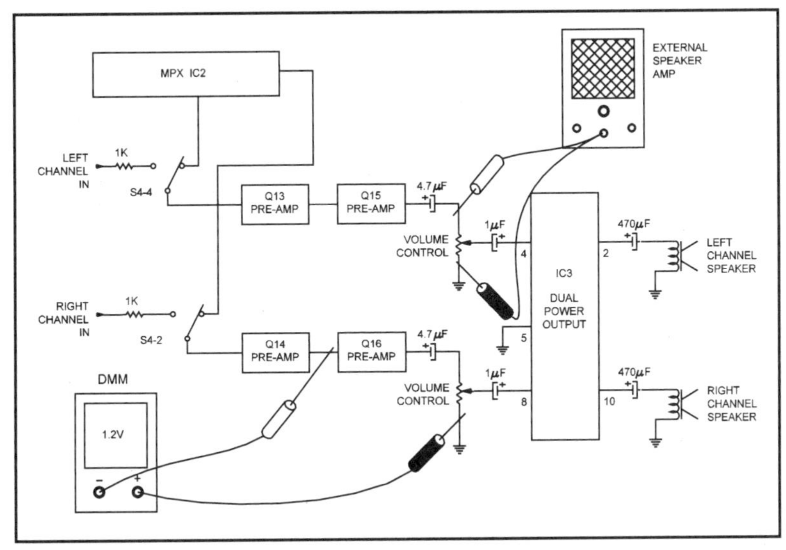

Figure 6-16. Signal trace the audio signal and take critical voltage measurements to locate the defective stage.

Start at the base terminal of the preamp (Q13) and compare the signal with Q14, when the audio is weak at the left volume control. You know the defective component is in the preamp circuits. Place the signal probe to the base terminal of the left preamp Q15 and compare the signal at the base of Q16. Likewise check both sides of the electrolytic capacitor for a weak left channel. When the signal appears to be weak, you have located the defective circuit.

For instance, if the signal becomes weak at the output collector terminal of Q15 and the base signal was normal, suspect Q15. Take critical voltage measurements upon Q13 and Q15. Now compare these measurements with the right channel. Check both transistors with in-circuit transistor tests or a diode-junction DMM test. Take critical resistance tests on all Q15 terminals to common ground, if the defective component is not already located. Remember, weak audio signals can be caused by defective transistors, ICs, electrolytic coupling capacitors, and bias resistors.

SERVICING TV STEREO OUTPUT CIRCUITS

The TV stereo output circuits might consist of transistors, ICs or both. Only three output transistors are found in the audio output circuits of an Emerson MS1980R model *(Figure 6-17)*. Each stereo channel has identical driver and sound output amplifier circuits. Notice the sound circuits have a much higher dc supply source then most solid-state sound output circuits.

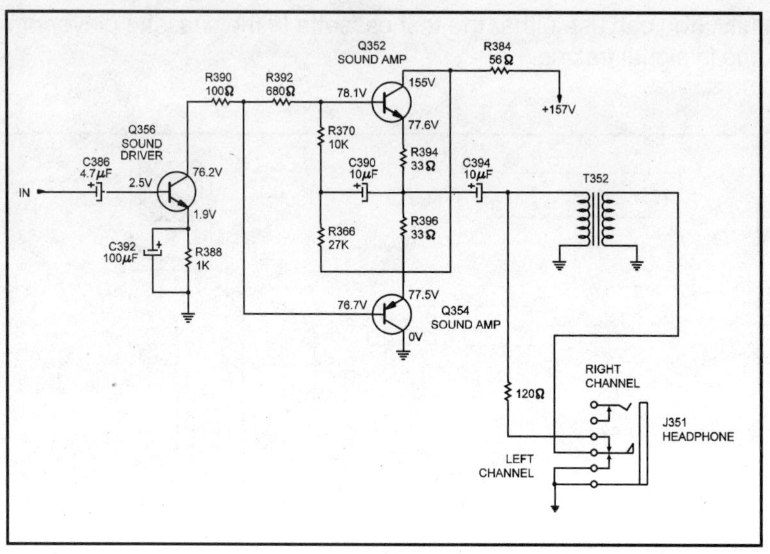

Figure 6-17. The left channel audio driver and output circuits in the Emerson MS1980R TV chassis.

C386 couples the audio from a tone control IC (IC371) to the base terminal of driver transistor Q356. A direct-resistor couples the audio at the collector terminal of Q356 to both base terminals of the sound amps (Q352 and Q354). The audio output is coupled from the emitter resistors to C394 (100 µF). A step-down transformer couples the power output to a 8 ohm PM speaker. The stereo headphone jack switches in the speaker or headphones when the plug is inserted into J351.

Service the TV sound circuits as any audio output stage. Signal trace the audio into the output circuits and at the speaker terminals. Check each transistor with in-circuit tests. A weak output circuit might be caused by an open driver transistor or electrolytic coupling capacitor. The dead output circuit can be caused by the open driver transistor, coupling capacitor, sound output transistors, or speaker coupling capacitor.

A leaky coupling capacitor, driver and sound output transistors can cause distortion in the speakers. Take critical voltage measurements on each transistor. Notice that these transistors operate from a much higher voltage source.

For a no audio symptom, signal trace the TV audio circuits with the scope or external audio amp. Check for an open transistor, IC or coupling capacitor for no audio in the speakers. Suspect a defective transistor or IC when the audio stops after operating for several hours. The no sound symptom can occur with a defective AF or preamp transistor or IC. A leaky or shorted output IC can cause no sound in both stereo channels. A defective switch or volume control IC can cause no audio in one or both audio channels. Suspect a shorted power output IC that might shutdown the whole chassis.

The no sound symptom might result from a blown power line fuse. Check the speaker fuse for open conditions when one channel is dead and no hum in the speaker. Suspect a defective speaker relay or poor relay points for no audio in the speaker. Check for poor or bad soldered board connections on power output transistors and IC components. A cold soldered connection upon coupling and bypass capacitors can cause a no sound symptom.

Do not overlook a missing power supply source for no sound in both channels. Check the voltage across the main or largest filter capacitor. Next check the supply voltage source for correct voltage. Small ohm resistors (1.2 to 2.7 ohms) in the power source may be cracked or open. Inspect low ohm resistor terminals for burned areas upon the PCB, indicating a poor soldered connection. Open bias resistors in the emitter circuit of the audio output transistor or upon a power output IC can cause a no sound symptom.

Check for shorted or open electrolytic capacitors that may be off of their terminal pins of the power output ICs, for no audio. Within a JVC C1950 TV chassis a no sound symptom with popping noises, resulted in a defective IC901. Most no sound problems result from shorted or leaky output transistors and ICs with a blown line fuse. Do not overlook the audio program set up procedures in the latest TVs for a no sound symptom.

After a couple of years it is possible to collect several different no sound problems in a certain TV chassis. In the RCA CTC145 chassis, on different occasions, a no sound symptom was caused by Q1202, Q1203, and the sound IF IC (U1001). Check bias resistors R1209 and R1210 with a shorted output transistor. A no sound problem can be caused by a defective C1207 (100 μF) and C1206 (680 μF) electrolytics *(Figure 6-18)*. Also check R1211, a 5.6 ohm resistor off of collector terminal Q1202 for open conditions. Sometimes this resistor will open with a shorted Q1202 audio output transistor.

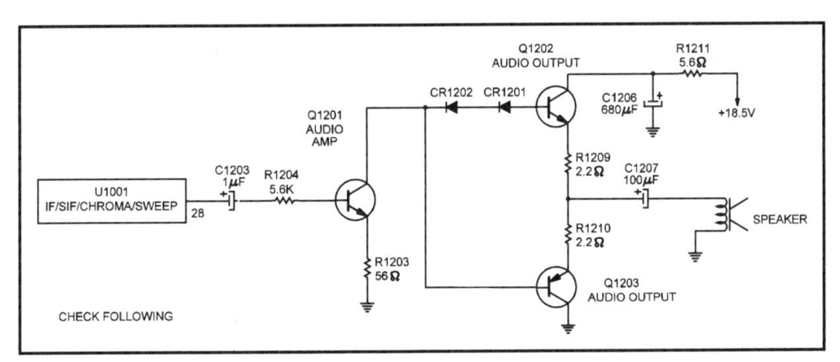

Figure 6-18. Check the following components for sound problems in the RCA CTC145 TV chassis.

REPAIRING THE CASSETTE STEREO CIRCUITS

The function switch of an AM/FM/MPX boom-box player, switches in the cassette stereo circuits instead of the radio circuits. The same audio circuits are used for both receiver and cassette player circuits. Of course, a separate play/record switch provides playback and recording features. The preamp or equalizer circuit might contain a transistor or IC tape head circuit *(Figure 6-19)*.

The tape head is switched into the base circuit through an electrolytic capacitor in play mode. The weak tape head signal is amplified and switched into the tape/radio function switch. Likewise the recording circuits are switched to the tape head to record the microphone or auxiliary music. At the same time, the bias oscillator circuits are switched in for better recordings.

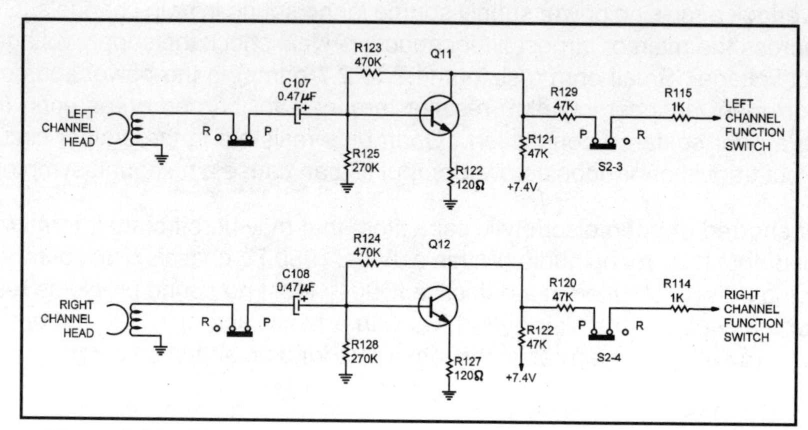

Figure 6-19. Start at the tape head and signal trace the audio with the scope or external audio amp in the cassette tape circuits.

Most problems found in the cassette player preamp circuits are related to the tape head. A dirty R/P switch can cause weak or intermittent stereo channels. The dirty tape head might produce a dead left channel and weak right channel during recording or playback modes. A loud rushing noise might be caused by an open tape head or wires torn from the tape head. Poor tape head connections can cause intermittent audio.

Insert a cassette and notice if either stereo channel is active. When the right channel is dead or weak, locate the right channel preamp transistor. Touch the right channel tape head terminals with the blade of a screwdriver and the volume wide open. You should hear a loud hum. If no hum, take critical voltage measurements upon Q12 and compare them with the left channel of Q11. Check the 7.45 voltage source on R122 (47K ohms). Test Q12 with in-circuit transistor tests. The cassette music can be traced from the tape head to function switch S4 with external amp or scope.

TROUBLESHOOTING CD PLAYER STEREO CIRCUITS

Start at the D/A converter when either stereo audio channel is weak, dead or intermittent. Signal trace the audio with the scope or external audio amp at the L and R channels of D/A converter IC. Locate the D/A converter upon the CD player chassis with a schematic layout drawing *(Figure 6-20)*.

For instance, if the right channel is weak compared to the left channel line output, start signal tracing at pin 8 of IC706. Suspect IC706 and corresponding circuits if the signal is weak on pin 8. When the signal is normal at pin 8, check the audio signal at pin 6 of IC702. Compare this audio signal with pin 2 of IC702 (*Figure 6-21*). Proceed through the audio stages and through the line output circuits until the weak right channel component is located. Take critical voltages upon each IC or transistors. Shunt electrolytic capacitors with the exact capacity and working voltage to locate the weak audio signal.

Figure 6-20. Locate the suspected D/A converter IC on CD player PCB.

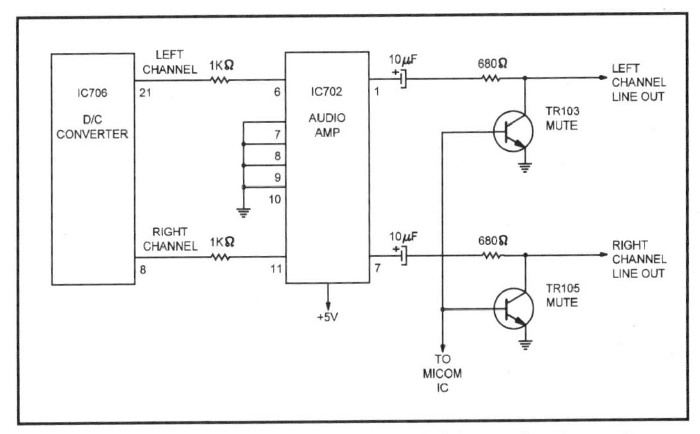

Figure 6-21. Signal trace the left and right audio channels on pins 8 and 21 of IC706.

A defective D/A converter IC has been noted to cause no audio, weak, intermittent, and distorted sound. Check the D/A IC for a background noise when the music is playing. Defective electrolytics have caused weak, dead, noisy, and distorted music. The defective Ram IC in the CD player can cause a high-pitched noise, distortion and a ticking noise in the sound when the disc is playing.

The defective line amplifier can produce a clicking noise, distortion after warm up, and intermittent static in the audio. Check for poor soldered pin connections for intermittent sound of the audio amp IC. Replace the IC amp for frying noises. Check the supply voltage source for low voltage with weak and distorted music symptoms. A defective power source voltage regulator can cause weak or no audio, intermittent audio, and a clicking relay with poor regulator transistor terminals in the CD player.

SERVICING THE AUTO STEREO CIRCUITS

In the early auto receivers, the auto front-end (AM/FM/MPX) circuits are switched into the stereo audio output circuits at the volume control. A cassette player IC preamp circuit amplifies the tape head circuits when the cassette is inserted into the radio. When the cassette is inserted, the receiver circuits are switched out with a function switch or with fixed diodes.

When the receiver circuits are functioning and no tape music can be heard, signal trace the tape head and preamp circuits. If both tape head and radio circuits are dead, weak or intermittent, signal trace the audio output circuits.

Check the audio at the input circuits at pin 6 of both audio output ICs (IC4 and IC5). Suspect IC4 when the left channel is weak or dead and the right channel is normal *(Figure 6-22).* Likewise, suspect IC5 when the right channel is noisy or distorted with a normal left channel. Check for low or improper voltage source when both channels are dead or weak. Check all components tied to the audio output ICs before removing and replacing the suspected IC.

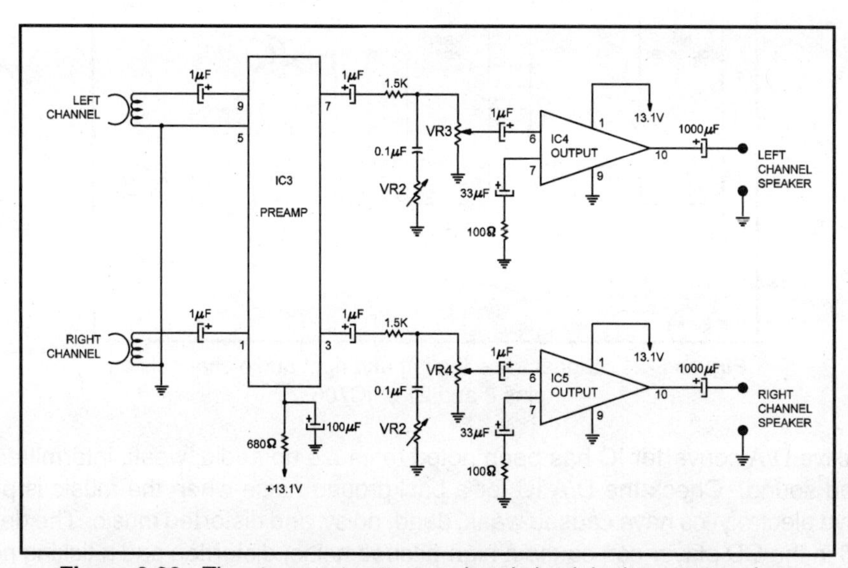

Figure 6-22. The stereo cassette tape head circuit in the auto receiver.

Suspect an open tape head with a loud rushing noise in one channel. Check for a broken tape head wire for no sound, when the cassette player is normal in the play and reverse mode. Suspect a dry supply reel when the audio seems to flutter during playback. Check for an open electrolytic coupling capacitor for motorboating in one channel. A loud popping noise can be caused by poor soldered connections on the output IC. Replace IC4 for a low rushing noise in the left channel with the volume turned way down.

The left channel was dead in an older Motorola FM481AX car radio. The right audio channel was normal. Q9 was found leaky and replaced. The left channel was still dead. Several voltage measurements were off on Q7 and Q9 *(Figure 6-23).* The output transistor Q9 measured 3.5 volts on the collector and should have been 1.8 volts. Driver transistor Q7 measured only 3 volts upon the collector terminal and should be around 11.4 volts. The emitter terminal of Q7 was quite high (2.9v). A shorted electrolytic C7 (100 µF) was replaced and solved the dead left channel.

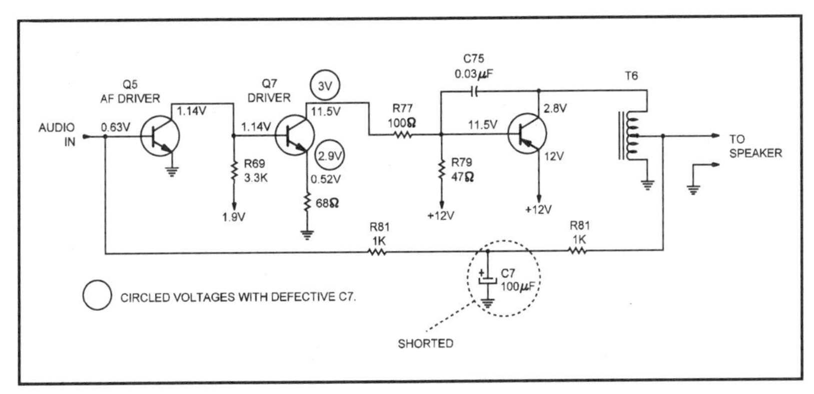

Figure 6-23. Capacitor C7 destroyed Q9 and produced improper voltages on Q7 and Q9.

REPLACEMENT OF POWER OUTPUT TRANSISTORS

After removing the suspected power output transistor and locating a correct replacement, test the new replacement before installation *(Figure 6-24).* Check the replacement with a transistor tester or the diode test of a DMM. If not, many hours can be wasted looking for other problems. Believe it or not, you can receive a new replacement transistor or component that is defective.

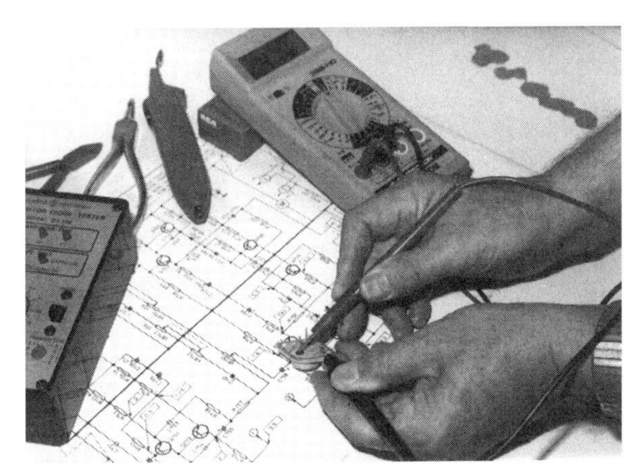

Figure 6-24. Check the new transistor replacement on transistor tester or diode-test of DMM.

Most audio output transistors are mounted upon a heat sink with a piece of insulation between transistor and metal heat sink; the heat sink might be the metal chassis *(Figure 6-25).* Apply a coat of silicone grease on both sides of the mica insulation. Since the emitter and base terminals are offset with a T0-43 type transistor, you cannot place it incorrectly within the socket.

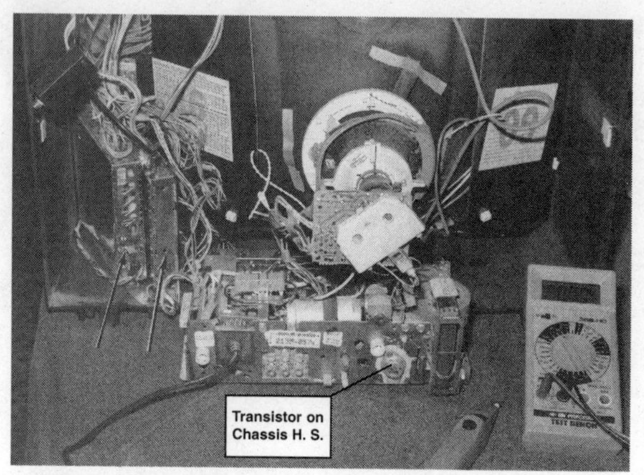

Figure 6-25. The metal chassis might serve as a heat sink in TV chassis.

Some heat sinks are mounted on top of the PC board. Here, the heat sink is insulated away from other components and does not require a piece of insulation between heat sink and transistor. Silicone grease should be placed upon the heat sink to provide greater heat dissipation from the transistor to the metal heat sink.

The metal chassis of the auto receiver might serve as a heat sink for the power output transistor or IC. Sometimes a large heat sink is mounted at one end of the radio and bolted to the metal chassis *(Figure 6-26)*. The output transistor or module are bolted to the metal heat sink. Make sure all metal mounting screws holding the transistor or IC are snug and tight against the heat sink. A loose mounting screw that connects the collector terminal to the output circuit might produce intermittent or erratic reception. Now solder the transistor or power IC terminals to the PC board.

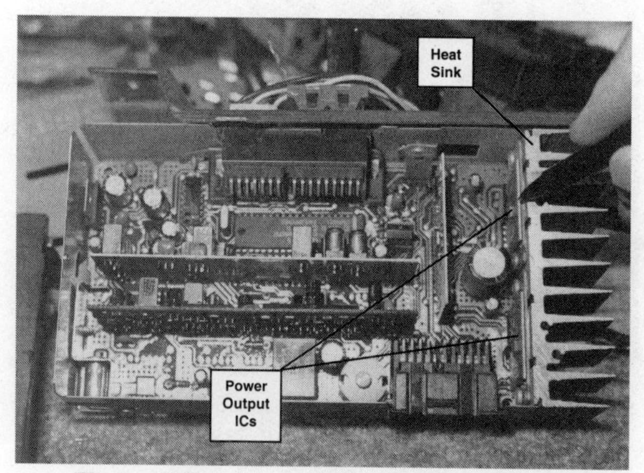

Figure 6-26. A large heat sink provides heat dissipation in the auto radio.

REPLACING POWER OUTPUT ICS

After determining the power output IC is leaky or shorted, remove it from the heat sink and chassis. Remove the many terminal pins with solder wick or a sucking tool. Lay the solder mesh material along side a row of IC terminals and heat with a large soldering iron or gun. Apply excessive heat upon wick material and move it alongside of the IC terminals, picking up excess solder. A de-soldering bulb, iron or de-soldering pump helps to remove solder around the pin terminals.

Make sure all solder is removed from each pin terminal and pc wiring. Flick each pin terminal with an insulated tool or small screwdriver to make sure the pin is loose and free. Remove the metal screws holding the IC component to the heat sink.

Replace the IC with original part number. If not available, look up the part number in a universal replacement manual for an replacement. Universal ICs and transistors work nicely in the audio circuits. Universal solid-state components can be found in RCA, Sylvania, NTE, Howard Sams cross reference, and Japanese universal replacement manuals.

The large IC sound output circuits are often mounted upon a large heat sink with many metal fins for cooling *(Figure 6-27)*. The heat sink is usually mounted directly on the PCB. Apply silicone grease on the heat sink and IC component for greater heat dissipation. Bolt the large IC to the heat sink with the same metal screws. You might find plastic screws on a few IC components that insulated the IC away from the heat sink. Snug up the screws or bolts, so the heat sink will make a very tight connection.

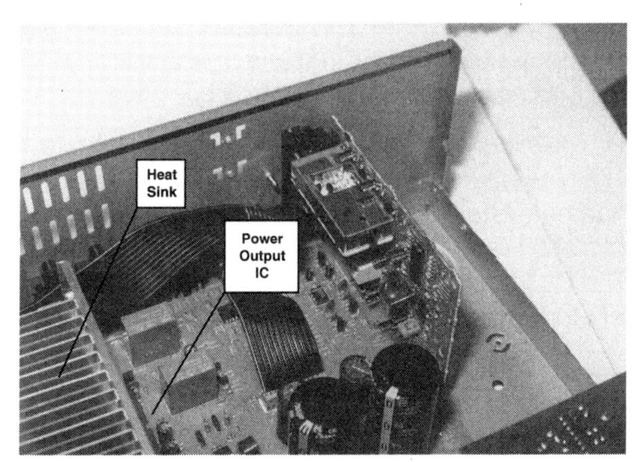

Figure 6-27. Many cooling fins are found in the power output IC heat sink.

Solder up all pin connections to the PCB. Check each pin terminal for excess solder that might bond two or more terminals together. Remove excess solder around the pin terminals and solder up again. Check the resistance between each succeeding pin terminal for possible leakage. Clean off all excess rosin around the terminals for a clean appearance, with cleaning spray or brush.

TUBE STEREO OUTPUT CIRCUITS

The high-wattage tube-amplifier might have dual-triode tubes as AF and driver stages. The output circuits might consist of four power output tubes in a push-pull operation. Two of the power output tubes have the plates tied together and feed into the primary winding of the output transformer. The secondary winding might provide a 4, 8, and 16 ohm speaker impedance taps *(Figure 6-28)*. You might find ultra wide-band toroidal output transformers in some tube circuits. The 100 watt amplifier can operate from 475 to 550 volts of low voltage power supply.

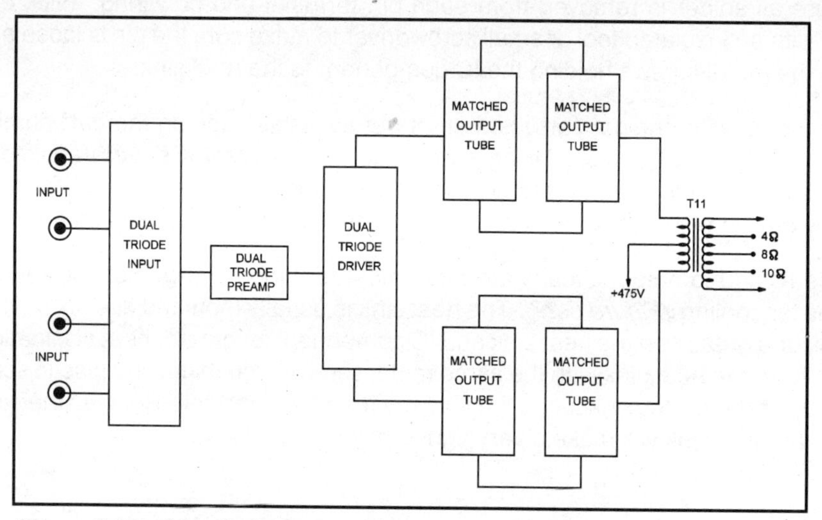

Figure 6-28. A block diagram of the 100 watt tube amplifier with four output tubes in parallel and push-pull operation.

The stereo tube circuits have identical high-powered circuits from 6 to 10 tubes with the AF and driver circuits containing dual triode tubes. Two separate single-channel tube amplifiers can serve in the stereo system. The 20 watt amplifier can contain four dual-triodes and four output tubes. A 50 watt stereo push-pull tube amplifier might have four 6550 or KT88 output tubes. The EL34 output tubes are found in 20 and 30 watt amplifiers. Some of these tube amplifiers today appear in kit form from $700 to $1400.

The distorted tube amplifier can be caused by a gassy, weak or shorted vacuum tube. A bluish-red color inside the tube might indicate a gassy tube. Check the suspected tube within the tube tester. Substitue another new tube when a tube tester is not available. A matched set of tubes should be installed in the audio output circuits. Although a complete set of tubes is rather expensive, keep one each of the preamp, driver and output tubes on hand.

A shorted output tube might destroy the screen grid resistor and damage the power output transformer. Improper negative bias applied to the grid circuits of the output tubes can destroy or damage them. Check for a negative bias supplied from a power transformer winding through a bias adjustment control. An increase in resistance of a grid or plate load resistor within the AF and preamp circuits can cause distortion and a loud hum symptom.

Suspect a burned or open cathode resistor of a shorted or leaky AF or preamp triode tube. A change of resistance in the grid circuits of a preamp or AF tube can cause slight distortion and hum in the sound. Poor soldered input grounds can result in pickup hum in the speakers. A dirty or worn volume control can cause erratic or noisy audio. Excessive hum can be found in the sound with dried-up decoupling or filter capacitors in either the positive or negative voltage sources.

Chapter 7

Repairing SMD Audio Circuits

Surface mounted devices (SMD) are found in many different consumer electronic products such as the TV, portable cassette player, compact disc, VCR, and camcorder. SMD parts are very miniature in size and are difficult to identify. Although, the SMD or SMT components are very tiny, they can be tested in the same manner as the standard or larger parts.

Because many surface-mounted devices have similar shapes and sizes, sometimes they are quite difficult to see upon the chassis. The SMD part might appear as a black, brown or silver part on the pc wiring. These surface-mounted components are so small as they might be mounted between a regular IC terminals upon the pc wiring side *(Figure 7-1)*. SMD components are mounted directly on the pc wiring, usually on the bottom side, and standard parts are found on the top side of the PCB.

Figure 7-1. SMD components might be found between standard IC's within the RV chassis.

Commercial SMD resistors might appear as round, flat and leadless devices. The fixed resistor might appear black or silver with several numbers stamped on top, indicating the correct resistance. The round SMD resistor might have color-coded rings around the component for correct resistance.

The SMD ceramic capacitor might be flat with outside terminal connections and tinned ends. The SMD part terminals are found at each end, except transistors and IC chips; either end can be mounted between the pc wiring pads without any difficulty.

The commercial surface-mounted transistor might appear in a chip form with flat or gullwing contacts at one side, top and bottom, or on both sides. You might find more than one transistor inside one chip.

The same applies to fixed diodes and LED SMD components. Two or more diodes might be found in one component or chip. Each SMD part can be tested like standard components. They are easily checked and tested when wired or soldered to the pc wiring.

SMD CONSTRUCTION

The SMD parts are now available for electronic project construction. They are marked and mounted, somewhat like the commercial SMD component. These SMD parts are very small in size and must be handled with care. Since these parts are so tiny, they can easily be flipped out of sight. For surfaced-mounted resistors and capacitors, select the largest physically, highest wattage, and working voltage; they are much easier to work with.

Choose SMD electrolytic capacitors with at least a 16 volt rating for small 9 volt electronic projects. Select thick film resistors with a 1/8th watt size. Select a ceramic capacitor chip with a 50 volt working voltage. The ceramic electrolytic capacitor chip can be selected with a 5, 10, 16 or 25 volt rating. A radial or stand up electrolytic might have a higher voltage rating. These surface-mounted parts are soldered directly to the pc wiring. They are ideal for building small electronic projects *(Figure 7-2)*.

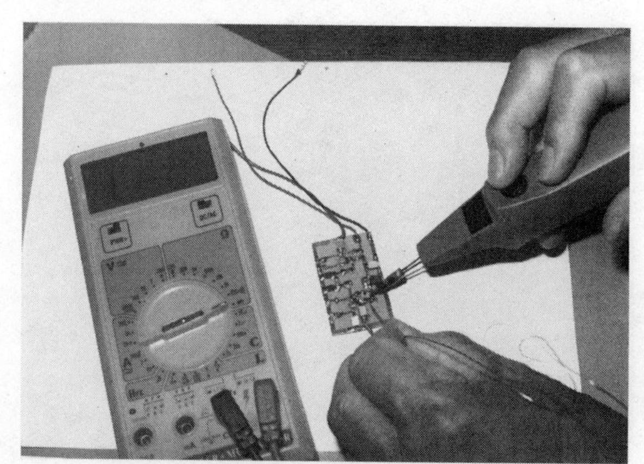

Figure 7-2. Surface-mounted components are now available for project construction.

The most common SMD parts available for electronic construction are capacitors, resistors, transistors, ICs, LEDs, diodes, and inductors. The SMD ceramic capacitor might be available in three or four different case sizes, 1210, 1206, 0805, and 0603. The 1210 and

1206 case sizes are physically the largest chips to work with. For instance, the SMD 1210 case is 3.05 in. long, 2.54 in wide, and 1.27 mm thick, while the 1206 is the same length (3.05 mm), 1.5 mm wide and 1.27 mm thick. The smaller the capacity in picofarads (pF), the smaller the case size of the capacitor.

Most ceramic chip capacitors are available with a 50 volt working voltage. The capacity can range from 0.5 pF to 0.068 µF. These ceramic chip capacitors are available from some mail order firms in single (1), 10, 100, 500, or 1000 lot prices. It's best to purchase SMD parts in a 10 lot price. The ceramic chip capacitors are used in bypass and coupling electronic circuits.

Remember the ceramic chip capacitor is a non-polarized capacitor. You can solder any end into the pc wiring without any problems. The ceramic capacitor might have a letter and number stamped on the top, indicating the actual capacitor value; while in other chips there are no markings or values with only end connections. Always keep these small SMD components inside the marked plastic envelopes, so they will not get lost or mixed up *(Figure 7-3)*.

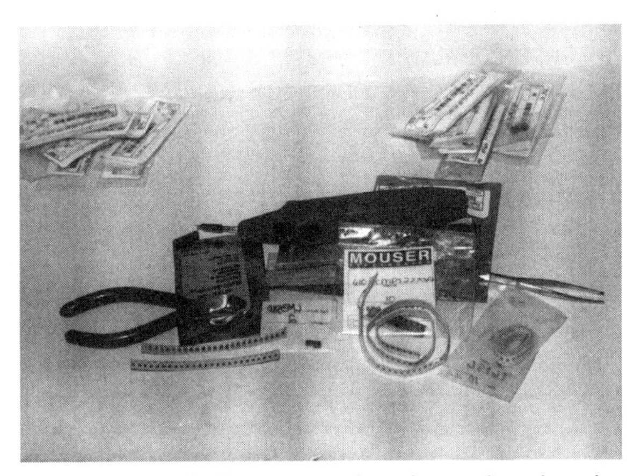

Figure 7-3. SMD parts are found on strips placed inside plastic envelopes.

The SMD aluminum electrolytic capacitors are polarized and should have at least a 16, 25, 35 or 50 volt working voltage. Do not use an SMD electrolytic capacitor of 10 volts or less working voltage for a 9 volt circuit, as they have a tendency to break down.

Often the voltage and capacity are stamped on top of the electrolytic capacitor. The top black marked area indicates common ground terminal. Observe the correct polarity of electrolytic capacitors or if installed backwards, they can run warm, and blow up in your face. They are available from 0.15 to 1000 microfarads (µF). The aluminum SMD electrolytic capacitor stands up while the solid chip lays down on the pc wiring.

Tantalum electrolytic chip capacitors are found in a smaller capacity and can be purchased in the 16, 20, 25, and 35 working voltage types. The black or white polarity bar on the top side and at one end is the positive terminal. Most standard electrolytic capacitors have a black line that indicates a negative or ground terminal. These small SMD chip electrolytic capacitors have the reverse, a positive (+) polarity with a black or white bar at one end. Place the bar at the positive voltage connection with a SMD electrolytic chip capacitor.

The Tantalum electrolytics are available from 0.47 to 47 microfarads (μF). The SMD aluminum electrolytic capacitor is used in B+, decoupling and power supply circuits, while the lower capacity Tantalum capacitors might be found in coupling and bypass circuits.

The SMD fixed resistor might be identified by numbers stamped on the top side of the chip. These SMD resistors appear in thick film chips of 0805, 1206, 1210, and 2512 case styles. Choose the 0805, 1206 and 1210 case styles for electronic construction. The 0805 resistor is 1/10th watt, 1206 is 1/8th watt, and 1210 is 1/4 watt. The 2512 case is a 1 watt SMD resistor. These fixed resistors appear in 0, 10 to 1.0 megohms. The zero (0) resistor might be used as a feed through or to tie two circuits together.

These SMD resistors can be purchased in 1, 10, or 100 lot prices. It's best to choose SMD resistors in a 10 lot price of each value.

Remember, either end of the resistor can be soldered into the circuit with the resistance value at the top. For instance, the SMD resistor might have 122 stamped on top, where the first two numbers equal the amount with the last number rated in zeros *(Figure 7-4)*. The number 1 and 2 equal 12 with two zeros at the end equals a 1200 ohm resistor.

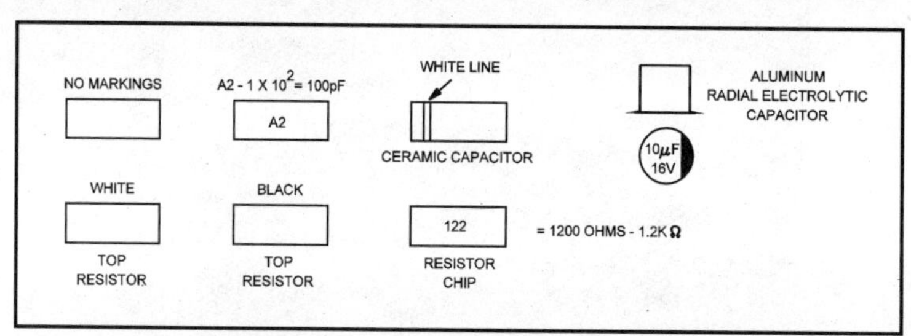

Figure 7-4. The SMD capacitor might have a letter and numbers, while the SMD resistors have numbers for the correct value on top of the component.

The surface-mounted transistor (SMT) might appear as a chip with flat contacts at one end, top and bottom, or both sides. You might find more than one transistor in one chip. The standard or conventional SMD transistor has a SOT-23 package outline, while the one watt power transistor has a SOT-89 outline with a heat sink *(Figure 7-5)*. The SOT-89 and SOT-223 might consist of two transistors in one chip or in a Darlington arrangement.

The conventional transistor SOT-23 is a general purpose transistor that you will find in electronic projects; such as the Digi-Key FMMT3904CT-ND (2N3904) NPN or the FMMT3906CT-ND (2N3906) PNP type transistor. The Mouser listing for the same type of transistor has a part number of MMBT3904 (2N3904) NPN or the MMBT3906 (2N3906) PNP type. The conventional 2N2222 general switching transistor in the same SMT types is listed as MMST2222 at Mouser and FMMT2222ACT-ND at Digi-Key Corporation. The Digi-Key corporation and Mouser Electronics are electronic mail order firms.

The SOT-23 general purpose transistor has the collector terminal on one side (at the top) with the base to the left and emitter terminal to the right on the bottom side of the package outline. Some have flat or gullwing type terminals. These transistors can be tested with the transistor tester or on the diode-test of the digital-multimeter (DMM).

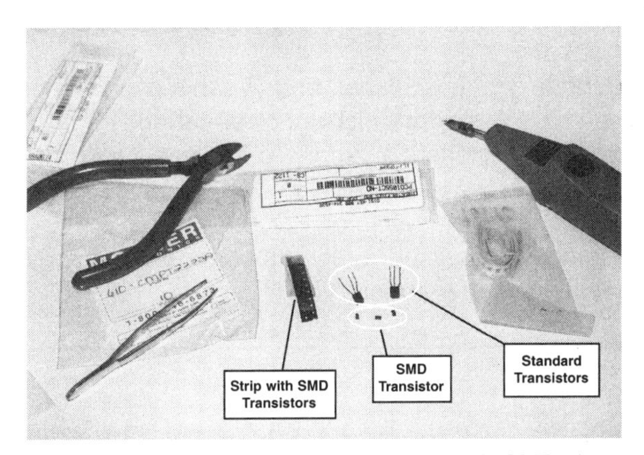

Figure 7-5. The standard transistor with SMD size transistors below.

The SOT-23 general purpose FMMT3904CT has a 1A stamped on top, the FMMT3906CT can be identified with a 2A on top, and the CMPT2222A switching transistor has a C1 printed on top (body) of chip.

The SMD IC chip is constructed like the standard IC with gullwing terminals. These sensitive-static devices appear in dark static-sensitive bags.

Terminal 1 is identified by a "U" or indentation circle on top and is located on the bottom left hand corner of the chip. The linear LM386 low power amp IC is a SMT (LM386-1-ND) and has an SO-8 outline. The numbers 386 are stamped on top with an indented circle at terminal 1 *(Figure 7-6)*.

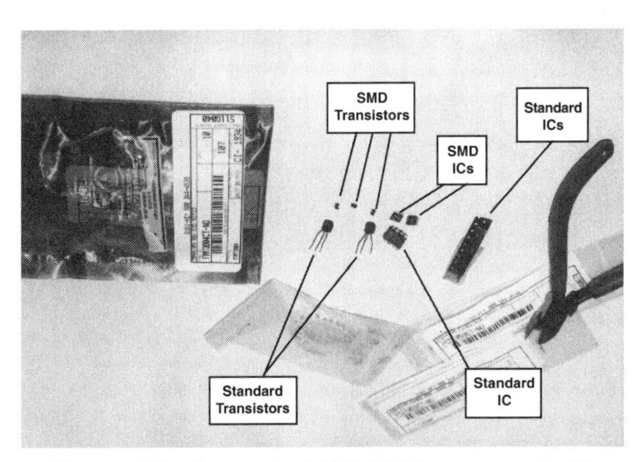

Figure 7-6. A standard LM386 IC compared with SMD gull wing audio output IC's.

These IC devices are made of ceramic or plastic moldings. IC chip devices are not heat-proof or shockproof; they should not be subject to a hot soldering iron or magnetic components. Do not apply unnecessary stress to the chip. Handle SMD semiconductors with extreme care. Keep the SMD IC chips inside the plastic container until ready to be mounted. Install the chip on the printed circuit board. Use a low wattage or battery iron to solder the small terminals.

The leadless, fixed, Schottsky barrier, zener, and variable capacitance diodes appear in SOT-23 packages. These diodes might look like a three-legged transistor. The SMD diode can be identified by two alphabet letters or numbers. You might find one or two fixed diodes in one chip *(Figure 7-7)*. Only two of the terminals are used with one fixed diode and all three terminals are used for two diodes in one SOT-23 chip. These diodes can be purchased in one (1), 10, and 100 lot pricing. SMD components can be obtained for construction projects from several mail order firms.

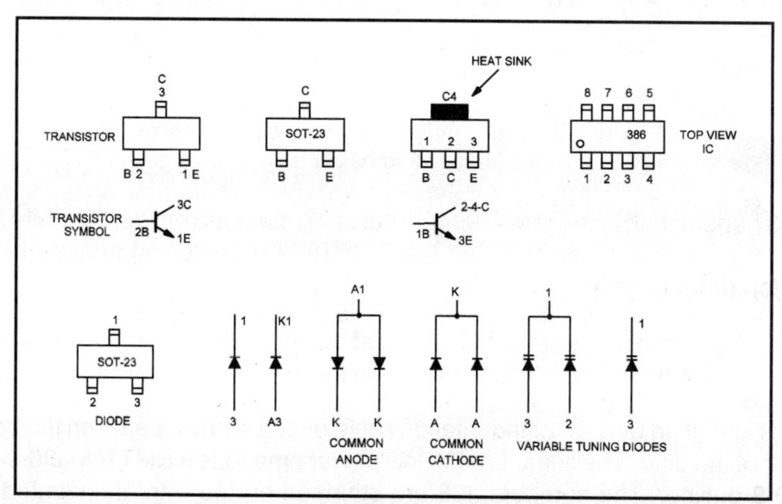

Figure 7-7. The SMD transistor and diode connections inside the SOT-23 package.

The SMD diode might appear in several different packages. The SMD signal diode SD914 or MMB914 might appear in a SOT-23 package, while a SMD silicon signal diode appears in a round LL-34 chip. The SMD zener diode might be found in a flat, round chip, or SOT-23 package. The case and power rating of a 6.1 volt zener diode rated at 200 mW, 300 mW, and 400 mW is found in a SOT-23 package. The LL-34 case zener diode is rated at 500 mW in a round package. The SMD on (1) watt zener diode appears in a round (SM-1) or PSM flat chip. You can test the SMD diode chip with the diode Test of the DMM.

WHICH SIDE IS UP

Mount the SMD resistor with the numbers on top and black side upward. A bypass or coupling chip capacitor might not have any markings on the body and should be mounted with the contact points downward on the end pads. Mount the small chip electrolytic capacitors with the capacity and voltage listed on top with contacts at the bottom. Make sure the white line on the top side of the SMD ceramic capacitor connects to the positive voltage. The top black edge of the aluminum electrolytic capacitor is connected to ground.

The transistor is mounted with the number and letter (example 1A) upward with the terminals over the three pc pads. Place the indented dot of the small IC at terminal one on the pc wiring. Make sure all terminals line up with the pc pads and connections.

There are only a few more tools needed than those found on the service bench. Select a hand-held magnifying glass to locate and identify each SMD component. A pair of eyebrow tweezers are required to hold the SMD part in position. Select a small screwdriver to help lineup the SMD component and a low wattage soldering iron. Select a 30 or 40 watt, or less, soldering iron, a battery operated soldering iron is ideal for miniature connections *(Figure 7-8)*. The bench-type temperature control soldering iron does a good job in soldering sensitive-static ICs and microprocessors.

Figure 7-8. Soldering up a round SMD resistor pin on the PC board with battery soldering iron.

IDENTIFY SMD PARTS

Within the latest TV chassis, surface-mounted parts are soldered directly to the pc wiring, while standard components are mounted on the top side of the board. The SMD part found on the electronic chassis might look like tiny brown, black and gray specks. The fixed resistor might be marked with white numbers on the black chip.

The SMD fixed component might be shiny and round with color marked rings to show the value of the part *(Figure 7-9)*. A ceramic capacitor chip might have a letter with a number along side to identify the value. Some bypass and coupling SMD chips are not marked at all. The electrolytic ceramic chip can be identified with a white line at one end, indicating the positive terminal. The electrolytic capacitor might have three numbers marked on top with a value (222) that equals 2200 μF. An aluminum electrolytic capacitor has a black line on top, indicating the negative or ground terminal. Besides the polarity marking, the SMD chip electrolytic capacitor might have the capacity value and working voltage stamped on top of the plastic body. A number on top such as 3R2 equals 3.3 μF, where R is considered a decimal point.

Figure 7-9. You can remove and replace SMD components with a small screwdriver, tweezers, battery iron, and magnifying glass.

The SMD transistor has three terminals with two on one side and one on the top side. The SMD diode might have the same form as a three-terminal transistor. The terminals might be marked 1, 2 and 3. Check for the correct part number printed in white on the TV PCB. The SMD transistor or IC might have the part marked on top or no markings at all.

The ceramic IC chip has many terminals on each side, while some microprocessors have gullwing-type terminals. You will find many ICs on the camcorder and CD player PCB. The IC chip has a indented terminal 1 or a white dot. You might find a white dot also upon the pc board, indicating terminal 1 *(Figure 7-10)*. The small IC might have 8, 16, or 18 terminals while the microprocessor has up to 80 soldered terminals. Of course, you must have a magnifying glass under a strong light, to identify the small numbers and letters stamped on the tiny SMD component.

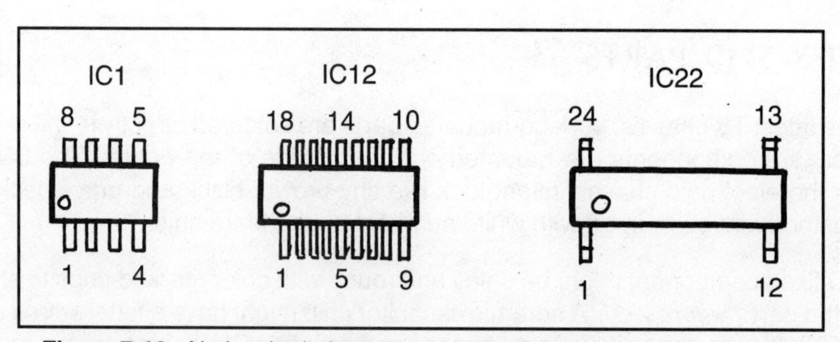

Figure 7-10. Notice the indented zero indicating pin terminal 1 on most SMD IC's and microprocessors.

There are many different surface-mounted devices found in the CD player. The D/A converter and audio amp ICs might be SMD components. These chip devices have many gullwing terminals. The chips are usually made of ceramic or plastic molding, and they should not be subject to a direct shock. Always wear a grounded wrist strap, when working with sensitive devices. Most SMD replacements are obtained from the manufacturer because they fit in the required space and replace the original part number.

REMOVING SMD COMPONENTS

After locating the defective component or damaging the new SMD part with too much heat, the component must be removed from the PCB. A cheap and easy method is to apply the iron at one end and slip a small screwdriver blade under the part as heat is applied at the other end. Solder-wick or a desoldering iron with sucking bulb can remove the excess solder. A desoldering station is ideal, but they are rather expensive. Throw the removed component away. Do not try to use it again.

Heat each individual terminal of the transistor and pry up each terminal with a pocket knife or small screwdriver. Likewise, remove each gullwing terminal from the defective IC until all terminals are removed. Make sure all terminals are loose on the SMD IC before prying them out of the mounting pads. Be careful, it is very easy to damage the PC wiring. Touch up the soldering pads with solder-wick or mesh-braid and soldering iron. Lift the excess solder up from the pc pads and wiring.

REPLACING THE SMD COMPONENT

Before replacing a defective SMD part, test out the fixed resistor or capacitor with a resistance test. Likewise, test all transistors, inductors and connections with the low ohm scale of a DMM. Place the DMM fine probe tip at each end of the resistor and check for correct resistance. The fixed ceramic capacitor can be checked with the 20K ohm scale by touching the two ends with ohmmeter test probes.

A low capacitance capacitor might show a resistance test for only a few seconds. The electrolytic capacitor will charge up and down according to the capacity measured. For instance, a 2200 μF electrolytic will charge up the numbers for several seconds. Then reverse the test probes and the capacitor will charge up again. Charging action without any leakage on the meter indicates the capacitor is normal and probably okay. Check the capacitor with a capacity tester if one is handy.

Check each SMD transistor or diode with the DMM diode-test as standard solid-state devices are tested. Be very careful as these miniature parts can fly or flip out of site while being tested. Place the SMD part on a white sheet of paper for replacement and tests. Double check the polarity of diodes and electrolytic capacitors before and after installation.

Take one component out of the package or off of a replacement card at a time *(Figure 7-11)*. Remove the part from the strip (usually capacitors and resistors) by placing a razor or knife blade under the piece of cellophane. Place the SMD device on a white sheet of paper, where you can easily see it.

Seal the remaining parts on the cut strip with a piece of cellophane tape. Return all other parts back into the original packet or bag. Seal up with tape or staple the plastic packet so the parts will not spill out. These parts can easily be mixed up as they all look somewhat alike.

Figure 7-11. Replace the strip of SMD components back into the plastic or sensitive-proof bags.

Grasp the tiny part with a pair of small tweezers and hold the ends to be soldered over the correct set of pads. Make sure the wiring pads are tinned with solder. Choose the smallest diameter of rosin core solder available for those tiny connections. Tack one end to the pad with the soldering iron. Only a dab will do. Then go to the opposite end and apply enough solder for a clean joint. Now you are soldering the component ends of the SMD part to the pc wiring pads. Go back and resolder or touch up the tacked-end side and make a good soldered joint. Notice that a good soldered bond upon each end of an SMD component will have a bright and clean connection.

Do not leave the iron on the joint too long to damage the SMD part or lift the pc pad and wiring. Double check the soldered connection with the magnifying glass. After installation, check for correct resistance across the fixed resistor and capacitors for leakage.

The semiconductors are the most difficult SMD components to solder into the circuits. They have such tiny connections and so many of them. Try to center the three transistor terminals over the right soldered pads or tabs with a pair of small tweezers. Tack in one terminal to hold the part in position. Then carefully solder all three terminals with the fine point of the soldering iron. Be careful not to apply too much heat from the iron to destroy the transistor.

Make sure terminal one (1) of an IC or microprocessor is at the right pad. Check for the indentation circle or dot on top of the IC that indicates terminal 1. A white dot on the PCB can indicate terminal 1. Double check to determine if all terminals are over each pc wiring tab. Like other SMD parts, tack in one terminal on each side of the IC so it will stay in place. Now solder all IC terminal connections to the pc wiring. Inspect each connection with the magnifying glass.

Take a resistance test between each IC element or terminal with the 200 ohm range of the DMM for leakage. Sometimes too much solder will lap over and cause leakage between two or more terminals. Make an in-circuit diode-transistor test of each diode and transistor. Make sure the transistor or IC are not damaged and has good soldered connections.

These tests can be made upon the PC wiring where the semiconductor terminals are connected. A sharp-pointed test probe does the trick.

The resistance and diode tests of resistors, capacitors, transistors, inductors, and ICs insure that no parts are damaged, the correct part is in the right place, and good soldered connections are made. Besides, the electronic chassis will perform after all parts are mounted. There are no greater rewards, when the electronic project or product chassis is fired up after repairs, and it begins to play.

LOCATING THE DEFECTIVE SMD IC

Signal trace the audio signal in and out of the preamp or output IC with the scope or external audio amplifier. If the signal is found at the input of the IC and no output, suspect a defective IC or surrounding components. Take critical voltage measurements of each IC terminal and compare them with those shown upon the schematic. Also compare the same voltages with the other stereo channel. When one or more terminal voltages are improper or low, check for a leaky component tied to that terminal or a leaky IC. A leaky output IC might have lower then normal voltage on the voltage supply pin (Vcc). Take critical resistance measurements from each terminal to common ground.

LOCATING AND REPAIRING DEFECTIVE POWER ICS

Locate the defective power output IC with signal in and out tests of the scope or external audio amp. Notice if the power IC is running extremely warm on the heat sink. A leaky IC might have a very low supply voltage *(Figure 7-12)*. The shorted IC can contain overheat marks with burned PC wiring. Often, the leaky or shorted power output IC can blow the main power or speaker fuse in the receiver or amplifier.

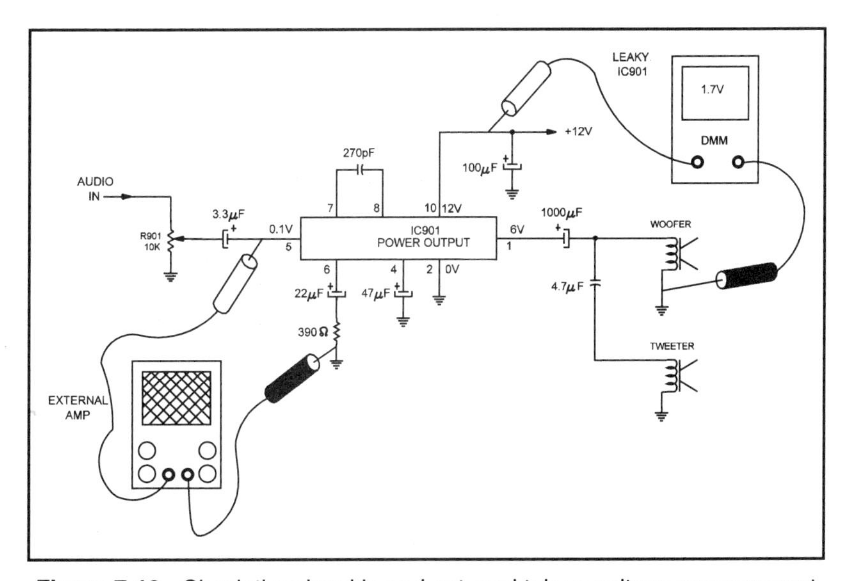

Figure 7-12. Check the signal in and out, and take a voltage measurement upon the supply pin to locate a defective output IC.

After locating and removing the defective power output IC, obtain the manufacturers correct replacement. If the correct replacement is not available, check the semiconductor replacement manual for a universal replacement. At the present time, some universal replacements are very difficult to locate for critical output ICs and processors.

Before mounting the new replacement, clean up all pc terminal pads. Make sure all pads are clean and bright. Smear the back side of the power IC with silicone grease before mounting. Snug up both metal screws to the heat sink. Now solder all terminal contacts to the power IC. Some smaller ICs and microprocessors have glue between the component and PCB. It is not necessary to apply glue to the new replacement, before mounting and soldering the component to the pc wiring.

INTERMITTENT PC BOARDS

Most intermittent problems found in the SMD audio output stages are related to intermittent components or poor PCB connections. The pc board might tend to warp and bend, breaking contacts between the tiny SMD parts and pc wiring. Most SMD symptoms are caused by poor board connections.

Try pressing up and down with an insulated tool or a long pencil with the eraser pressed against the board connections *(Figure 7-13)*. Sometimes by moving the board up and down around the various audio components, the intermittent connection can act up. Spraying the suspected SMD component with coolant can make some parts become intermittent. Solder all SMD part connections after locating a section of the board that appears intermittent, to try and locate the defective connection. A cracked board in a portable cassette player or hand-held TV might result from the unit being accidentally dropped on a hard surface.

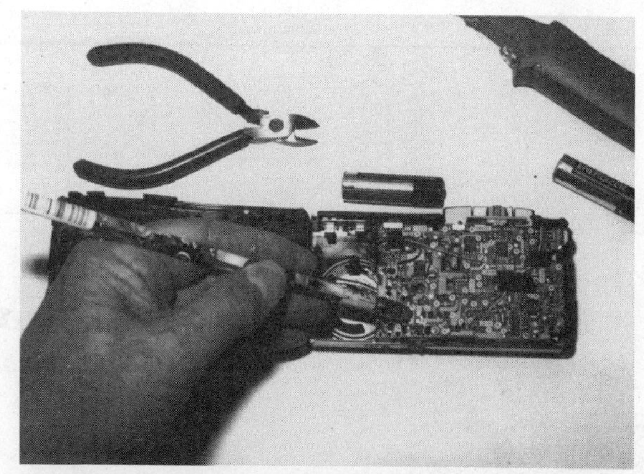

Figure 7-13. Locate the intermittent SMD component with insulated tool or pencil.

Besides poor SMD part connections, the intermittent sound might result from poor socket connections. Move the connections around as the unit is playing to locate a poor socket connection after several years of operation. Pressed socket wires have a tendency to make poor connections after a few years. While flexing the wire cable, take critical low ohm resistance measurements between connecting wires and socket. A bright light and magnifying glass can help to spot poor board connections.

REPAIRING SMD CIRCUITS IN THE CASSETTE PLAYER

The SMD components are usually found in the pocket, mini or micro cassette player and recorders. A few standard parts might be found on the top side of the pc board and tiny SMD components mounted directly on the pc wiring *(Figure 7-14)*. A trace of the pc wiring and part numbers might be found upon the top side of board. Most problems related to the various cassette players are defective components, dirty switches, worn parts and damaged boards.

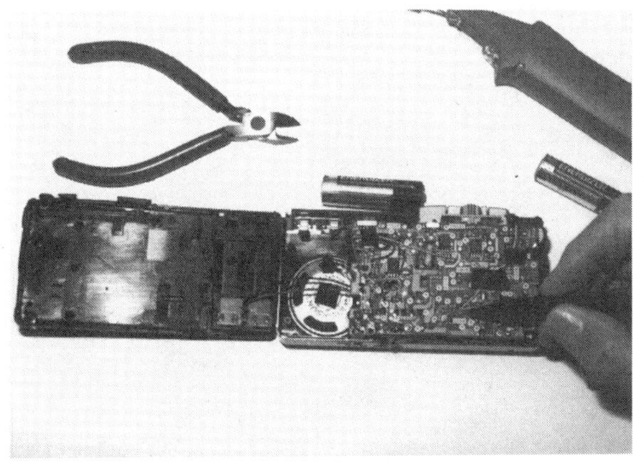

Figure 7-14. The SMD components found on the PC wiring of a personal cassette player.

The stereo cassette mini-amplifier might consist of a dual-preamp IC and a dual-power output IC. Some small cassette players might have all of the audio amplifier circuits in one IC. The stereo tape head circuits are switched into the preamp circuits. A noisy volume control can be cleaned up with cleaning fluid. Replace the control if excessively worn. One channel might be noisy, intermittent, or distorted, pointing to a defective output IC.

Although the small cassette audio circuits are very simple to follow, the biggest problem is locating the defective SMD part on the PCB, with many parts cramped together. Locate the defective or distorted circuit by inserting a test cassette and signal tracing with the scope or external audio amplifier.

Suspect a small coupling capacitor between volume control and input terminal of the output IC for a weak and frying noise. Suspect a defective power output IC with a low rushing noise in the left channel with the volume turned down. A weak audio signal might result from a leaky preamp transistor or IC. The dead or distorted speaker might be caused by a leaky output IC.

For a very weak recording or playback, check for weak batteries. A noisy right channel might be caused by a dirty headphone jack. Clean the male plug and female jack with cleaning fluid. Spray the function switch with cleaning fluid when a bad or noisy recording is noted. A jumbled recording might result from a ground wire coming off at the erase head. Replace a defective function switch with intermittent recording and playback. First solder all function switch terminals for intermittent reception.

SERVICING SMD AUDIO CIRCUITS IN THE PORTABLE CD PLAYER

The portable CD player might contain a SMD D/A converter processor with SMD audio preamp ICs, mute line and earphone SMD transistors. SMD resistors and capacitors might be found in these circuits with a few standard components *(Figure 7-15)*. The left and right audio channels are taken from pins 4 and 7 of the D/A converter (IC7). The weak audio signal is amplified by a dual-SMD IC8-1 and IC8-2. Q101 and Q102 provide muting of the left and right output line jacks.

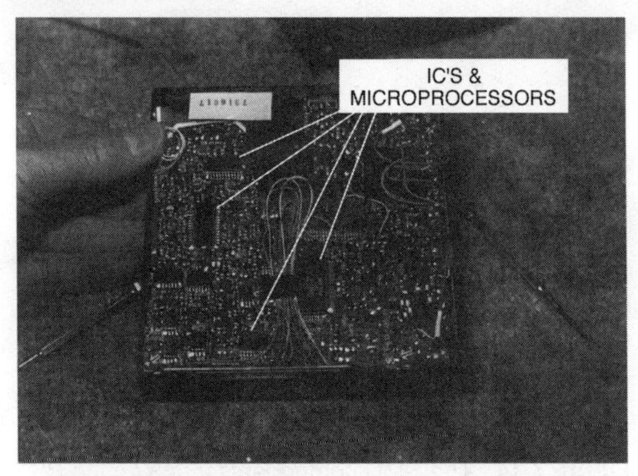

Figure 7-15. The bottom view of a portable CD player with SMD IC's and components.

Signal trace the left and right audio pin terminals 4 and 7 from the D/A converter with the scope or external audio amp. Check the audio input signal at pins 3 and 6 of IC8. Likewise check the amplified audio signal at pins 1 and 7 of the IC amplifier. The CD player audio signal can be traced to each audio line output jacks or stereo headphone jack.

A no left audio channel can be caused by a defective Low Pass Filter network (LPF101) or D/A converter (IC7). Replace IC7 for static or frying noise in each channel. Check IC8-1 or IC8-2 for distortion in both channels *(Figure 7-16)*. Replace IC7 for a background noise. Resolder all terminals of IC8 for intermittent sound in the right channel. Suspect a leaky 10 µF electrolytic capacitor for no sound while the disc is playing. Replace the D/A converter (IC7) for distorted audio. Check Q101 for a leaky SMD transistor with a muted left channel.

REPAIRING SMD COMPONENTS IN THE STEREO AMPLIFIER

SMD audio components are found in the portable cassette and compact disc players, portable and auto receivers, and power amplifiers. The portable compact disc might have SMD parts throughout the audio circuits. A portable cassette player can have both input SMD audio components and standard power output parts. The preamp audio circuits might consist of some SMD parts with standard components in the output circuits. A dual-preamp IC and dual-output ICs are found in the low wattage amplifier circuits. A separate SMD audio board might be located in the cassette and CD players.

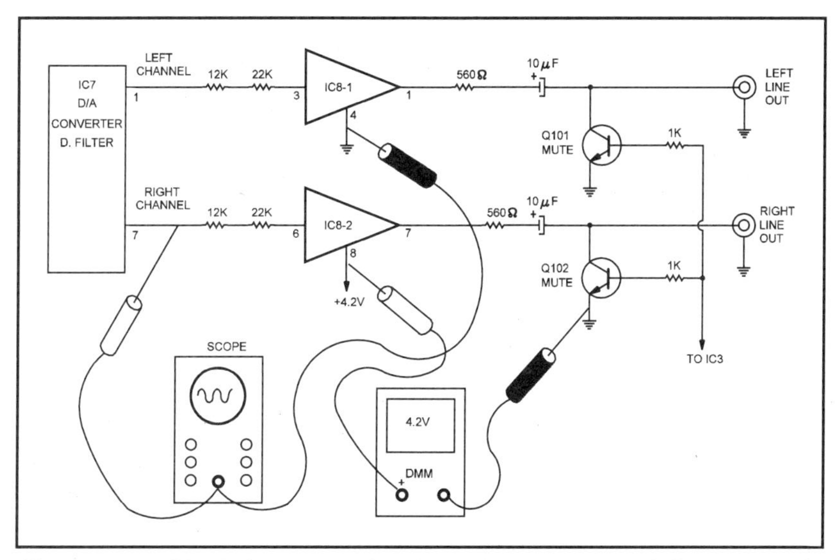

Figure 7-16. Check the normal signal in and distortion out of IC8-1 and 2 of the analog amplifier CD section.

Signal trace the SMD audio circuits like any standard audio circuit. Start at the D/A analog output audio circuits within the portable CD player. Check the audio signal from the tape head terminals through each stereo speaker in the cassette player.

Begin at the signal diode detector within the radio circuits to the volume control and small PM speakers. Compare the audio signal at the volume control to determine if the preamp or audio output circuits are defective. Then compare the audio signal from base to collector of each audio transistor, from input to output of the dual-preamp or dual-output IC, or at any given point in the circuit.

TROUBLESHOOTING SMD TV AUDIO CIRCUITS

The portable hand-held TV and small screen receivers might have SMD parts in the audio circuits. The SMD parts and component layouts are found in some service manuals. You cannot tell by looking at the schematic which parts are SMD components. By checking the schematic, you can see what ICs or processors have gullwing terminal connections, the part list will tell you this, as well.

All semiconductors such as diodes, transistors, ICs and microprocessors will be listed as all chip types. The SMD resistors might be listed as metal glass chips. A star next to a capacitor in the part list indicates a chip type capacitor.

You might find both SMD components and standard parts within the audio output circuits. The standard parts are mounted on top and the SMD parts are found underneath the PC wiring. The low wattage resistors (1/10 watt) can be SMD components. The ceramic low voltage (under 50v) capacitors are used in coupling and bypass circuits. Often, low voltage Tantalum capacitors under 16 volts are found in SMD bypass, coupling and decoupling circuits. Most high-capacity electrolytic capacitors are standard components in the audio circuits.

Although, SMD components are found in very low voltage circuits, they do fail and breakdown. Besides semiconductors, coupling capacitors and SMD resistors produce most of the SMD TV sound problems. Check for cracked pc wiring around heavy heat sinks and intermittent terminals on IC components. Suspect cracked SMD resistors for distorted audio. High ohm SMD resistors have a tendency to increase in resistance, causing weak and distorted audio. In a Sharp 19 inch portable TV, a feedback resistor R3206 (100K) produced distortion in the speaker with increased resistance *(Figure 7-17)*.

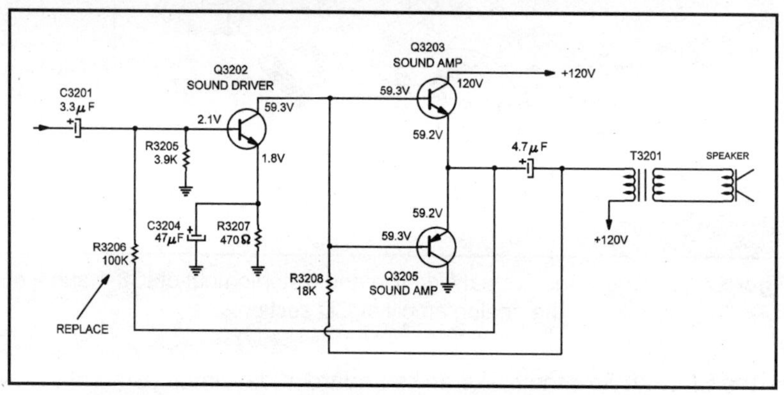

Figure 7-17. Feedback SMD resistor (R3206) had increased in resistance producing distortion in a Sharp portable TV.

SERVICING SMD COMPONENTS IN HIGH-POWERED AUDIO CIRCUITS

You might find SMD components in the front-end audio circuits where low voltage is fed to preamp transistors or ICs. They are often located under large standard parts upon the pc wiring side.

The preamp stereo circuits might contain two op amps in one SMD component. The op amp part might be found in the input treble and bass tone circuits. Most of the audio circuits within the auto high-powered amplifier have mylar, polypropylene, and ceramic capacitors with carbon film or metal oxide standard resistors. Standard high-wattage components might be located on a double-sided PCB in the high-powered auto amplifier.

The high-powered auto amplifier can be checked with signal in and out tests. Check the chassis for overheated components and PC wiring *(Figure 7-18)*. Connect a 4 or 8 ohm-100 watt load resistor across the left and right output speaker terminals; test speakers can be connected if no voltage is found at the output terminals. Keep the gain or level controls low as possible, if found at one-end of the high-powered amplifier.

Figure 7-18. Look the chassis over for possible burned SMD parts and PC wiring.

Inject a 1 kHz signal from a function or audio signal generator to both input terminals. Connect a bench power supply (13.6v) to the amplifier. Scope each circuit and compare the audio signal stage to stage in the high-powered amplifier. When the audio drops out or becomes distorted, you have located the defective stage. When possible, replace all critical audio components with original part numbers.

Chapter 8

Servicing High-Powered Audio Circuits

The audio components within the high-powered amplifier circuits consist of many transistors and IC components. A high-powered amplifier might operate with an output power of 35 to 1000 watts.

The early home receiver output might be less than 1 watt, while today's receiver-amplifier might average between 35 to 100 watts *(Figure 8-1)*.

Figure 8-1. The high-powered circuits are found in a separate auto amplifier.

The early auto-cassette receiver had less than 10 watts of output power and now might average 25 to 40X4 watts peak power output. Today's auto CD receiver might have an average of 35 to 50X4 watts peak power output with a preamp output voltage of 0.5 to 4 volts, with the average around 2 volts output.

The average high-powered auto amplifier might have a 50X2 to 400X2 RMS at 2 ohm output. The average RMS power (watts X channels) 35X2 up to 300X2 watts with 14.4 volts of battery input voltage. The high-powered amp might have a built-in low pass (LP) or a high and low pass crossover network. A signal-to-noise ratio might be from 95 to 110 dB.

The fuse rating of the auto amplifier might be 15 to 60 amps with 2X20 up to 2X40 amps of power.

The Root-Mean-Square (RMS) measures the power in watts the auto amplifier can produce continually. The measured output power of an amplifier depends on the automobile's input voltage. The average voltage from a battery, with the car operating, can be around 14 volts and 12 to 13 volts when the engine is shut off. Likewise, the amplifier will have a higher wattage output at 14.4 volts than at 12.6 volts of input battery power. For instance, a Jensen KA2500 2-channel auto amplifier has a bridged RMS power output of 300X1 watts compared to 250X2 watts of peak power.

HIGH AND LOW LEVEL INPUTS

Most commercial audio amps have both a high and low impedance input. The high level jacks are to be used with the auto-cassette speaker output. A low-level input might match the separate CD line output jacks. Usually, the high and low level inputs are found in the auto stereo frequency equalizer/booster and high-powered audio amplifier. A low-level input might connect the preamp output circuits to the low-level input of the auto amplifier.

The high-level radio and cassette player outputs might be switched into the IC1 preamp circuits of an equalizer-booster circuit or low-powered auto amplifier *(Figure 8-2)*. The high and low level inputs in a resistance network is switched by SW-1 into the input of IC1. Both left and right stereo channels are switched with dual SW-1 into the preamp circuits.

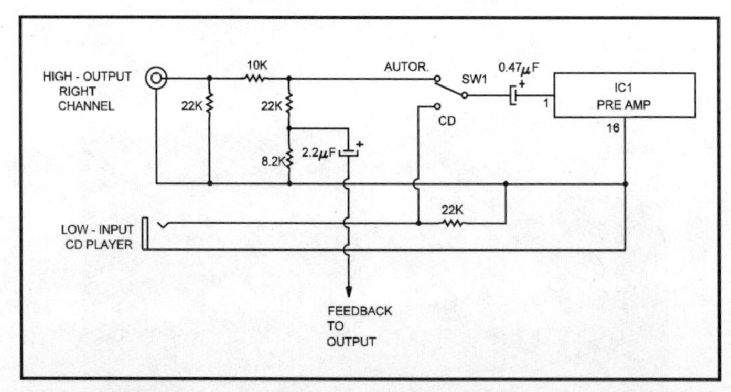

Figure 8-2. The high and low level audio signal is switched into the input of IC1 preamp.

The high-level input might match the input impedance with transformer coupling. A transformer input is found on both left and right audio channels *(Figure 8-3)*. Some auto audio amplifiers have speaker level inputs from the radio receiver, CD and cassette player. The multi channel amplifiers might have both speaker level input and preamp output jacks.

HIGH-POWER TRANSISTOR CIRCUITS

Besides the high-powered auto amplifier, a complete home theater audio system of a shelf or rack system might have a 3-CD to 24-CD player, dual cassette, AM/FM/MPX receiver, VCR, and system remote control features. These stereo receivers are found in the 50 to

100 watt systems. The dual-cassette player features normal and high-speed synchro dubbing, auto reverse, continuous play, and Dolby-BNR for clean, dynamic sound. The speaker system might include two 10 inch 3-way main speakers, two 4 inch full-range speakers for rear channel, and a shielded 4 inch full-range speaker for center channel sound.

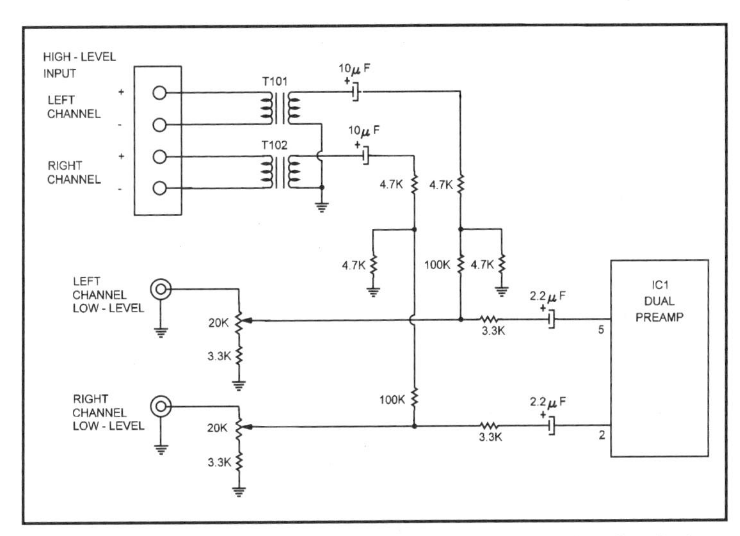

Figure 8-3. The high level output audio is coupled to the dual preamp through T101 and T102.

The high-powered amplifier might appear in 50 to over 1000 watts in the auto amp, auto shelf or rack system, auto receiver, CD player and PA systems. The lower voltage amplifiers might have power output ICs while the high-wattage amplifiers might contain many transistors in directly-coupled power output circuits. A typical 150 to 200 watt amplifier might contain two ICs in the preamp circuits, one transistor in the muting system, ten power amp transistors, and a single transistor in the overload and shutdown circuits, within each stereo channel.

The 150 watt amplifier might consist of two directly-coupled AF transistors, four driver transistors and four output transistors in a push-pull directly-coupled power output circuit *(Figure 8-4)*. Q109 and Q110 are NPN power output transistors connected to the positive (+) 35 volt source. Q111 and Q112 are PNP power output transistors connected to the minus (-) 35 volt source. Q109 is a 2SC3421 or 2SC600 transistor, while Q110 is a 2SC3907 or 2SC3519 NPN transistor. Q111 is a 2SA1358 or 2SB631, while Q112 is a 2SA1516 or 2SA1386 PNP transistor. Replace all power output transistors with original part numbers, when available.

OVERLOAD OUTPUT CIRCUIT

The overload protection output circuit is designed to shutdown the DC to DC power supply when the output circuits become unbalanced. The directly-coupled output circuit to the left and right channels is balanced at zero volts. No dc voltage is found on the voice coil of a connected speaker (usually a woofer type speaker).

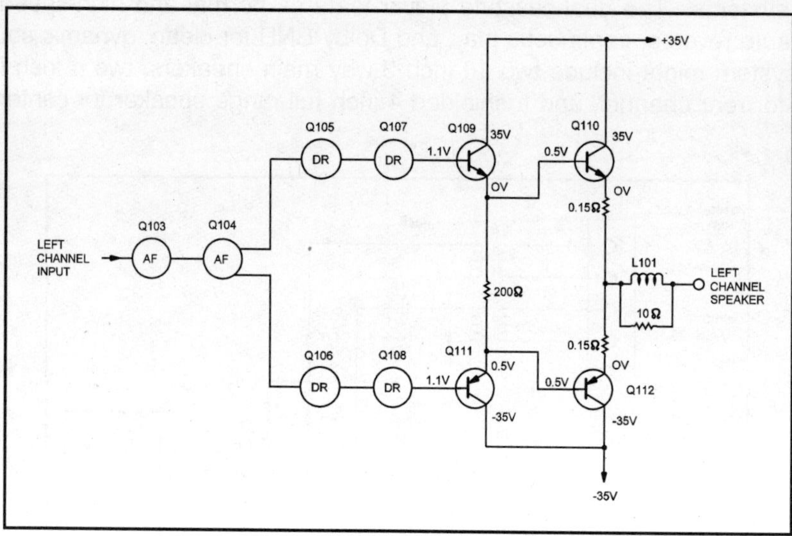

Figure 8-4. A directly-coupled power output circuit of a transistorized 150 watt amplifier.

When a driver or directly-coupled power output transistor becomes open or leaky, the output voltage of Q109 and Q111 change, resulting in a dc voltage at the speaker output terminals.

If a dc voltage is left too long on the voice coil of a speaker, the winding of the voice coil will become warm and drag. The voice coil might result in a frozen or burned coil on the speaker magnet. Now the speaker is damaged and must be replaced. The overload output circuit is suppose to protect the woofer speaker or the speaker connected to the output terminals.

The dc voltage on the emitter terminals of Q109 and Q111 is zero. This same voltage is connected to the base and emitter circuit of the overload transistor Q113 *(Figure 8-5)*. D101 isolates the DC to DC power circuits from the overload transistor. When the power audio output circuits fail, a dc voltage is found at the base and emitter of Q113, upsetting the balanced output circuits. The overload transistor conducts, applying a signal voltage to the protection circuits in the dc power supply.

The protection circuit shuts down the power supply, disconnecting the positive and negative 35 volts to the output transistors, and removing the dc voltage from the speaker terminals. Each stereo channel has the same type of overload protection circuit.

The defective audio output circuits might come up and then shut down at once. Sometimes the audio output transistor or IC might become leaky and blow the main power fuse. The dc voltage on the speaker fuse might cause the fuse to open. When a blown fuse is found or the audio chassis fires up and shuts down, check the dc voltage at the speaker output jacks or terminals.

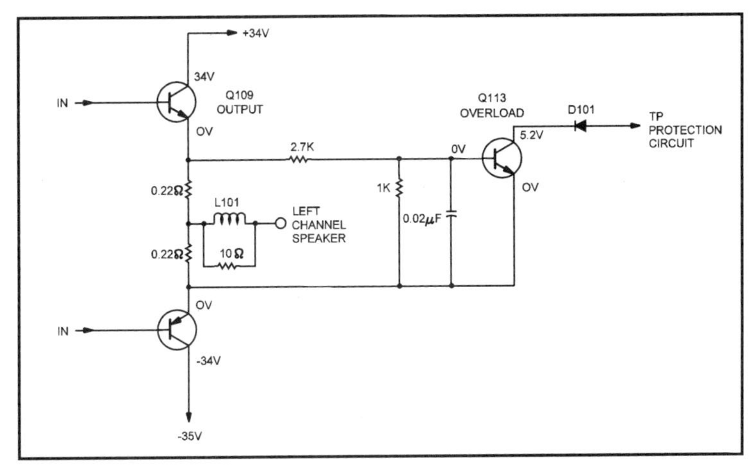

Figure 8-5. Q113 provides overload protection to power output transistors Q109 and Q111.

A test speaker should never be connected to the amplifier that has a dc output voltage at the speaker terminals. Connect high-powered output speaker load resistors instead. Then take a low voltage measurement at the left and right output speaker terminals.

KEEPS DAMAGING OUTPUT TRANSISTORS

The high wattage audio amplifier symptom was a loud hum and distortion, then quit operating. The chassis was dead. The line fuse was blown with shorted and open power output transistors. The woofer speaker voice coil was found open. The left channel loading resistor was running warm, while the right speaker load resistor was quite cool. A -24.1 volts was measured at the left channel speaker.

A quick in-circuit transistor test of both output transistors indicated Q109 was open and Q112 appeared to be leaky. Both transistors were removed from the circuit and both proved to be defective. Before replacing the power output transistors, the emitter bias resistors were tested on the low ohm scale of a DMM. Although, emitter bias resistor (R137) had correct resistance, the 5 watt resistor's ceramic body was cracked and showed burn marks. R137 (0.15 ohms) was replaced with another 5 watt replacement. Both bias resistors were replaced.

The amplifier was powered up, the fuse blew and the chassis shut down. Q112 was damaged with a dead short between emitter and collector terminals. All transistors were checked in the left channel and appeared normal *(Figure 8-6)*. Although voltage measurements were off at the output transistors, it was difficult to check voltage on the other transistors when the chassis would shut down. The voltages were way off with Q109 and Q112 out of the circuit.

Starting at the output speaker terminals, a resistance test was taken from the left speaker terminal to common ground. This same reading was compared with the normal right speaker terminal.

Figure 8-6. Check all transistors within the audio circuits with the diode-tests of DMM.

It's best to use an analog ohmmeter for these resistance measurements, as the DMM numbers will charge up and down, and take a long time to recover. Although, the analog meter is not as accurate as the DMM, a good comparison resistance test can be made. Naturally, the defective channel output resistance was off compared to the normal right channel.

The resistance measurements were made on each audio transistor terminal and compared to the same transistor in the right channel. When the emitter terminal of Q107 was checked, no resistance was found, even on the RX10K scale *(Figure 8-7)*. Upon checking the Q107 emitter circuit, R127 (47 ohms) appeared open. No resistance was measured across the 47 ohm resistor. One end of R127 was removed from the PCB and checked again. Underneath the small resistor, a small crack was found; which could not be seen by looking down at the resistor.

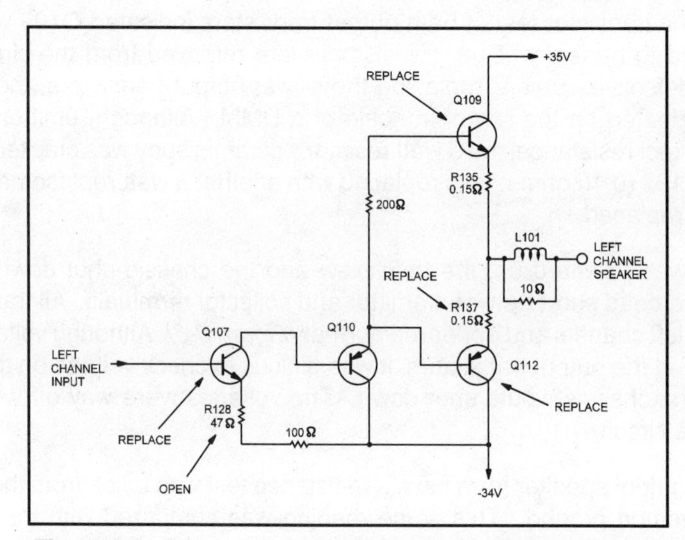

Figure 8-7. Power output transistors Q107, Q109, Q112, and resistors R128 and R137 were replaced in the output circuits.

Since the emitter resistor was open, Q107 might have arced-over or become leaky, damaging the emitter resistor. Q107 was replaced with a 2SC2229 replacement. Once again, Q112 was replaced and a resistance check was made. The left and right output speakers resistance were within 10 ohms of each other to common ground. Resistance measurements of Q109, Q111, and Q107 were quite close in both channels.

The amplifier was plugged in and the chassis remained alive. A quick voltage measurement on the output transistors was fairly normal. By replacing Q107, Q109, Q112, R137, and R128, the music was restored in the high-wattage amplifier.

When the chassis shuts down with damaged outputs and voltage measurements cannot be made accurately, try resistance measurements to common ground to locate a defective audio circuit. Before taking resistance measurements, discharge the large filter capacitors in the low voltage power supply with the amp turned off.

HIGH-POWERED TRANSISTOR AMP PROBLEMS

The high-powered amplifier symptoms are the same found in regular home receiver amps, except more extensive component damage. A leaky AF or driver transistor can cause distortion in the output circuits. Most distortion problems are found in the audio output circuits. The leaky bias diode in the base circuits of the output can cause distortion and sometimes a whistling noise. A distorted right channel might be caused by burned bias resistors and leaky output transistors. A leaky directly-coupled driver transistor can destroy the output transistors and result in an extremely distorted channel *(Figure 8-8)*. Check for an intermittent voltage regulated transistor when the audio becomes distorted after operating for several hours.

Figure 8-8. Go directly to the audio output transistors with a main fuse blown in the high-powered audio amplifier.

Go directly to the output transistors when the main fuse is blown. You might end up replacing a leaky driver transistor and four power output transistors in the high powered amplifier. The dead amp with a blown fuse might result from a leaky 2200 μF filter or coupling capacitor.

The left channel might be dead with a blown fuse caused by open emitter resistors in AF or driver circuits. Check for ring and board cracks on the PCB when the main fuse blows. Suspect leaky driver, output transistors, and leaky zener diodes for a blown fuse symptom.

Suspect a defective component on the volume control board when the amplifier cuts out at low volume. Check for poorly soldered component leads or a cracked board when the audio cuts out. Replace defective volume control with intermittent volume in one channel. When one channel cuts out as volume is increased, check for a defective speaker relay. Intermittent audio with some noise might result from a small electrolytic capacitor (0.47 mF to 1 mF) connected to the volume control. Clean up volume and tone controls with cleaning fluid sprayed down inside the lug terminals.

HIGH-POWERED IC CIRCUITS

The high-powered IC output circuits might consist of one large IC on a large heat sink. The entire audio circuit might be found in one large IC component. The audio circuits might be included from the volume control to the speaker terminals. The power output IC can cause many different sound problems. A high-powered IC might be found in home receivers, auto radios, and separate amplifiers.

Suspect a leaky output IC when the main fuse is blown with dead receiver and amplifier circuits. The dead left channel might be caused by a leaky output IC. Inspect the body of the large IC for blown out areas with a dead chassis. Check the body for overheated marks. Check for a change in resistance or burned bias resistor on the IC terminals, when the IC appears to be overheated. Suspect the output IC when both audio channels are dead. Replace the power output IC when the amplifier warms up and cuts out.

Check the signal in and out of the power output IC when both channels are distorted. Test out each electrolytic capacitor (47 to 2200 μF) that is connected to the IC terminals for leaky or open conditions. The left channel might be distorted and the right channel weak with a warm or red hot output IC. Suspect the power output IC for only a hum in the speaker and no sound. Both channels were distorted in a Fisher SUV7 amplifier with a leaky power output IC (STK805IC).

Check all hot resistors tied to the power output IC terminals with hum in the speakers. Replace the power output IC when running hot with a loud howl or no sound in both channels. Check and resolder all output IC terminals with a buzzing and motorboating symptom.

A noisy and motorboating sound in the left channel can be caused by a defective output IC. Suspect the power output IC when both channels are noisy, crackle, and pops in the speaker. Replace power output IC when the right channel pop and music is distorted. Retighten power IC mounting screws to the heat sink for hum and noisy reception.

PIONEER SX-780 TOUGH DOG

The woofer speaker in a Pioneer amplifier was damaged with -24.7 volts found on the left channel speaker terminals. To prevent damage to the test speaker, a 100 watt 10 ohm resistor was clipped to the left speaker output terminals. Of course, after a few minutes of operation the resistor began to heat up. The left output IC (STK-0050A) also ran warm *(Figure 8-9)*.

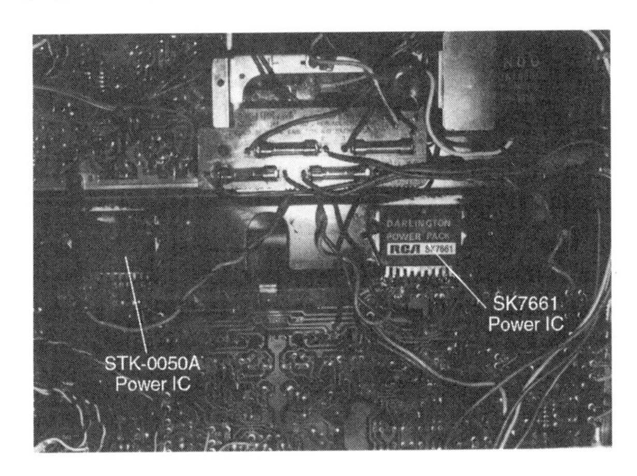

Figure 8-9. The audio output IC STK-0050A was replaced with an RCA SK7661 universal replacement.

The voltages found in pins 8,2,1,and 0 were highly negative. The normal voltage on terminals 0 and 1 should be a +1.4v and -1.4v respectively. The same comparable voltages found on the right channel output IC indicated a leaky left output IC.

Since Q11, Q13 and Q15 were tied directly to the output IC, they were checked in-circuit for open and leakage with the DMM. All transistors tested normal, except Q15 was installed backwards. Someone had worked on the amplifier, as both Q15 and the Darlington transistor Q7 were replaced. Replacing Q15 did not solve the problem. Sometimes additional problems might be found when working without a schematic. Critical voltage tests were made on each transistor and recorded upon the schematic.

Servicing the amplifier chassis was a little more difficult since no parts were marked on the bottom pc wiring. The correct transistor terminals were located with the diode-test of the DMM. The base terminal of a transistor is always common to both the emitter and collector terminals. For instance, with the diode transistor test, the resistance between base and collector terminals will be a few ohms lower (.722 ohms) than from base to emitter (.731 ohms) on a normal transistor. Remember that all transistors might have a different voltage measurement, except the comparable voltage will be quite close in resistance. Both measurements are quite close with the base to emitter test, a little higher in resistance.

Since the voltage was high on pins 0 and 1 of output IC (Q1), maybe output IC was leaky. Someone had replaced Q1 with an RCA SK7661 universal replacement. Perhaps, Q1 had overheated and was destroyed once again. Naturally, the output IC was not on hand or locally available. So the next best thing was to exchange the two output ICs (Q1 and Q2). After the exchange of parts, the sound was normal in the right channel indicating both ICs were normal.

All components were checked on Q15 with accurate voltage measurements. The collector voltage was off (-26.9v) and should have been a positive 1.4 volts at pin 0 of output IC (Q1) *(Figure 8-10)*. Diode D1 and all corresponding resistors were normal when one terminal lead was removed from the circuit. The wiring was traced and double checked with the RX1 scale of the DMM. Q15 circuits appeared normal.

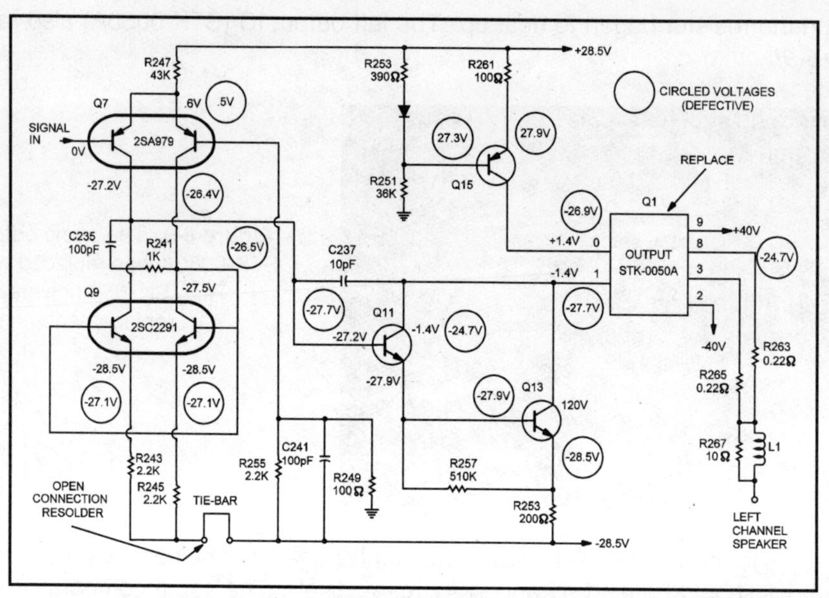

Figure 8-10. A negative -26.9 volts and -24.7 volts was found on power output IC (Q1).

Next the Q11 and Q13 voltages and components were checked. The collector voltage on Q11 and Q13 was quite high (-24.7v) and should be -1.4 volts. All components tested normal within the Q11 and Q13 circuits.

Transistors Q5, Q7 and Q9 were tested with voltage and resistance measurements. Since the Darlington transistors (Q7 and Q9) looked like flat IC components with five terminal leads, no leakage was found between Q7 and Q9 terminals. The voltage on Q9 was negative except at the collector terminal. All resistors on the terminals of Q7 and Q9 were fairly close in value.

When the wiring was checked and traced in the emitter circuits, R245 and R243 to the -28.5 volt source, indicated an open circuit at Q13. The -28.5 volt source was traced out again since there were no markings on the PCB. Both resistors were quite close in value (2.2k ohms). Since there are about five different voltage sources within the amplifier, the -28.5 volts was located at emitter terminal of Q13. Again a resistance measurement was made from connection of R245 and R245 to the -28.5 volt source with no resistance measurement.

Another method to double check the resistance or wiring between the emitter terminals and power supply source (-28.5v) is to take a resistance measurement between the emitter terminals of Q9 and Q13. The resistance measurement should be less than 2.4K ohms. Double check this resistance measurement by checking the same terminals in the normal right channel. No resistance was found even on the RX200K scale. Perhaps there was a break in the pc wiring of the emitter circuits. Remember, the voltage on both emitter terminals of Q9 were -27.1 volts and only off 1.4 volts from the schematic.

A quick resistance measurement between the emitter terminals of Q9 and chassis ground were made to determine if the wiring was open. The 20K ohm range was made for these resistance measurements. Another resistance measurement from the emitter of Q13 to common ground should be quite close. These two different resistance measurements are made through the -28.5 volt power source to common ground. No measurement was noted.

When tracing out the return of R243, R245 and R253 to the -28.5 volt source, one end of a wire tie bar was poorly soldered. Both sides of the tie bar were resoldered, the voltage returned to normal, and the music was restored in the amplifier. And to think, the poorly soldered terminals of a small tie bar, caused all these servicing problems in the left channel circuits.

FOUR CHANNEL AMPLIFIERS

The four channel or quad amplifier circuits were found in early auto, home, phono and cassette players. Four separate tape head directly-coupled preamp circuits were switched into a 4 or 2-channel cassette circuit. After the four channel input circuits, a function switch placed the phonograph, AM/FM/MPX and tape head circuits into the AF amp section.

The four identical auto output circuits contained a right and left audio output to four different speaker circuits. In the high-powered auto amp circuits, the 30X4 up to 80X4 watt multi channel amplifiers are located. These amplifiers can be bridged from 70X2 up to 240X2 watts.

The tape head circuit is coupled through a 1 μF electrolytic capacitor to the base terminal of Q10. The base terminal of Q11 was directly-coupled to the collector terminal of Q10. The output preamp circuits are coupled with a 33 μF capacitor to a selector output switch. All four tape head circuits are identical and each channel can be checked against the others for correct voltage, resistance, and component tests *(Figure 8-11)*.

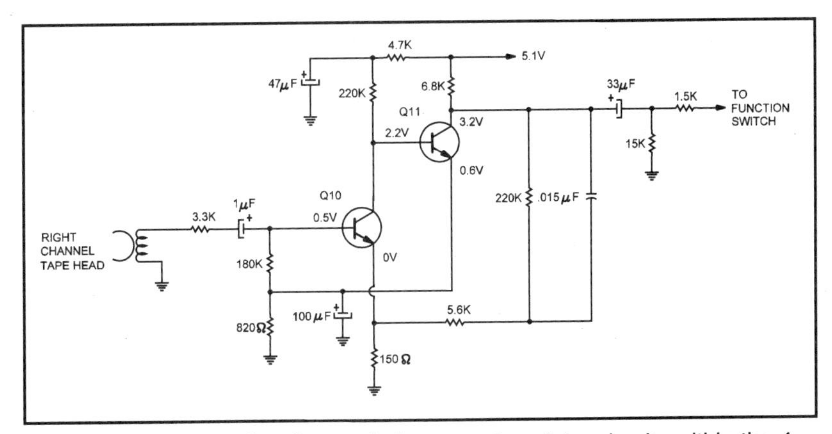

Figure 8-11. The directly-coupled preamp transistor circuits within the 4 channel amplifier.

The multi channel amplifier might contain preamp circuits. Check the preamp circuits with signal of a test cassette, from tape head to the output of each preamp circuit with the scope or external amplifier. Compare each stage with one of the normal circuits.

The output transistor circuits are switched from the AF amp transistor with the function switch. An AF amp transistor is directly-coupled to the function switch and into the balance, bass and treble control circuits. A volume control provides audio signal to the driver and push-pull audio output transistor circuits. Two separate left and right channel output circuits are found at the output of the four audio channels (two-left and two-right) *(Figure 8-12)*.

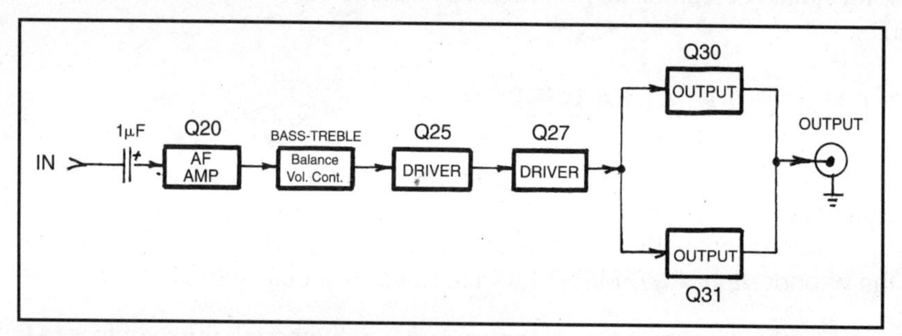

Figure 8-12. There are four identical transistor audio output circuits found in the 4-channel amplifier.

BRIDGED POWER OUTPUTS

The 2-channel auto amplifier can be bridged for a greater wattage output. For instance, a Kenwood PS-300T 2-channel amplifier that has a peak power output of 200X2 can be bridged to a single output of 400 watts. The outputs of the 2-channel amplifier (L and R) are tied together to achieve high wattage output into a 2 ohm load. In mono-mode, you combine the stereo outputs to power a subwoofer pm speaker. A bridged amplifier cuts in half the output impedance to the speaker. A 2-channel 4 ohm output amplifier, when bridged, cuts the output impedance to 2 ohms and doubles the wattage output.

Some auto amplifiers have multi channels like the Kenwood model PS500F (75X4) with a peak output power of 600X2 or when bridged (300X2). The bridged output provides a left and right channel of 300 total output watts. The RMS at 2 ohms equals 150X4. Most of these type amplifiers have built-in high-pass and low-pass filter crossover networks.

A Tri-way speaker output hookup powers a pair of stereo speakers and subwoofers. This is a cheap way to drive two main speakers and one subwoofer from a 2-channel amplifier. A Tri-way crossover network might have a low-pass filter to send frequencies below 100 Hz to the subwoofer, while the high-pass filter sends frequencies above 100 Hz to the main speaker. Of course, the amplifier that the Tri-way crossover connects to must have Tri-way capabilities.

In a 3-channel speaker hookup, a 4-channel amplifier is built with enclosed crossover networks. You bridge the 4-channel amp to run a subwoofer while the front channels drive a pair of regular stereo speakers. The 3-channel output provides greater control over the sound output *(Figure 8-13)*. The amplifier with preamp outputs allows a non-amplified signal to pass through the amplifier and into another amp in the system without separate crossovers.

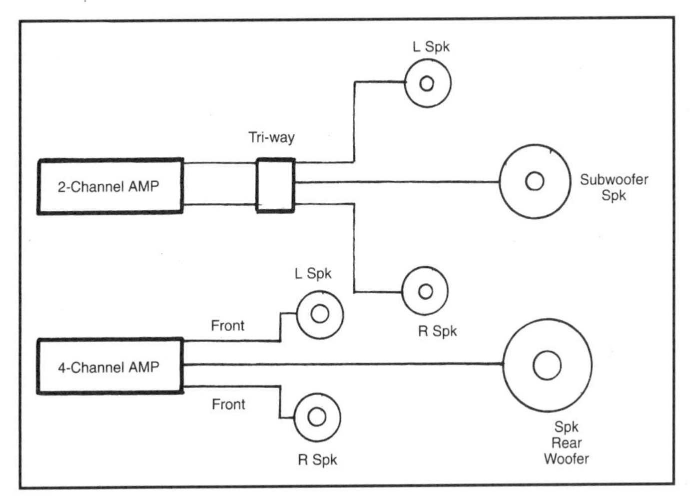

Figure 8-13. A 2-channel amp output with a tri-way crossover and 4 channel amp hookup connected to the speakers.

HIGH-POWER SPEAKER HOOKUPS

Most high-powered amplifiers provide speaker hookup information with a new unit. A simple auto amp and speaker hookup might contain the AM/FM/MPX receiver and cassette player, equalizer, 100 watt amplifier, and two mid-range with large rear subwoofer speakers *(Figure 8-14A)*. The average auto installation might consist of an AM/FM CD receiver, 6 disc changer, center control unit, equalizer, 120 watt amp for two mid-range tweeters, and a 250X2 watts amplifier for two woofer speakers *(Figure 8-14B)*.

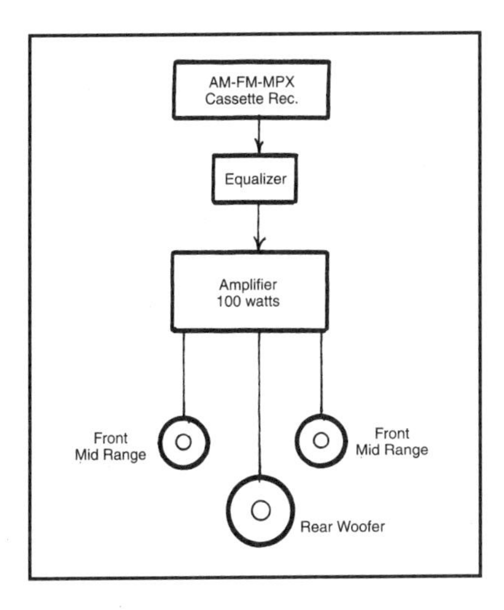

Figure 8-14A. A typical 100 watt amplifier hookup with a front midrange and sub woofer speakers.

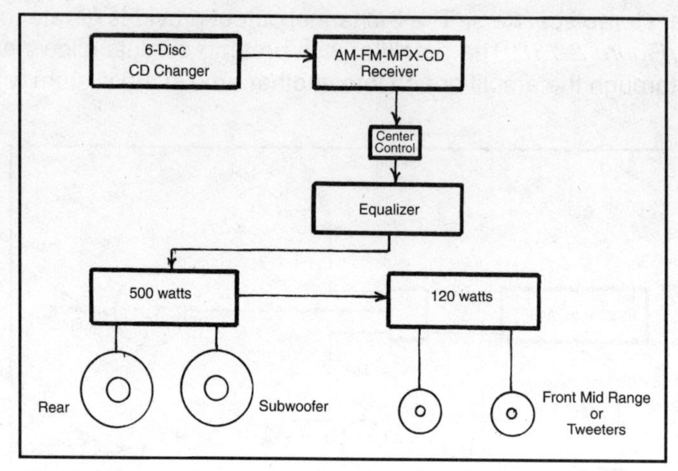

Figure 8-14B. Two separate amplifiers are found to drive two rear sub woofers and two front midrange speakers.

A deluxe high-powered system might include the AM/FM-CD receiver, CD changer, two auto-control equalizers, two audio control units, and a 50X2 watt amplifier feeding four tweeter speakers. The 100 watt amplifier provides all other frequencies to four mid-range speakers. The bass frequencies contain a 1000 watt amplifier connected to six large subwoofer speakers *(Figure 8-14C)*.

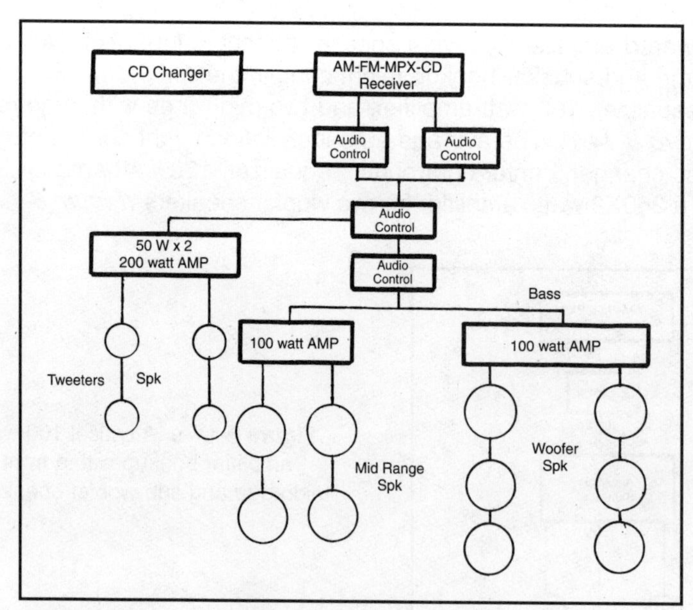

Figure 8-14C. A deluxe high-powered tweeter, midrange and woofer speaker hookup.

HI-POWERED SPEAKERS

The high-powered amplifier provides audio to multiple-driver, enclosed, and raw speakers. The multiple-driver speakers might include a tweeter, mid-range, mid-bass, woofer or sub-woofer. A typical multi-driver might consist of a compact 2-way speaker system with a soft-dome tweeter and a 6.5 inch woofer. The multiple-driver system is more than one speaker on one speaker channel. The driver system might contain two or three speakers.

The multiple-driver speaker channel might have 2, 4, and 8 ohms voice coil impedance and from 15 to 300 watt power range. The woofer speaker cone material can be copolymer, tuff-kote, polyprop, graphite, kraft pulp, pulp, carbon fiber, carbon filled polyprop, fiber, polymica, polypropylene, paper, laminate fiber, poly, mica, black resin, carbonic, acrylic resin, neodymium, injected graphite, poloygraphite, carbon mica, copolymer, treated paper, coated paper, mylar, polylaminate, silicone treated, injection w/butyl, injection cone, kelvar, kelvar-normex, poloymica cone, mica polymer, cellulose, polyglass, polylam, polymineral, graphite/copolymer, pmp, pmt, and polytex.

The tweeter cone material might be made from copolymer, silk, poly dome, polymide, polymer, titanium, poly-ethyl-imide, mylar, paper, textile, aluminum, poly, cloth, rubber-dome, fabric, strontium, whizzer cone, polycarbonite, soft dome, neodymium, treated silk, kelvar, tioxid, glass fiber, phenol, polymineral, polycell, ceramic, cellulose, dynamic cone, whizzer, piezo, pure titanium, titanium laminate, hybrid composite, soft linear, polyneo 11, supronyl dome, silk-cotton, crystal fiber, crystal paper, polyester, and trilaminate.

The raw speaker impedance might be 2, 4, and 8 ohms with a power range up to 1600 watts and a frequency response of 20 Hz to 20 kHz *(Figure 8-15)*. The driver material is made from copolymer, polyprop, kapton, cellouse, graphite, paper, polycarbonate, plastic pulp, epoxy pulp, poly, polycarb, polymer, carbon fiber, organic fiber, fiber, titanium, treated NCP, silk, treated paper, fiber laminate, mica graphic, glass fiber, PVA paper, treated linen, resin lam, polycoated paper, coated silk, poly/butyl, polyproplene, pcp, aluminum, mica, kraft pulp, polymineral, propylene-mica, and polyprop coated.

Figure 8-15. The large 12 or 15 inch PM speaker can handle high-powered auto amplifiers.

A typical Hyper-Throw model HT156D subwoofer 15 inch speaker has a 400 to 800 watt power range, a 3 to12 ohm impedance, and a 15 to 150 Hz frequency response. Another resonance model RW915 subwoofer is 15 inches in diameter, with a power range of 1000 watts, 4 ohm impedance voice coil and a frequency range of 20 to 300 Hz. The raw woofer or subwoofer speaker might vary between 8 to 15 inches in diameter.

The high-powered amp speaker damage results from too much power applied to blow out the voice coil, have an open winding, and a dragging cone. Of course, the open voice coil has no or infinite resistance measurement. A voice coil can be damaged by placing a dc voltage across the winding with speakers connected to a directly-coupled leaky or open power output transistors or IC within the amplifier.

When an electrolytic coupling capacitor is found between the output circuit and speaker terminals, the speaker seldom was damaged. The auto speaker might become warped, with the voice coil dragging or frozen against the magnet pole, and might contain water damage. A mushy speaker sound can be caused by a dropped speaker cone.

HIGH-POWERED SPEAKER RELAY PROBLEMS

In the large power amp receivers or amplifiers, the left and right speaker relay switches in the left and right channel speakers. The transistor-relay circuits are controlled by a microprocessor *(Figure 8-16)*. A separate transistor controls the left and right relays, providing audio output to each speaker. The defective relay circuits might be dead or inoperative, intermittent, or after the amp warms up the speaker relay cuts out. If the power output IC is normal and the speaker relay kicks out, check for a poorly wire soldered wire connection and jumper wire bridges.

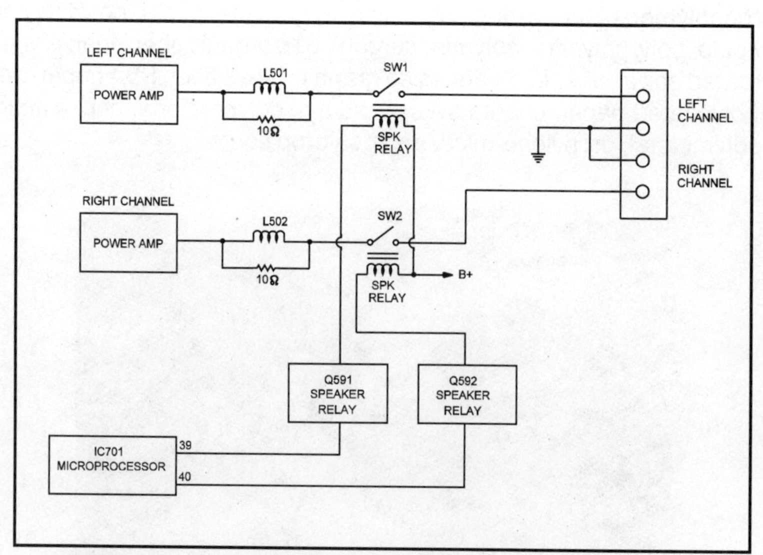

Figure 8-16. The speaker relay transistors are controlled by microprocessor IC701.

When one channel cuts out at low volume, suspect a defective relay. Check the relay for poor contacts when the volume cuts in and out on one channel speaker. Suspect a defective relay when one channel cuts out at low volume and as the volume is turned up, the sound cuts in.

Suspect the speaker relay when the right or left channel goes out with a loud static noise. Check the control microprocessor or output IC when the relay will not entergize. Check the muting switch transistor if one channel is intermittent and cuts out. Suspect an open solenoid or speaker relay when dc voltage is found across the solenoid winding. A dead, no power, loud arcing noise symptom might be caused by a defective ac power switch and not the speaker relay.

TROUBLESHOOTING HOME RECEIVER CIRCUITS

The tabletop AM/FM/MPX receiver might have several transistors in each audio channel or one large IC. The audio transistor circuits might consist of an AF, driver and push-pull output transistor circuits. In the latest stereo receiver amplifiers, only one large IC is found.

Just take the receiver symptoms and apply them to the schematic and components on the pc board. Check the line fuse if the receiver is dead. Measure the voltage on the largest electrolytic capacitor in the power supply with a dead chassis. If correct voltage is found in the power supply source, go directly to the power output transistors or IC. Scan the chassis for any burned components. Suspect leaky output transistors or IC with very low voltage at the output components.

Suspect a leaky or shorted transistor and IC when the fuse keeps blowing. A leaky output IC and directly-coupled transistors can blow the main line fuse. A dead receiver can result from a defective power switch. In a Pioneer SX737 receiver, the symptom was no sound, no relay click with a defective transistor regulator and diode lowering the voltage source.

Check for a weak left channel with open resistors and zener diodes in the audio voltage sources. Suspect an open or leaky driver transistor when the right channel becomes weak and then cuts out. Signal trace both sides of a small electrolytic coupling capacitor for weak sound. A change in emitter bias resistors and electrolytic bypass can cause a weak audio stage.

When one channel is out or dead, check the power output IC or transistors. Suspect the power output, driver transistors, power source, and emitter resistors when one channel is dead. Check for a leaky electrolytic coupling capacitor (1 to 2.2 μF) with a dead right channel. Replace the open driver transistor for no left channel.

Resolder plated feed-through holes with a weak or distorted left channel. Replace the power output IC when the sound in the right channel is distorted, pops and cracks in the speaker. Check for poorly soldered joints on the output IC and defective relay with intermittent audio. When the left channel shuts down at once, check the output and driver transistors, open or burned low ohm emitter resistors, and burned pc wiring.

REPAIRING THE MID-RANGE AUTO RECEIVER OUTPUT CIRCUITS

Today the in-dash auto cassette-receivers have a greater wattage output than yesterday's receivers. The peak power output might be around 25X2 up to 40X4 watts, while the RMS wattage range is 15X2 up to 22X4 watts. The auto receiver might have four speaker outlets and 1 or 2 sets of preamp outputs. The preamp jacks at the rear of the radio can be connected with an RCA patch cable to another outside amplifier. Some auto receivers have two sets of preamp outputs, to which an external 4-channel amp or a second 2-channel amp can be added for a higher power output.

In the early auto cassette receivers, you might find a directly-coupled transistor or IC preamp stage. The power output amplifiers might contain transistors or separate IC components. Each channel output IC was separate and fed one or more pm speakers *(Figure 8-17)*. The AM and FM stereo radio signal was tied to the base terminal of each left and right preamp transistor. When the cassette was inserted, the dc battery source was switched out of the radio circuits.

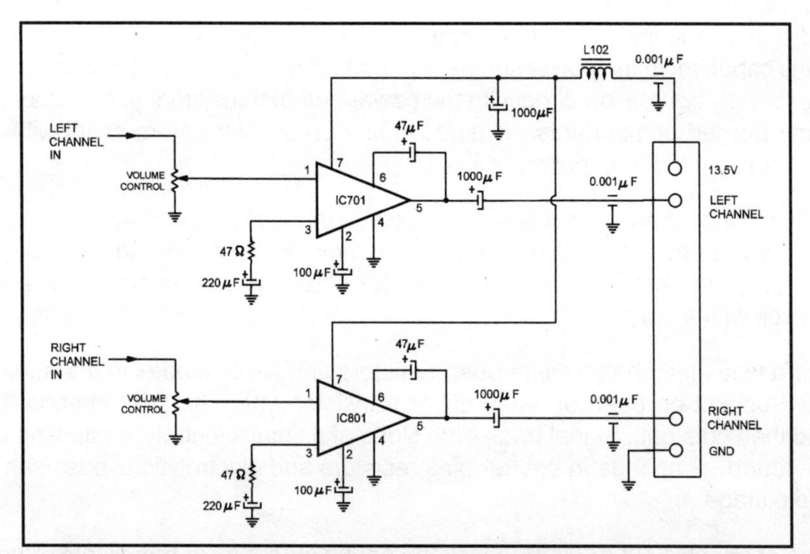

Figure 8-17. The early auto receiver audio output IC's were separate components.

A dead cassette receiver might have a blown fuse caused by a leaky or shorted output IC. Check for a leaky filter or in-line capacitor that can damage the line fuse. Suspect a shorted output IC when the fuse will not hold. Check the audio signal with the external audio amp or scope at the volume control for weak or distorted sound. The audio can be checked at each preamp transistor and output IC. Compare the defective left channel to the normal right channel or vice-versa.

The IC preamp circuits can easily be serviced by inserting a tape or cassette and checking the audio at the input terminal (1) of IC205. If no sound at pin 8 of the left channel IC205, suspect a defective IC or improper voltage source. Measure the voltage on pin 4 (9.1v). A leaky IC205 might lower the supply voltage at pin 4. Remove pin 4 from the pc wiring and notice if the voltage source increase, indicating a leaky output IC. Check capacitors C205, C213 and C211 before removing the suspected preamp IC *(Figure 8-18)*.

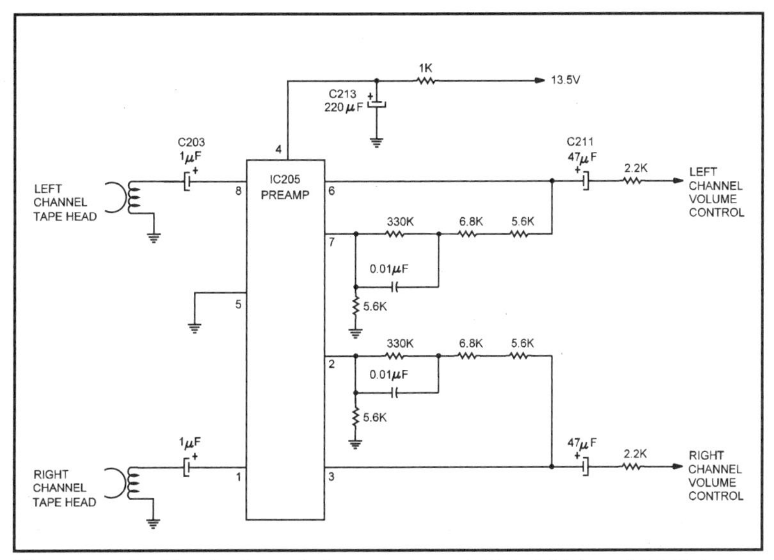

Figure 8-18. Check electrolytic capacitors C205, C211 and C213 to see if they are open or leaky before replacing IC205.

The medium-wattage output cassette-receiver-player might have an AF, driver and a separate power output IC in the 4-channel audio circuits. There are four different audio output circuits and IC components. All four output circuits are identical with the same components and voltage measurements. When a schematic is not available, compare the audio signal and voltages of the normal channels to the distorted or weak channel.

Determine if the radio receiver or cassette circuits are functioning. If the cassette circuits are not heard, suspect the tape head preamp circuits or the tape head itself. When the radio circuits are dead, proceed to the radio input circuits. If one channel is dead, distorted or weak, check that audio channel.

When the front left speaker is dead, start at the speaker output and check the audio at pin 9 (output) and pin 4 (input) of IC201 terminals *(Figure 8-19)*. Check the audio at the volume control to determine if the driver circuits are defective. When the signal is normal at any given point in the circuit, you have located the defective audio circuit.

4-CHANNEL AM/FM STEREO AMPLIFIER CIRCUITS

The AM/FM 4-channel stereo amplifier has identical audio amplifier circuits. Besides the AM/FM receiver circuits, the 4-channel receiver might have cassette and phono inputs. The phono and cassette inputs might contain directly-coupled transistor or IC preamp circuits. After the audio function switch, the audio circuits contain identical AF, driver and IC output circuits. A treble and bass controls are found in the driver input circuits. Each channel has identical volume controls at the input of each power output transistor *(Figure 8-20)*. Isolate the defective sound at each volume control and proceed either way until the bad stage is located.

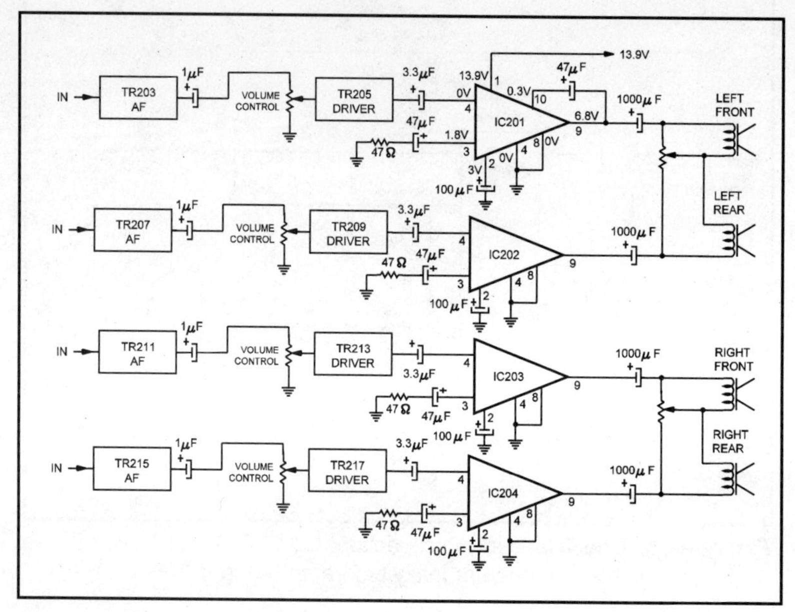

Figure 8-19. Signal trace the audio into pin 4 and out pin 9 of IC201 for weak, dead or distorted music.

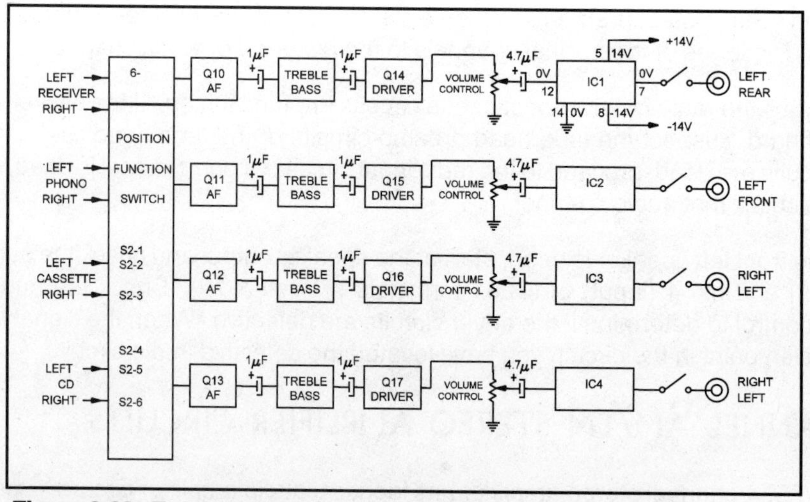

Figure 8-20. To locate the defective output IC, signal trace the signal in and out and compare critical voltage measurements.

Check small electrolytic coupling capacitors (1 to 4.7 μF), diodes, and output ICs or transistors for a noisy weak channel. A loud pop and arcing noise might result from a bad power switch. The noisy right channel might be caused by the audio IC function switch. A constant buzzing in the audio might result from loose IC mounting screws. Check the output ICs for a pop, crackle or hum noises. Suspect the output IC for a crackling noisy sound in the speaker. Clean up the volume control for noisy and intermittent sound. Replace volume control if excessively worn. Resolder the driver and output IC terminals for a low hum in one channel.

Suspect leaky (1 μF) coupling capacitors on the driver (Q15) input terminal for no audio in the left front speaker channel. Replace Q10 for a no audio symptom in the left rear channel. For no audio and a blown fuse, check for a leaky or shorted power output IC4. Suspect poor speaker relay terminals on the right rear speaker for intermittent audio. Measure terminal 8 of IC1 for a low -6.7 volts or improper voltage source that results in distortion at the left rear speaker. Check Q17, when the audio quits after several minutes of warm up. Suspect power output IC3 or small electrolytic capacitors (47 to 220 μF) connected to the output IC terminals for extreme distortion.

Chapter 9

TROUBLESHOOTING INEXPENSIVE ELECTRONICS AUDIO CIRCUITS

The early table model radio and portable cassette player contained a mono or one source of audio. Today, the clock radio and portable cassette player might have AM/FM/MPX reception. The monochrome TV provides a black and white picture. A mono-amplifier consists of only one channel of amplification with a speaker or headphones as indicator. The simple mono-amplifier might have only one audio IC or two and three audio transistors.

EARLY AUDIO CIRCUITS

In the early transistor table model radio, the audio was detected with a germanium diode detector (1N34A) from the 2nd IF transformer. This weak audio signal was controlled by a 5K ohm audio tapered volume control. A 5 μF electrolytic capacitor coupled the audio from the volume control to the base terminal of an NPN driver transistor (TR5). Interstage transformer T1 coupled the audio from TR5 to the base terminals of push-pull output transistors TR6 and TR7. The output transformer T2 stepped down the amplified audio to match the impedance of an 8 ohm pm speaker *(Figure 9-1)*. The low voltage power supply consisted of two silicon rectifiers in a fullwave ac circuit from a step-down power transformer.

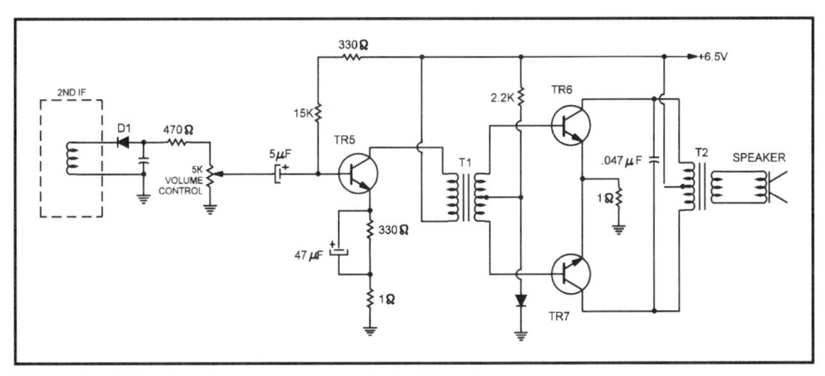

Figure 9-1. The early transistor radio audio circuits with interstage and output transformer circuits.

The early transistorized audio circuits within the table model cassette player might consist of two preamps, driver, and two push-pull output transistors. A 3.3 μF electrolytic capacitor coupled the tape head signal to a driver transistor. The driver transistor was transformer coupled to two output transistors. Instead of an output transformer, a 220 μF capacitor couples the amplified audio to the 8 ohm speaker *(Figure 9-2)*.

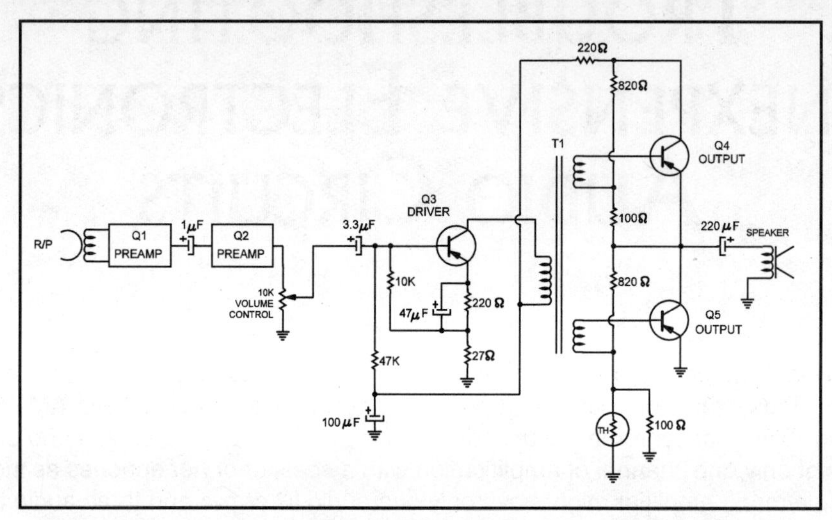

Figure 9-2. The audio output is coupled to the pm speaker through 220 μF electrolytic capacitor.

The portable tape player operated from batteries or the ac power line. A fullwave or bridge rectifier circuit provided a 6 volt dc source with a 2000 μF filter capacitor.

A SIMPLE IC AUDIO AMP

The most common IC audio amplifier found in many different electronic projects were designed around the 8-terminal IC chip (LM386N-1). This small low voltage audio IC only provided 250 mW with a 6 volt supply and the LM386N-3 provided 500 mW with a 9 volt source.

The early mono cassette player-recorder might require greater audio output with an IC component that was equivalent to 15 transistors within one IC. The simple cassette recorder had one transistor stage ahead of the power output IC. A separate ALC (automatic level control) transistor controlled the level of recording.

When in playback mode, the tape head was switched into the base circuit of Q1 through a 1 μF electrolytic capacitor. The weak audio signal was amplified and capacity-coupled to the 50K ohm volume control. VR1 adjust the correct audio applied to the input terminal of IC1 *(Figure 9-3)*. A 1 μF electrolytic couples the controlled audio to pin 9 and the output terminal 1 is coupled to an 8 ohm pm speaker. The cassette amplifier might be powered by five 1.5 volt "C" batteries or from a fullwave rectifier circuit of a step-down power line transformer.

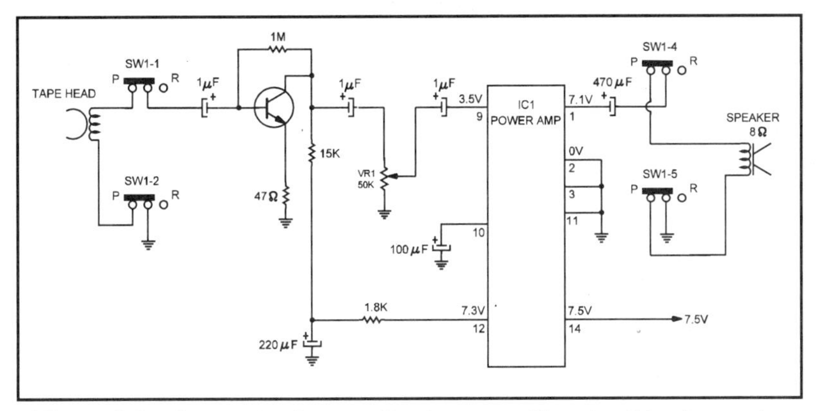

Figure 9-3. A very simple cassette player amplifier circuit in play mode.

A TYPICAL STEREO AUDIO AMP CIRCUIT

A typical inexpensive audio stereo output circuit might consist of only one dual-output IC. The left channel input is taken from a 100K volume control, coupled to pin terminal 4 of IC1 with coupling capacitor of 4.7 µF. Likewise, the left channel amplified audio output is coupled to pin 2, through a 470 µF electrolytic capacitor to the pm speaker *(Figure 9-4)*. The right channel is identical to the left stereo channel. This amplifier circuit might be powered by "D" cells or a step-down transformer rectifier circuit.

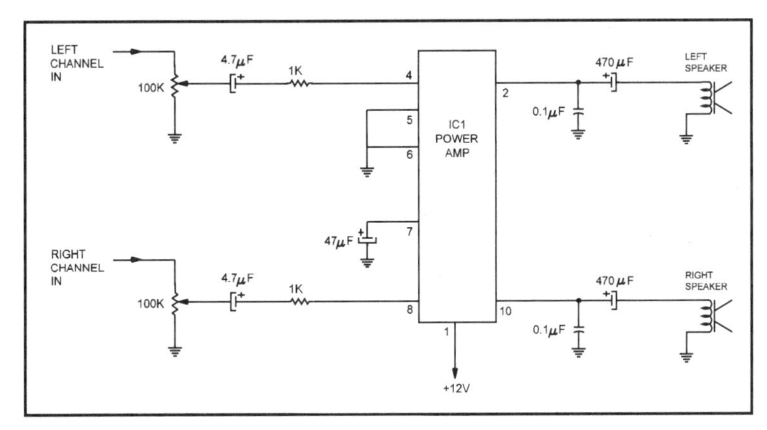

Figure 9-4. Large electrolytic capacitors couple the audio from dual power IC to speaker terminals.

THE EARLY TV AUDIO CIRCUITS

Before audio output ICs were included in the TV audio chassis, the early solid-state audio circuit consisted of a sound IF/Detector/audio preamp IC and a single-ended output transistor. The weak detected IF audio was amplified by a preamp audio stage and the volume was controlled inside the same IF-IC. IC201 was directly-coupled to the audio output transistor with a 680 ohm resistor *(Figure 9-5)*.

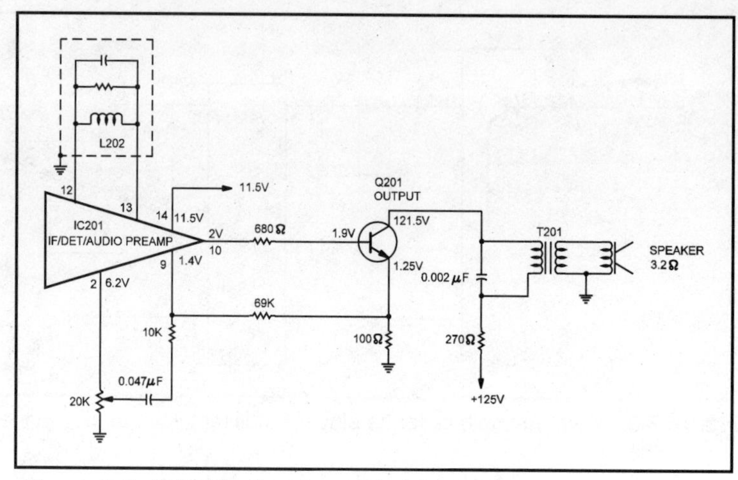

Figure 9-5. IC201 is directly coupled through a 680 ohm resistor to base of single-ended output transistor.

The audio output transistor (Q201) operates at a very high dc voltage through a step-down coupling transformer to the 3.2 ohm pm speaker. A distorted sound symptom can be caused by a leaky IC201 or Q201.

When the output transistor shorted or became leaky, the emitter resistor was found burned or overheated and was replaced.

Sometimes only a touch-up of the discriminator coil can cure the mushy-distorted speaker. Only a slight turn of the 4.5 mHz discriminator coil (L202) was needed. Slowly adjust the coil in one direction and then reverse to clear the distortion and obtain a clear sound.

EARLY TV IC OUTPUT CIRCUITS

The early output circuits consisted of one large IC with a heat sink glued to the top to provide audio to a pm speaker. IC201 consisted of the sound-IF input, preamp and audio output circuits. The volume was controlled by applying a dc voltage to pin 6 of IC201. The audio output at pin 8 is coupled to the 8 ohm speaker *(Figure 9-6)*.

Excessive distortion can be caused by improper adjustment of coil (T202) and leaky IC201. When a touch-up of T202 does not entirely erase the mushy sound, replace the small bypass capacitors inside the shielded coil. The noisy volume control can be cured by spraying cleaning fluid down inside the volume control lugs. Rotate the control knob back and forth to clean up the carbon-control area.

Intermittent or no sound can result from the defective 470 μF speaker coupling capacitor. Sometimes, just touching the capacitor can make the sound pop in and out. A defective IC201 or open 470 μF electrolytic can cause a weak sound symptom. Take critical voltage measurements on IC201 terminals to locate a leaky or defective output IC.

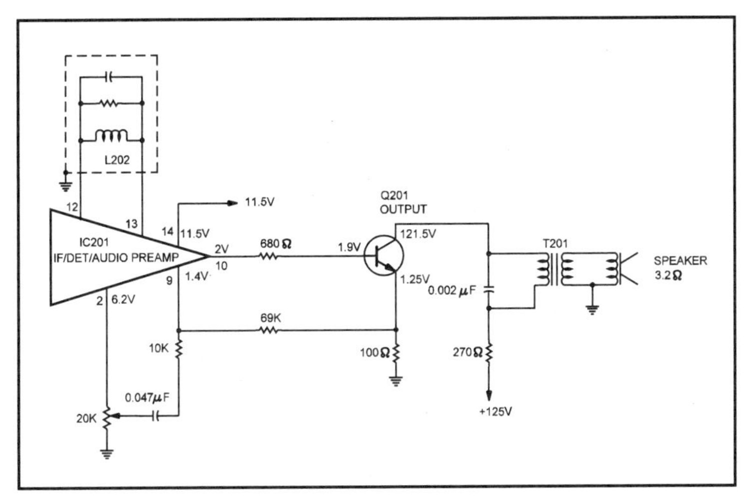

Figure 9-6. The early TV audio IC output circuits.

SERVICING COMPACT CASSETTE AUDIO CIRCUITS

The low-priced compact or portable cassette audio circuits might consist of only one large IC component to power an 8 ohm speaker. The audio output might have only 700 mW to 750 mW of audio power. The record/play tape head is coupled through SW1-A through a 1 μF electrolytic capacitor. SW1-B switches the other head lead to common ground. The preamp audio from pin 3 is coupled to the volume control through a 1 μF capacitor and switched into play circuit at pin 6 of IC-1 *(Figure 9-7).*

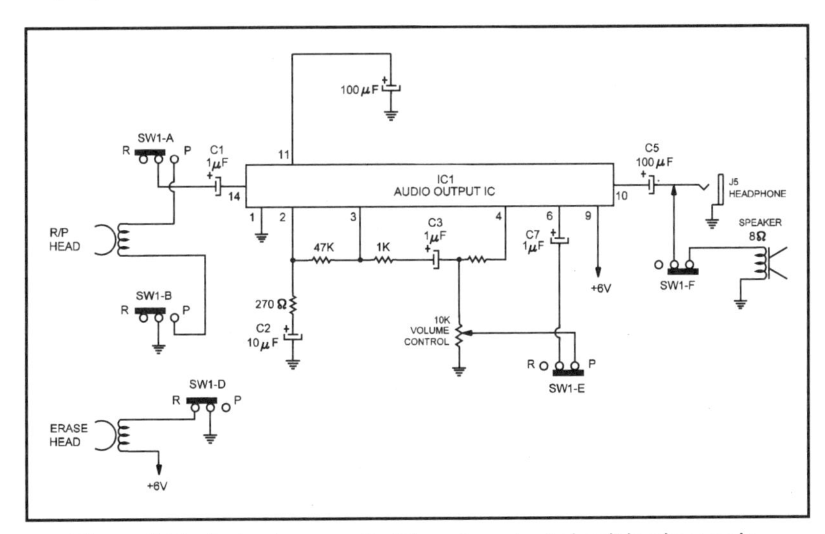

Figure 9-7. A simple cassette IC audio output circuit in play mode.

A 100 μF capacitor couples the audio signal to the earphone jack J5 and 8 ohm speaker. The erase head is excited by the +6 volt source and switched into the circuit in record mode with SW1-D.

For a weak sound symptom, clean up the tape head with alcohol and cleaning stick. Check C1, C3 and C7 for open conditions. Suspect a leaky C3, C5, C7 and IC-1 for extreme distortion; do not overlook a mushy speaker. If the sound is weak and distorted after head cleanup, suspect a defective IC-1. Take a resistance measurement to common ground of all IC terminals to locate a leaky capacitor or a change in bias resistors. Take an input signal test at pin 14 and output at terminal 10 with an external audio amp, to determine if IC4 is defective or there is connecting component failure.

TROUBLESHOOTING PHONO INPUT CIRCUITS

The simple single-play or portable phonograph with changer might include 3 or 4 separate transistors. A Darlington transistor or IC might serve as AF or driver transistor to two small output transistors in push-pull operation. The mono-crystal cartridge feeds into a volume control and directly-coupled the amplified signal to the output transistors. A tone control is found in the directly-coupled input audio circuits. C5 (100 μF) couples the output audio to a higher-ohm impedance speaker *(Figure 9-8)*.

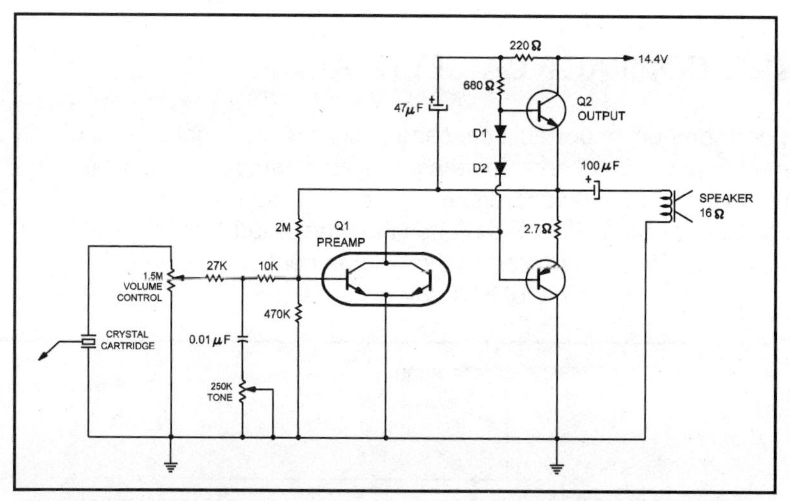

Figure 9-8. A simple phono crystal audio amplifier with only three audio transistors.

The crystal cartridge in this small record player might have a 0.5 to 3 volt output. A defective crystal cartridge might result from dropping the pickup arm or being left out in the sun too long, melting down the crystal cartridge. Dust and dirt collected from the record can cause a weak and mushy sound. Brush out the stylus with a small artists brush. A mushy-weak audio sound in the speaker indicates a defective crystal cartridge.

Besides a crystal cartridge, a leaky driver or output transistor can cause distortion in the speaker. A weak and distorted sound might result from leaky Q2 and open Q3 transistors. Suspect D1 and D2 with a low hum and distorted audio in the sound. Check each transistor with an in-circuit test for open or leaky conditions. Take critical voltage measurements upon each transistor. An open Q3 results in high dc voltage on the collector and emitter terminals of Q3.

PHONO INPUT CIRCUITS

The crystal phono input circuits in a large console or compact table model feed into phono input jacks. The stereo cartridge circuits are connected to a function switch that switches in a microphone, radio, tape, and phono input signal *(Figure 9-9)*. Check the defective crystal cartridge for weak, intermittent, or distorted sound. One stereo channel might be dead or intermittent with a broken wire connection at the cartridge.

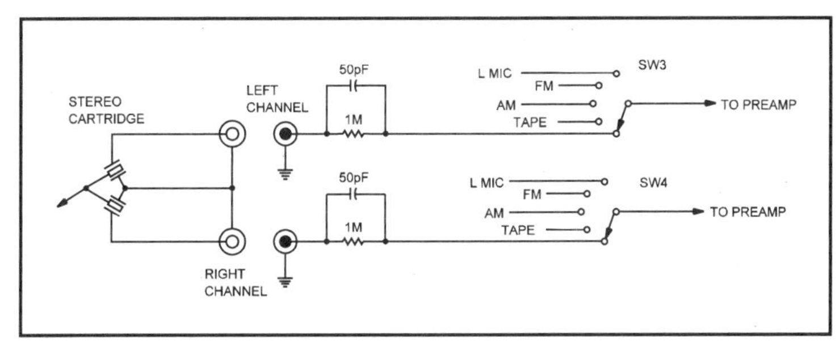

Figure 9-9. Stereo cartridge circuit connected to a function switch.

The phono player might have a magnetic cartridge instead of a crystal cartridge for better frequency response. A crystal cartridge is a transducer made up of piezoelectric crystal material. The stylus or needle is directly connected to the crystal element which follows the groove of the record. The vibration or movement upon the crystal surface provides an output voltage that is amplified into music production. The crystal cartridge has a much higher output voltage than the magnetic pickup.

The magnetic pickup is called a variable-reluctance pickup where the stylus or needle causes a metal vane or armature to move within a magnetic field. The coils on each side of the metal piece are fed into a preamp or equalizer stage. Since the signal is so weak in a magnetic cartridge compared to the crystal cartridge, both stereo channels must be amplified with a directly-coupled transistor preamp or IC input circuit *(Figure 9-10)*.

Figure 9-10. A single play stereo phonograph turntable.

A separate stereo preamp, equalizer or tone directly-coupled transistor circuit is coupled to the phono input jacks. Sometimes the inexpensive units have a preamp circuit that is switched in for the radio, tape and phono functions. The higher-powered output amplifier circuits usually contains a separate preamp stage for the magnetic stereo cartridge.

The magnetic input jacks are coupled with a electrolytic capacitor to the preamp Dual-IC. The same Compact combination receiver, tape, and phono circuits might have separate transistors as the preamp stage in the tape circuits of the high-powered output circuits. The weak magnetic phono audio is amplified by IC2 and capacity coupled to the function switch.

Check the signal in and out at pins 5 and 7 of the left channel and 3 and 1 of the right channel with an external audio amplifier. The audio signal might be quite weak at pins 3 and 5. Check for an open coil in the magnetic pickup with the low ohm scale of the DMM. Take critical voltage measurements on the supply terminal pin 8 (22.5v) of IC2 to determine if IC2 or the power source are defective *(Figure 9-11)*.

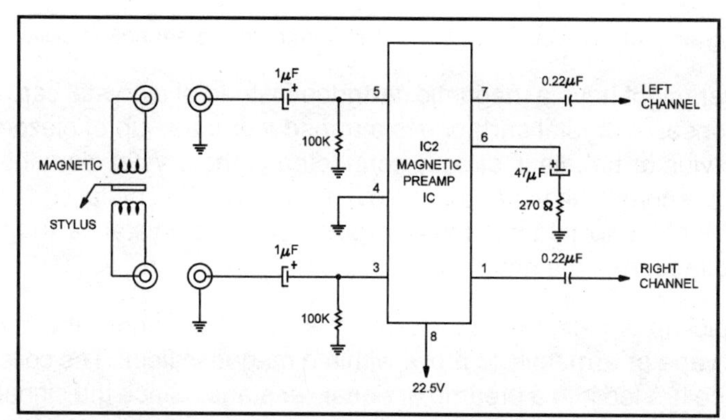

Figure 9-11. The magnetic phono pickup and preamp IC circuit.

PHONO PREAMP TRANSISTOR CIRCUITS

The phono stereo preamp circuits might consist of two transistors that are directly-coupled together. The left and right channel phono jacks are coupled to the base of the first NPN transistor. The collector terminal of Q101 is tied directly to the base of Q102 in the left channel. The audio phono signal is amplified by Q101 and Q102, and coupled with a 1 μF electrolytic capacitor to the function switch SW-1 *(Figure 9-12)*. Both right and left channels are identical.

When the left channel is weak or dead, signal trace the phono audio at the base of Q101 and the output at the collector terminal of Q102. Check the voltage on each transistor and compare to the schematic. If one transistor becomes open or leaky, the other directly-coupled transistor voltage will also change. If a diagram is not handy, compare the voltage and resistance readings to the normal channel.

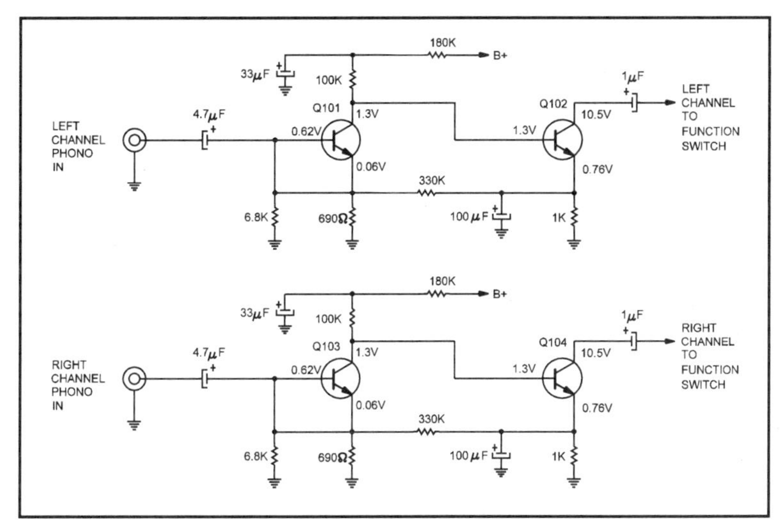

Figure 9-12. A directly-coupled L & R phono preamp transistor circuit.

A leaky preamp transistor can cause weak and distorted music. The weak phono audio might result from an open coupling electrolytic capacitor or transistor. The intermittent preamp transistor might restore to normal when tested in circuit or voltage measurements are taken at each terminal. Simply replace the suspected open or intermittent transistor. Universal transistors can be replaced in the phono circuits without any problems.

SERVICING THE SIMPLE BOOM-BOX AUDIO CIRCUITS

The boom-box player with radio, cassette and CD functions might have a dual-output IC circuit. A special function switch applies each function to the output audio circuits. A muting system might be found in the input or output circuits. The left and right channels are switched to the headphones or external speaker terminals. Some of the larger boom-box players have additional left and right speaker output jacks.

The preamp circuits might contain transistors, IC or the preamp circuit can be included within one large IC output component. Bass and treble tone controls are usually located between the preamp and volume control. The right and left channels are coupled through an electrolytic capacitor (1 μF) from the volume control to the input terminal of IC204 *(Figure 9-13)*. The amplified output of IC204 is found at pins 2 and 11.

Suspect the audio output circuits, when the sound is intermittent, dead or weak within the CD, radio or cassette functions. A dead left or right channel might result from a defective IC204, 1000 μF electrolytic capacitor, or improper supply voltage at pins 1. The intermittent left channel can result from a bad 1 μF or 1000 μF coupling capacitor and IC204. The weak right channel can be caused with open 1 μF, 1000 μF, and improper voltage at pins 10 and 11.

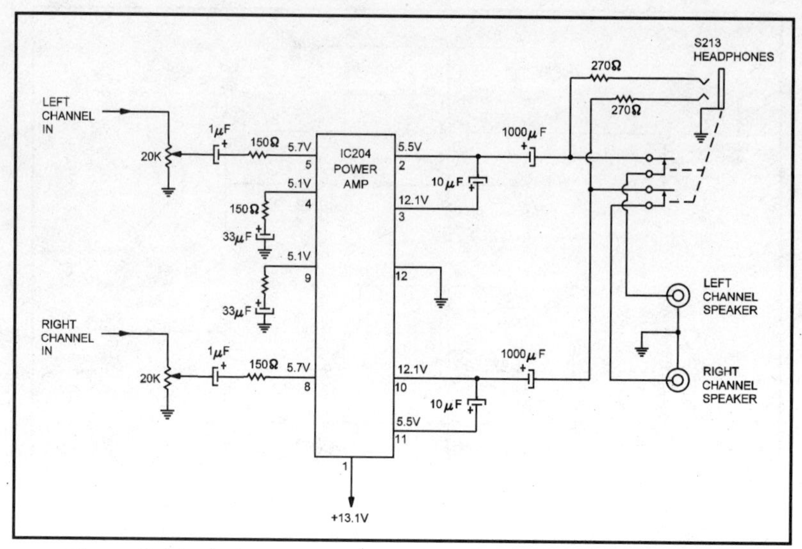

Figure 9-13. A dual-power output IC circuit in a boom-box player.

Check the switching contacts of earphone jack for noisy, erratic or cutting-out of sound with dirty switching contacts. A noisy or worn volume control can be silenced with cleaning fluid sprayed down inside the control terminals. Rotate the volume control back and forth to smooth out the noisy audio.

SERVICING THE PORTABLE AUDIO CD CIRCUITS

The portable CD player audio circuits begin at the D/A (digital-analog) converter stage. The D/A converter and audio amplifier might be included in one large IC or microprocessor.

In the earlier chassis, the audio amps after the D/A circuits might be separate transistors or IC components. The portable CD player is chock-full of SMD processors with gullwing terminals. The portable CD audio output components might consist of a separate headphone IC amp or a dual-IC audio amp. The audio amp can be attached to the line output and headphone jacks *(Figure 9-14)*.

Scope the D/A converter audio output terminals when either channel is weak or dead. The external audio amp can be used to signal trace the audio from the D/A converter to the line output jacks. Suspect the headphone amp IC or improper voltage source when the headphones are weak, distorted or dead. A dirty or worn headphone jack can cause intermittent or erratic sound in the headphones. Determine if a low frying noise in either channel is caused by the D/A converter, headphone or line amp IC with an external audio amplifier. Take critical voltage measurements on each IC terminal to determine if the IC is leaky or if there is a defective voltage regulator transistor in the power supply source.

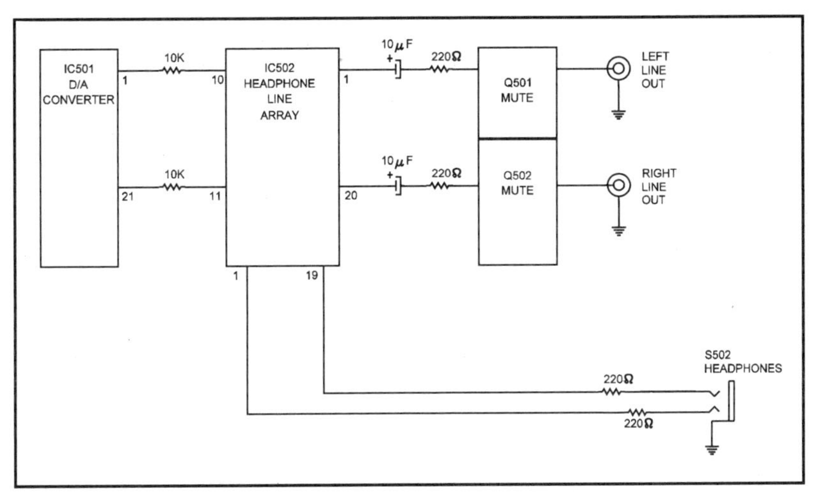

Figure 9-14. The portable CD player line and headphone output circuits.

TROUBLESHOOTING THE AUTO CASSETTE AUDIO CIRCUITS

The deluxe car receiver-cassette player might have a normal reverse tape deck, audio muting, IC preamp or equalizer, transistor preamp, Dolby, switching, transistor AF amp, line output amp, and power output IC circuits. The dual-right and left cassette tape heads are switched into the input circuits of a preamp or equalizer dual-IC component. The tape heads are coupled to the input terminals with a 4.7 μF electrolytic capacitor *(Figure 9-15)*. The output terminals of IC501 are pins 3 (left) and 6 (right). These input circuits are tied to the radio/tape function switch.

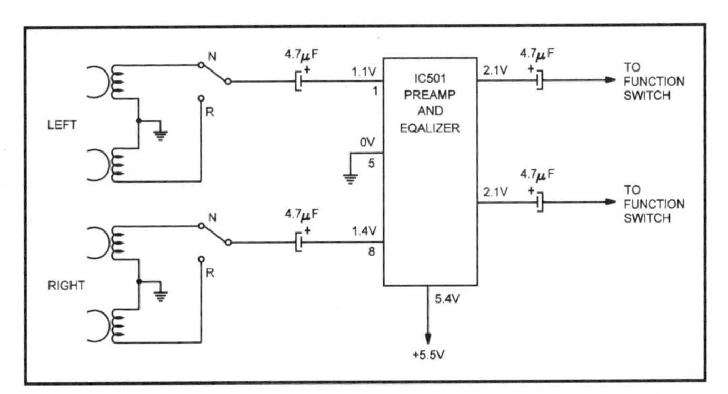

Figure 9-15. The auto cassette input preamp and equalizer circuits.

Signal trace the head circuits at pin terminals 1 and 8 of IC501 with the external audio amp. Insert a test cassette for the audio signal. If signal is found at the input terminals and not at the output, suspect a defective IC501 or improper voltage source at pin 4. Check all voltages on each IC terminal to locate a leaky IC501. Suspect IC501 is leaky when the voltage at pin 4 is very low compared to the schematic. Remove pin 4 from the pc wiring with soldering iron and solderwick. Now take another voltage test at pc wiring. Replace leaky IC501 if the voltage rises above 5.5 volts.

The power output IC might amplify both right and left channels to the respective speaker terminals. An input switch can switch the preamp audio to the external amp or line output jacks. A separate line output IC amplifies the output jacks, while another power amplifier provides from 5 to 30 watts to several different speakers *(Figure 9-16)*.

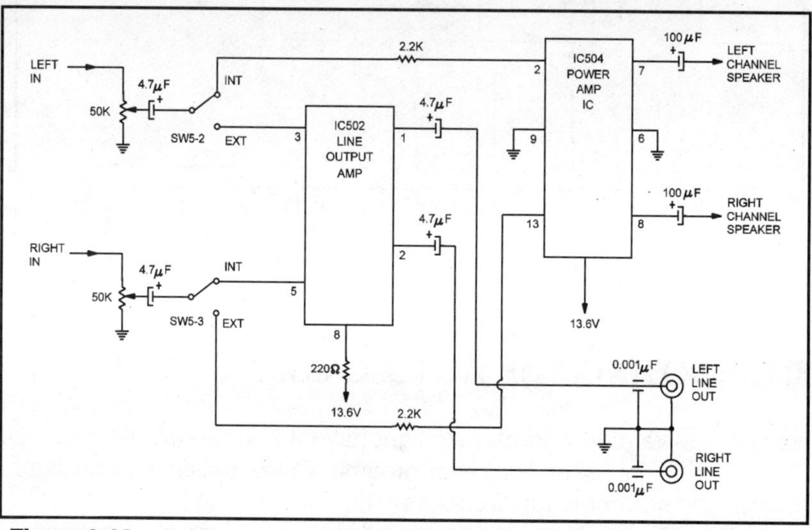

Figure 9-16. A deluxe auto-cassette receiver line and power output circuits.

Suspect the 4.7 μF coupling capacitor or IC502 when the left line output channel is weak, dead or distorted. Check the supply voltage at pin 8 of IC502 for a leaky line output IC. Check SW5-3 and IC502 with a noisy right line output sound at IC502 when both line output jacks are dead or distorted. Check the audio at pin 2 with the external audio amp when the left channel is noisy or weak.

Take voltage measurements on all IC504 terminals with a dead symptom in both speaker channels. Spray each transistor or IC with coolant when one channel is intermittent *(Figure 9-17)*. Apply several coats of coolant before going to the next component. Likewise, spray electrolytic coupling capacitors for weak or intermittent symptoms.

Figure 9-17. Spray each intermittent transistor, IC or audio component with coolant.

REPAIRING AUTO CD PLAYER AUDIO CIRCUITS

The auto CD player might have a few more audio circuits than those found in the portable CD player. Signal trace the audio from pin 17 and 18 of D/A converter (IC301), with the external amp or scope. Check the IC circuits where the audio stops. The left and right channels can be signal traced from the D/A converter to the line output jacks. Most auto CD player audio symptoms can be repaired with signal tracing methods and critical voltage measurements.

Signal trace IC301 and IC401 for noisy or garbled audio *(Figure 9-18)*. Check the D/A converter (IC301) for distorted audio. No audio in the left channel might result from a defective low pass filter (LPF) circuit. Suspect a voltage regulator IC with a defective right channel and normal left channel.

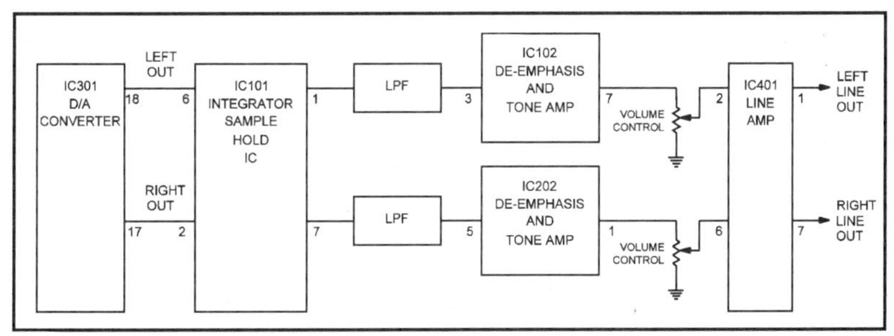

Figure 9-18. A block diagram of the auto CD player audio line output circuits.

Check IC401 with a frying or static noise in both channels. Suspect a dead, noisy or intermittent music with an open or defective electrolytic coupling capacitor. Solder all pin terminal connections upon IC301 and IC401 for intermittent, noisy or weak sound symptoms.

REPAIRING AUDIO HEADPHONE CIRCUITS

A dirty earphone jack or connecting wires can cause intermittent or erratic music in the headphones. Clean around the ground jack input area for erratic audio. Check the cord for breakage where it enters the earphone or at the male plug. A dirty male plug might not make a good connection where it enters the headphone jack.

Check the audio output components when there is no audio from either the speakers or headphones. The mono headphone has only two connections at the male plug. A stereo male plug has a common ground, left and right stereo connections. The low impedance headphones might have an impedance from 8 to 55 ohms, while the high impedance phones might be 1000 to 2000 ohms.

The stereo headphone jack might switch the speakers out of the audio circuit and place the headphones in the output circuit. A small 100 or 220 ohm resistor is placed in series with the audio output to lower the level for headphone reception *(Figure 9-19)*. Of course, the volume control should be lowered when the headphones are plugged into the output jack.

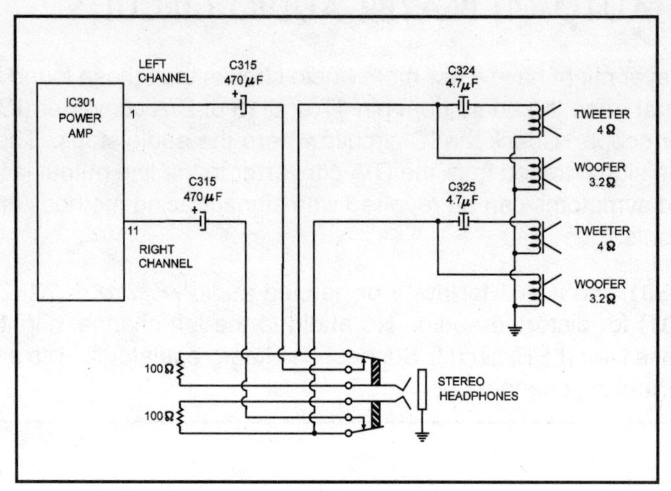

Figure 9-19. The stereo heaphone output jack coupled to the PM speaker circuits.

Clip the RX1 ohm range of the DMM across the tip and ground area of the earphone plug for continuity tests. Both right and left resistance measurements should be quite close. An 8 ohm impedance headphone might have a 7.5 ohm measurement on the DMM. Flex the cord while clipped to the DMM. Notice if the resistance changes, indicating a break within the headphone cord.

Clean up the male plug with cleaning fluid for erratic or intermittent reception. Spray cleaning fluid down inside the headphone jack and work the plug in and out to help clean up the contacts. Check for a cord break at the earphone plug. Most electronic stores stock male headphone plug replacements. Remove the rubber headphone cover or cap and check for a break in the cord or a torn-loose wire connection. The small, low impedance headphones might have miniature 8 ohm pm speakers inside the covers.

CD PLAYER HEADPHONE CIRCUITS

The headphone circuits within the portable CD player might have a separate IC amplifier or be contained within the line output circuits. The headphone amplifier connects directly to the line output jack of each stereo channel. A separate volume control might be used to control the audio to each stereo headphone. The input and output audio is coupled through an electrolytic capacitor at the dual-headphone IC amplifier. Some headphone circuits have mute transistors at the input or output amp circuits.

A defective volume control might cause no sound, intermittent or erratic music with a worn control. A leaky coupling capacitor (C111) and IC11 can cause distortion in the headphones *(Figure 9-20)*. The leaky or open IC11 might cause distortion in one or both stereo channels. C113 can cause weak, intermittent and no sound in the headphones. A dead right channel might result from a defective IC11, with a normal left channel.

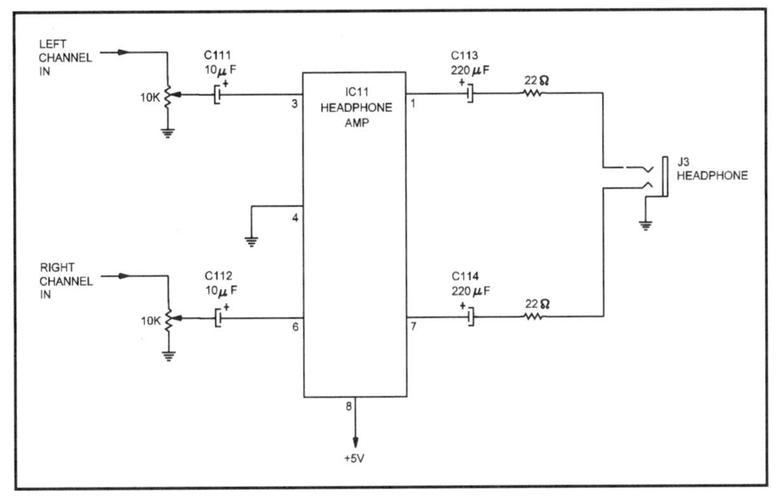

Figure 9-20. A portable CD player headphone output circuits.

CHECKING RECORDING CIRCUITS

In the most simple cassette players, the recording and playback circuits are the same. Sometimes only the preamp circuits are used for playback and recording circuits. In larger or deluxe recording circuits, a separate transistor or ICs are used for only recording features. The stereo tape heads are switched into the input amplifier circuits for playback, while in recording mode the tape heads are switched into the output circuits.

The electret or condenser microphone is switched into the preamp circuits of IC401 with the tape/radio switch and record/play switch S203 in record mode *(Figure 9-21)*. The microphone audio is amplified through IC401 and switched back to the record/play (R/P) head so the audio can be recorded on the cassette tape. The bias oscillator is switched into the tape head circuit by S2-1 for high-fidelity recordings.

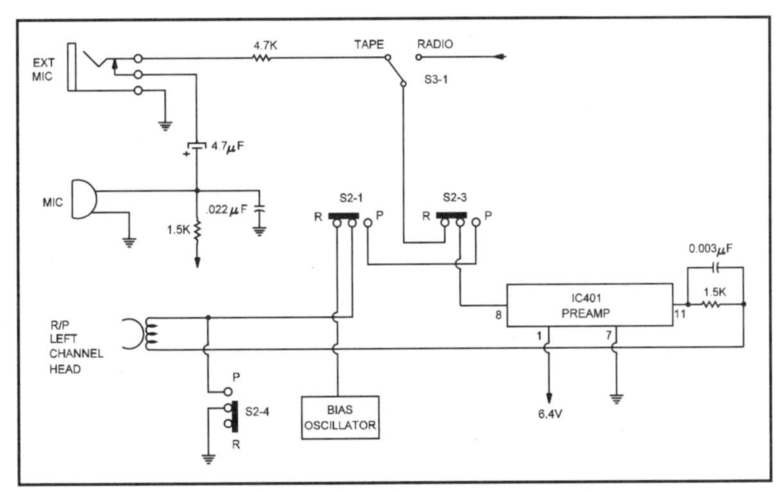

Figure 9-21. A portable cassette recording circuit.

Often, when problems occur in the recording circuits, the same conditions exist in the same playback circuits. If the cassette player operates normally in the play mode and not in the record mode, check for dirty or poor switching contacts. First, Clean up the function R/P switches. Clean up the tape heads with alcohol and a cleaning stick. Check the microphone circuits by inserting another external microphone.

Now check the bias oscillator circuits for no, poor or distorted recording. Scope the output of the tape head in record mode. A sine waveform should be noted when the bias oscillator circuits are performing. The jumbled-noisy recording might be caused by a defective erase head or circuit. Both the R/P and erase heads are excited with the bias oscillator, when the bias oscillator circuits are switched in the recording mode.

BIAS OSCILLATOR CIRCUITS

A bias oscillator circuit provides a 30 to 100 kHz frequency to the tape heads to erase and bias the system for linear recordings. The bias oscillator circuit operates only when in record mode. Some cassette recorders switch in the bias oscillator circuits with a positive voltage source or grounds the entire oscillator circuit. A separate waveform is sent to the erase head to erase the previous recordings and to both R/P heads for high-fidelity and linear recordings.

The bias oscillator circuit might contain one or two different transistors. A bias oscillator circuit might not be found in inexpensive cassette recorders. When a supply voltage is switched into the bias oscillator, the circuit provides a waveform to the erase and tape heads *(Figure 9-22)*.

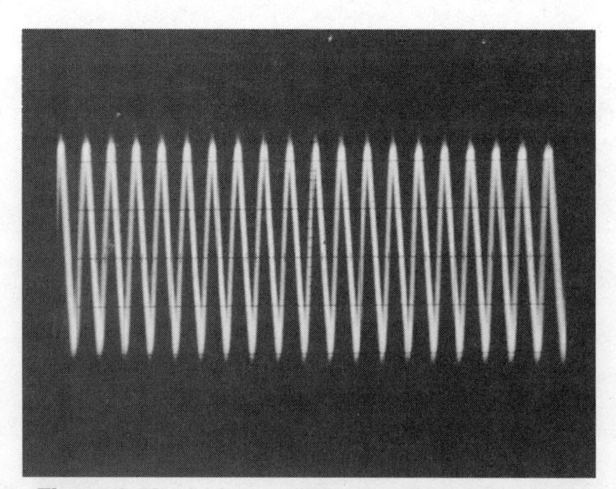

Figure 9-22. A bias oscillator waveform at the tape heads.

Check the bias oscillator circuits with no sound or jumbled recordings. Scope the ungrounded erase and R/P heads for a waveform in record mode. Go directly to the bias oscillator circuits when no waveform is located on the tape heads. Suspect a dirty record switch or power source with low or no voltage at the bias oscillator circuits. Test the bias transistor with in-circuit diode tests of the DMM or transistor tester.

Take a continuity measurement of T507, if the bias transistor is normal. Sometimes the oscillator transformer might have voltage going into it and no voltage to the collector terminal, indicating an open connection or winding *(Figure 9-23)*. Take a voltage measurement upon the collector terminal of bias transistor for open or leaky conditions. Check and shunt the small bypass capacitors for leakage, shutting down the bias oscillator circuits. Solder up all bias oscillator connections for intermittent recordings. Monitor the tape head windings when the recordings appear intermittent or erratic.

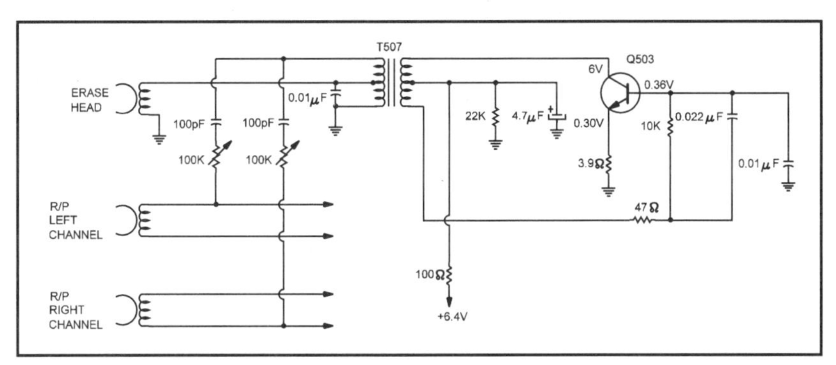

Figure 9-23. The bias oscillator circuit connected to R/P tape and erase heads.

SERVICING DOUBLE-CASSETTE HEAD CIRCUITS

The double-cassette recorder might appear in a portable boom-box AM/FM/MPX receiver or integrated stereo component systems. Two different sets of stereo tape heads are found with one set to record and playback, while the other set is used for only playback features. Tape head 1 is connected directly into a preamp IC and tape head 2 is switched into the record and playback modes. The erase head and R/P tape head (2) are excited with a transistor bias oscillator circuit.

Cleanup the tape heads for packed oxide and R/P switch with dirty contacts. A good clean-up might solve many tape sound and speed problems. The dead right channel can result from an open tape head winding. Check and compare each stereo head resistance with the other normal channel. The tape head should be okay if quite close to the other head resistance measurement. A high resistance measurement can be caused by a poor internal tape head connection. Intermittent and erratic playback can be caused by a poorly soldered or loose cable wire head connection. Push on the head terminals with a pencil eraser and notice if the resistance changes, indicating a poor internal connection.

Check tape head (2) for intermittent recording and playback modes. Make sure the R/P function switch is clean and not worn. Solder all function switch contacts on the pc board. Notice if the tape head (2) is switched into the preamp IC at different pin terminals than playback head (1) *(Figure 9-24)*. Tape head number 2 is switched into the record mode from a separate record IC. The record or playback audio signal can be traced from the tape heads through IC5 with the scope or external audio amp.

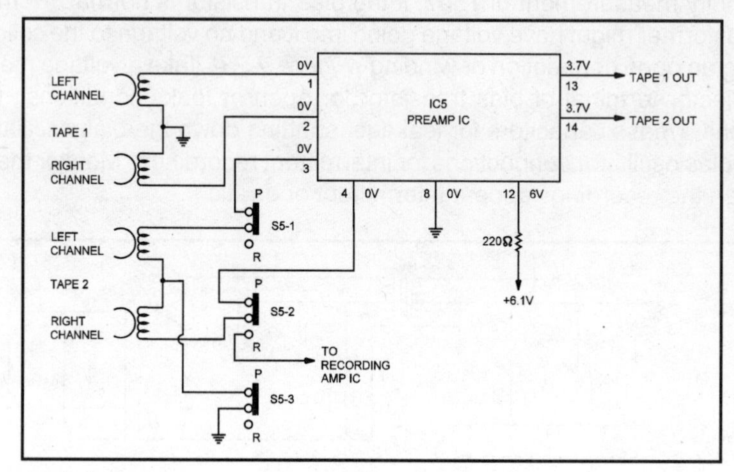

Figure 9-24. The double playback and R/P heads are switched into a preamp IC circuit.

The leaky preamp IC5 can cause a dead, weak or distorted record or playback symptom. Take critical voltage measurements on all IC terminals. Check the supply voltage at pin 12 to determine if the IC is leaky or shorted. When the recorded audio is found at the input terminals of IC5 and not at terminals 13 and 14, suspect a defective IC. Sometimes the IC voltages test close or are only off a fraction of a volt compared to the schematic or the other channel and still the IC might be open. Simply replace the suspected IC when voltages are close and no audio is found at either output terminal.

REPAIRING THE ERASE HEAD CIRCUITS

The erase head removes the recorded music from the tape. The erase head might contain an erase magnet or a coil with erase current. The erase head is always mounted ahead of the R/P tape head to remove the previous recordings. The cassette could not be used over and over again, to record music or information, without erasing the previous recording with the erase head. A low-priced tape recorder might have a dc voltage from batteries or power supply to provide a magnetic field to erase the magnetic tape, instead of a bias oscillator circuit.

The bias oscillator circuits provide a magnetic erase current to the erase head and also excites the R/P tape heads. The dc voltage is switched to the erase head to demagnetize the tape in record mode. Likewise, a dc voltage is applied to the bias oscillator circuits when making a recording *(Figure 9-25)*.

The erase head is usually mounted ahead of the R/P head within the tape path. In some cassette players, the erase head might be pivoted out of the way in playback and dubbing modes. The erase head is mounted on a swivel type assembly and in record mode is moved up alongside the R/P tape head. When in play mode, the erase head is pivoted down away from the tape path.

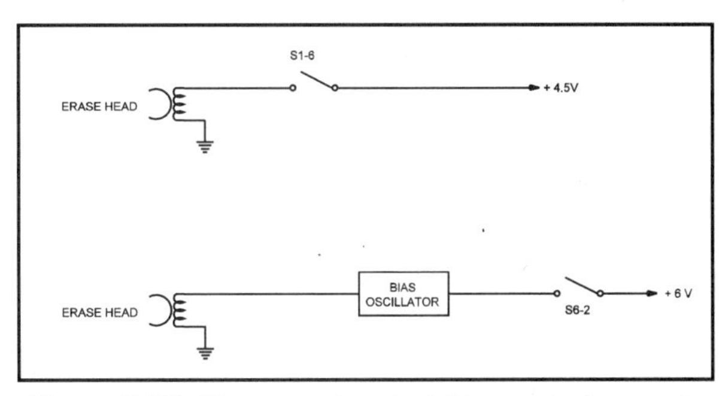

Figure 9-25. The erase head might operate from a dc source or bias oscillator circuit.

The defective erase head might appear open or may not touch the magnetic tape to erase the previous recording. Inspect the mounting screws for improper mounting of the erase head. Check the continuity of the erase head on the 2K ohm scale of the DMM. Often the erase head is normal with a continuity measurement. Check for a dc voltage across the erase head winding in record mode with a dc voltage excited erase head circuit. No voltage might be caused by dirty or poor recording switch contacts or low voltage from a voltage regulator transistor.

Scope the erase head for a bias waveform on the ungrounded head terminal in a bias oscillator circuit. Check the recording switch or defective bias oscillator circuit with no waveform.

MICROPHONE CIRCUITS

The microphones found within the cassette recorder are usually the electret-condenser type. Two identical microphones are found in a stereo cassette recorder. An external microphone self-shorting jack is located in larger players. The electret microphone is made from a dielectric disc and when sound waves strike the disc, a small voltage is developed. Most electret mikes operate with a low external dc voltage (1.5-10 volts DC). The dc voltage is provided from batteries or a dc source with a voltage dropping resistor to provide correct operating voltage.

The electret microphone is capacity coupled to the self-shorting external mike jack and switched into the input terminal of the preamp IC *(Figure 9-26)*. The small microphone can be checked by inserting another microphone in the external jack J101. If the subbed microphone operates normally, you know the amplifier circuits are good. The preamp circuits are normal if the playback cassette can be heard. An infinite measurement should be found across this type of microphone.

First clean the R/P function and tape/radio switches with cleaning fluid. Clean up the external mic jack (J101). Check the dc voltage across the microphone terminals in record mode. No voltage might indicate an open switch, poor voltage source or an increase in resistance of the voltage dropping resistor. Shunt the coupling capacitor (47 μF) with another electrolytic capacitor. Sometimes sharp objects might be stuck into the mike holes, damaging the microphone element.

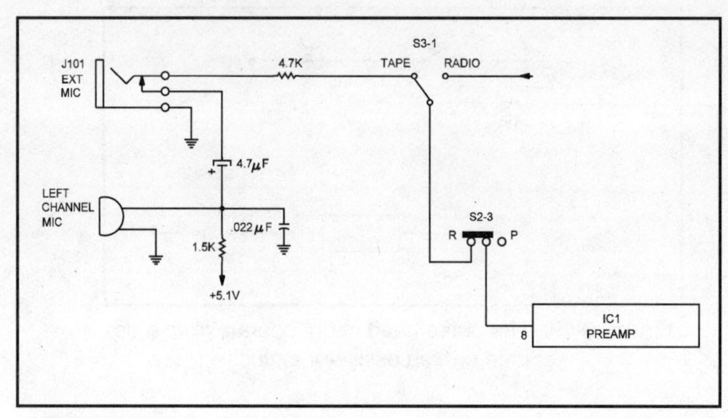

Figure 9-26. The left channel electret microphone input switched into preamp IC1.

SERVICING INTERCOM CIRCUITS

The early tube intercom might have 2 or 3 tubes *(Figure 9-27)*. Most solid-state intercoms have simple circuits with three or more transistors or one large IC. Most failures in tube intercoms are weak or shorted tubes. Check the tubes in a tube tester or substitute another one.

Figure 9-27. The early tube intercom unit operating from the AC power line.

A dried-up filter capacitor in the tube intercom can cause a constant hum in the speaker. Shunting the electrolytic capacitor with a known filter capacitor can solve the loud hum problem. Sometimes a dirty talk-listen switch produces noisy, static or no sound when in talk position. Clean all switches with cleaning spray.

Check all transistors in the solid-state intercom with a transistor tester. Place a connecting speaker close to the main station and notice the loud feedback-sound when the amp is normal. Suspect a break in the connecting wires with no sound in the remote unit. Check all wire connections *(Figure 9-28)*. To locate broken intercom wires, twist the wire ends together at the remote and take a continuity measurement with the ohmmeter at the main amp.

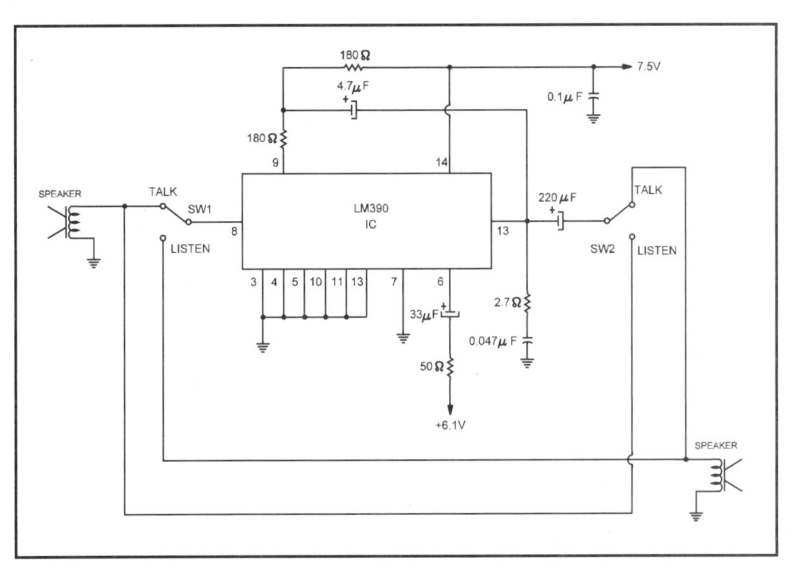

Figure 9-28. A simple intercom unit with one IC as audio amplifier.

Remove the top cover from the intercom unit and take critical voltage tests on transistors or IC components. In most cases, no intercom voltages or schematics are found, and you must troubleshoot the unit without a schematic. Clean up the function switch with cleaning spray for noisy or erratic conversations. Remember the small intercom unit might have only a few transistors or a couple of IC parts. Signal trace with a 1 kHz injected signal at the input terminals and scope the waveforms through the audio circuits.

Chapter 10

Servicing Special Consumer Electronics Audio Circuits

The audio circuits in special consumer electronics might consist of the telephone answering machine, speaker relay circuits, remote control, and muting circuits. Servicing the tube amplifier of an ac/dc radio, musical and high-powered amplifier might be considered special circuits. Since the vacuum tube is finding it's way back into the consumer electronic field with Hi-Fi audio gear, that might include amplifiers, preamps, phono stages, and tuners. Troubleshooting the tube chassis can introduce different service problems *(Figure 10-1)*.

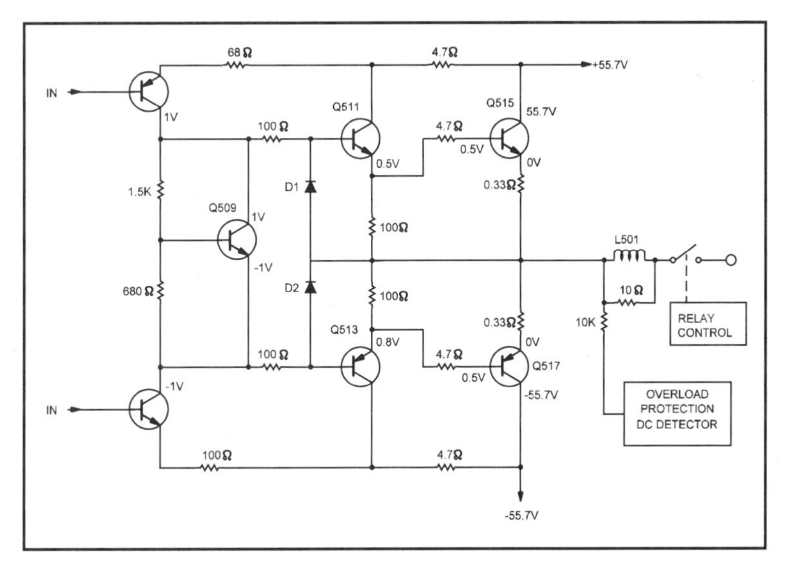

Figure 10-1. The telephone answering machine takes a telephone message when one is not home or available.

TROUBLESHOOTING AUDIO CIRCUITS IN THE ANSWERING MACHINE

The audio circuits in telephone answering machines are quite common to most cassette audio circuits. The audio circuits might consist of a recording amp, preamp, or main amplifier. The microphone (condenser or electret) circuit is fed into the recording amp, to outgoing and incoming R/P heads, audio preamp, and finally to the power output IC *(Figure 10-2)*.

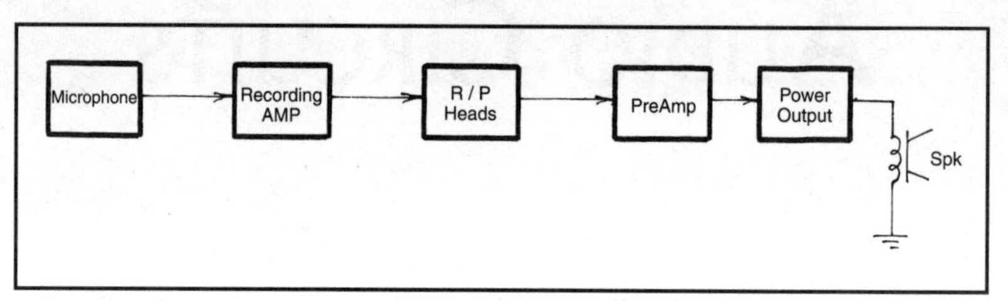

Figure 10-2. A block diagram of the audio system within the telephone answering machine.

The audio from the telephone line is amplified and recorded on a miniature cassette. When the message is recorded, the cassette can be played back by pressing the play button. The electret microphone is used to change or make different recordings of outgoing messages. The recording amplifier is used in only recording messages and disabled during audio playback. The audio preamp is not used during incoming messages. Both the preamp and power amplifier are in force to repeat the incoming message.

The recorded message is taken from a recording transistor or IC amplifier circuit and capacity coupled to a volume control. The preset volume control audio is fed to the input of a transistor or IC preamp stage. The voltage preamp IC amplifies the weak recorded message and is directly-coupled to a power output IC or transistors. An electrolytic capacitor (33 μF) couples the audio to a 32 ohm pm speaker *(Figure 10-3)*. The audio output circuits can be two push-pull transistors or one large IC.The ac power supply might consist of a fullwave or bridge rectifier circuit with a large electrolytic (2000-3000 μF) filter capacitor. The function switch or push buttons apply voltage to the various answering machine circuits. Very little current is drawn when in the off position. Most power supply problems are related to improper voltages with dried-up filter capacitors and leaky silicon diodes *(Figure 10-4)*.

A no erase or jumbled recording might result from an improperly seated cassette or one with a dirty or open tape head. The erase head circuit might be excited with a dc voltage switched to the erase head. The no-playback message might be caused from a defective transistor, IC or amplifier component.

A weak or distorted message can be caused by a dirty tape head, defective microphone and wrinkled cassette tape. Check the microphone and input circuits when the announcements and incoming messages do not record. Suspect the input circuits when the message is not recorded. Clean up the R/P head for a weak and distorted message. A no sound system might result from a defective amplifier or power supply circuits. Clean up the function switch, shunt filter capacitors, and check leaky blocking diodes, when the amplifier oscillates in the play mode.

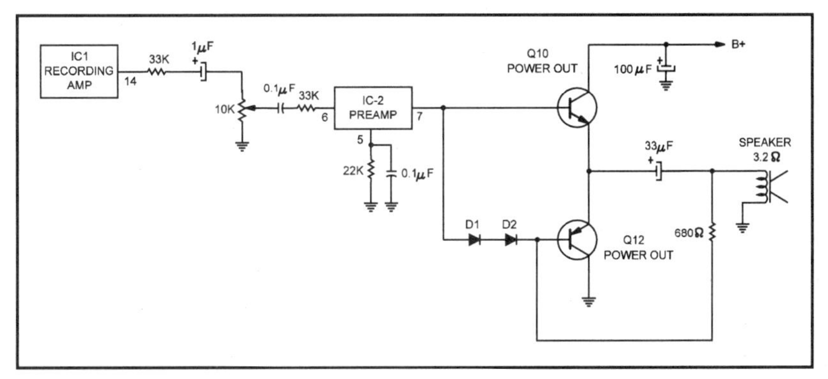

Figure 10-3. Typical audio stages within the telephone answering chassis.

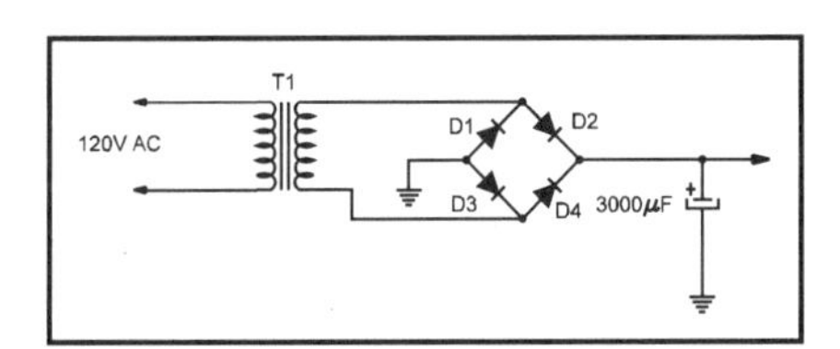

Figure 10-4. A simple ac power supply in the answering machine.

Intermittent and erratic sounds might result from a dirty function switch or push buttons. A push button may not seat properly, appear erratically, indicating poor switching contacts. Spray cleaning fluid down inside the switch area. Rotate function switch or move push buttons up and down to help clean the switch contacts. Be careful not to apply too much cleaning spray that might drip down on the moving tape, idlers and belt area.

PHONO EQUALIZER AMP CIRCUITS

The phono equalizer amp might also be called a tone or phono head amp with a single transistor or two transistors in a directly-coupled amp circuit. The equalizer phono amp circuit is ahead of the stereo volume controls. The auxiliary and playback jacks might be switched into the volume control in the high-wattage amplifier circuits. A magnetic phono pickup connects directly into the equalizer circuits. Both stereo equalizer circuits are identical.

Within the high-powered amplifier, the stereo equalizer circuits might have a phono 1 and 2 as phono input jacks. Either of the stereo phono circuits can be switched into the input of the equalizer phono amp circuits *(Figure 10-5)*.

A 3.3 μF electrolytic capacitor couples the input phono audio to the base terminal of Q501. Q502 is directly-coupled to Q501 and the signal is amplified to a function switch (S1-1). This same switch assembly switches in the AM/FM/MPX, tape playback and phono circuits.

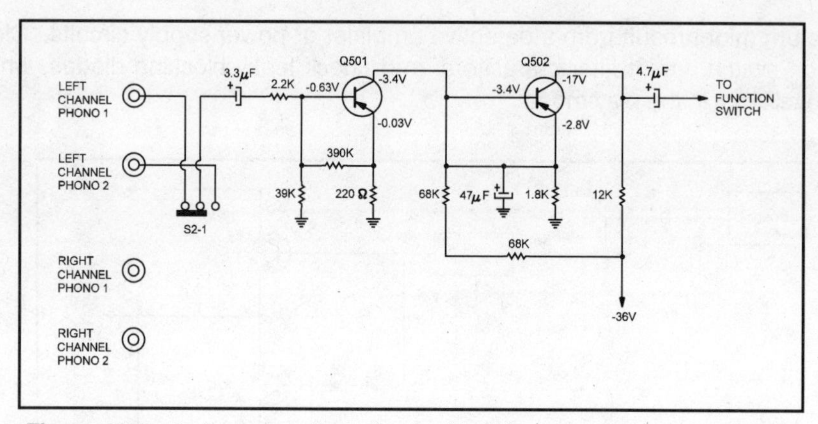

Figure 10-5. A phono equalizer circuit with PNP audio preamp transistors.

Check each transistor with in-circuit transistor tests with the transistor tester or diode-transistor test of a DMM. A leaky transistor can cause a weak and distorted phono audio signal. A leaky or open transistor can cause different voltage measurements on the other directly-coupled transistors. The open electrolytic coupling capacitor provides a dead phono audio circuit from the pickup cartridge. Take critical voltage measurements on each transistor to determine if open or leaky. A change in an emitter resistor can produce a distorted audio signal.

REMOTE CONTROL AUDIO CIRCUITS

The hand-held transmitter can change the volume within the latest TV chassis. The remote can increase or decrease the volume in the table-top AM/FM/MPX receiver. The volume control IC driver might operate a motor that controls the volume in a digitally synthesized audio/video surround receiver or an integrated stereo component system.

The audio within the RCA CTC157 TV chassis is controlled by signal from the AIU (U3300) processor that is connected to the input circuits of a volume control IC. Besides controlling the sound, U3300 also controls the bass, treble, and balance of the audio output circuits. The volume control IC (U1801) internally controls the audio signal fed to the sound output IC. Both the left and right audio stereo volume is controlled by U1801 *(Figure 10-6)*.

The left channel audio is fed into pin 4 and the right audio is at pin 6 of U1801. The microprocessor AIU (U3300) signal controls the volume internally for U1801 at pins 2 and 9. The controlled volume of U1801 is found at pins 3 and 7. Thus, the audio is coupled to the power IC (U1900) through a 1 μF electrolytic capacitor at pins 4 and 8. U1900 amplifies the controlled audio from pins 1 and 13 to each stereo speaker.

Scope the system control volume signal at pins 37 of AIU (U3300). Suspect IC (U1801) when the volume cannot be controlled with remote transmitter. Take critical voltage measurements at each IC terminal.

Replace U1900 for no or distorted audio. Solder all pins on U1900 for distorted audio, especially ground pins 6 and 14. Intermittent sound can result from poorly soldered connections of U1900. Solder the ground terminal of IC1900 for a hissing and popping noise in the speakers. Replace leaky C3313 (0.22 µF) capacitor from system control U1801 for noisy or no sound in the speaker. Take critical voltage measurements to locate a defective U1801 or U1900.

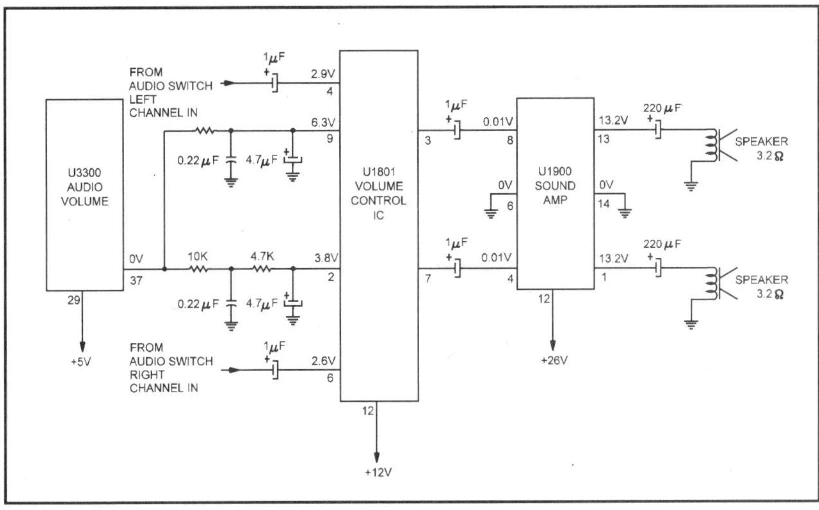

Figure 10-6. The volume is automatically controlled by U1801 at inputs 2 and 9.

SERVICING RECEIVER VOLUME CONTROL CIRCUITS

The rotation of the volume control in remote controlled receivers are both electronic and mechanical. The mechanical rotation of the motor turns the volume control up and down. A motor controlled IC signal is fed from the system control IC.

When holding or pressing the remote volume control buttons, the positive button raises the volume as long as the button is held. Likewise, the motor is rotating the volume upwards, increasing the volume level. The negative button reverses the motor direction or lowers the volume setting when the button is pressed *(Figure 10-7)*.

Figure 10-7. A close-up of a remote transmitter that controls the volume of an AM/FM/MPX receiver.

The volume control motor is controlled directly by a motor control IC. The motor volume-driver IC might be controlled by a PLL Controller IC in a stereo receiver. A volume indicator light lights up, indicating the volume is being rotated or changed. An LED is often controlled by a single transistor when the volume is rotated. The voltage out of (out 1) rotates the dc motor in one direction and out of (out 2) turns the motor in the reverse direction. A change in polarity of the dc voltage applied to the motor controls the direction (up & down) of the volume control.

Most schematics do not show voltages for the motor control IC. Once these correct voltages are taken, they should be marked on the schematic. Some separate voltage listings might be found on a voltage chart. Mark the correct voltages on the IC terminals before troubleshooting the motor circuits. Measure all voltages on the driver IC terminals. Press the volume up button to apply voltage to the motor and then check for correct voltage. Likewise, measure the voltage on the down terminal of motor control IC to reverse the motor.

No motor rotation might be caused by a leaky or defective motor control IC371 *(Figure 10-8)*. Suspect a leaky motor drive IC if the voltage is low at pin terminal 2. Erratic or intermittent rotation of the volume control knob can result from a defective motor control IC or poor board terminal connections. Suspect a defective zener diode when the volume will go way up or down, once pressed and when the button is released.

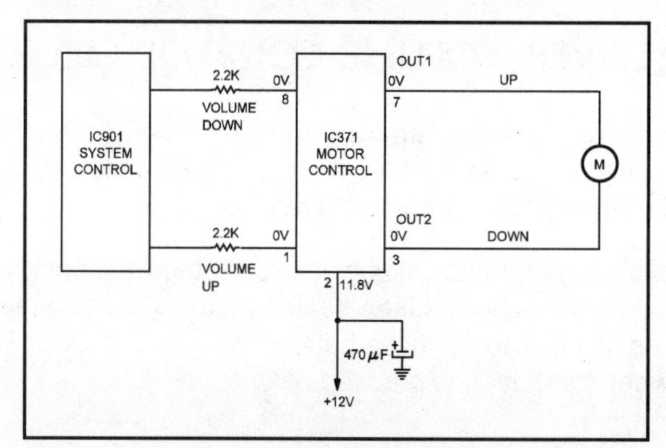

Figure 10-8. A volume control driver IC circuit in a stereo receiver.

Suspect a defective volume control or speaker relay when the sound is intermittent or volume is increased. Check for open volume control when one channel has no volume. Suspect a defective system control IC, motor control IC, and transistors on the synthesizer PCB for no control of volume. Suspect a leaky capacitor off of the volume control terminal or system control IC for no volume control action.

When the volume is changed and a squeal is heard in the speaker, suspect a defective electrolytic coupling capacitor off of the volume control IC in the TV chassis. Readjust the audio bias control or when the volume is increased, check for open control at high volume level.

REGULATED MOTOR CIRCUITS

The motor circuits in the cassette player might operate directly from a fixed or regulated voltage source. The cassette player might have two motors for high-speed dubbing and normal record-play modes. In lower-priced models, one motor might drive two different tape decks (play only and record/play decks). A long belt arrangement with one motor rotates both capstan drives in each tape deck *(Figure 10-9)*. When two separate motors are found in two separate tape compartments, one motor operates the record/play head unit in play, fast-forward and reverse modes. The play only deck is operated by another motor with a single play head and one erase head.

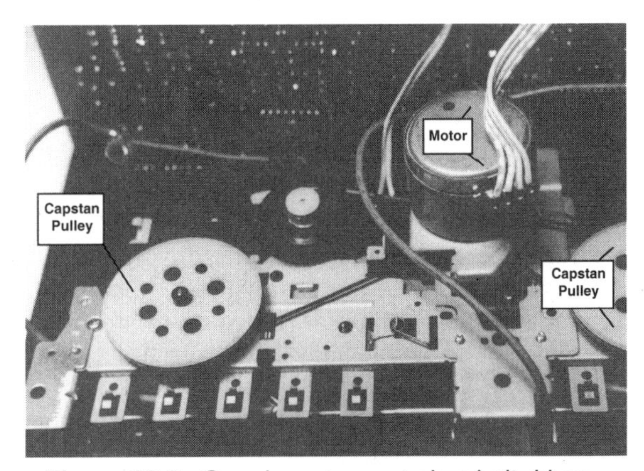

Figure 10-9. One dc motor controls a belt driven capstan in each cassette department.

The early cassette motor circuits were quite simple with a motor leaf switch (SW1-1), electrolytic bypass capacitor and dc motor. The motor operated from several batteries in series or from a step-down power transformer in the ac power supply. No voltage regulator circuits were found in the play only cassette player *(Figure 10-10)*.

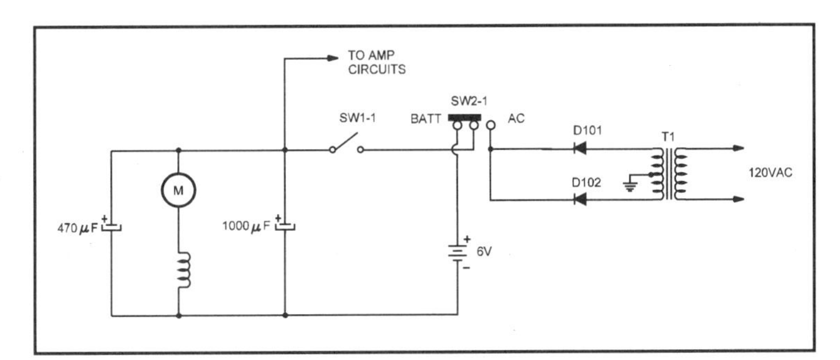

Figure 10-10. The cassette motor transistor regulator circuit.

A cassette motor might operate from a fullwave or half wave ac power supply. The cassette motor can operate from a dc regulated source or within an auto-stop motor circuit. When the cassette tape reaches the end of rotation, the motor stops, and is shut down by a transistor auto-stop circuit. The typical motor regulated circuit consists of one or two separate voltage regulators, controlling the voltage source applied to the motor winding. Q101 regulates the 11.1v dc voltage applied to a cassette motor (M). The zener diode (ZD101) helps to set the regulated voltage source *(Figure 10-11)*.

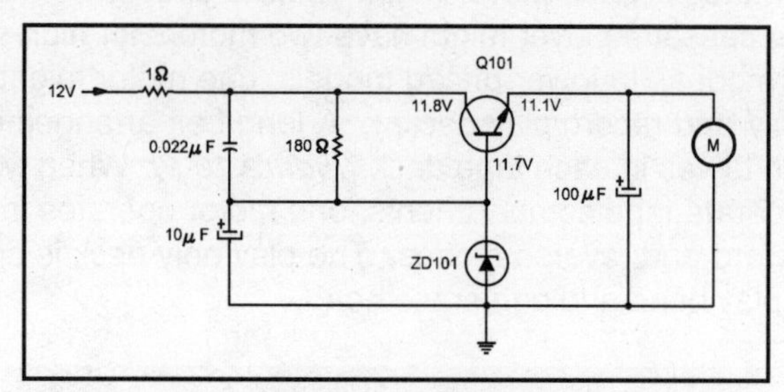

Figure 10-11. The cassette motor transistor regulator circuit.

When the motor will not rotate, measure the dc voltage across the motor terminals. An open regulator transistor provides no or very low voltage to the motor terminals. When the regulator transistor becomes leaky, zener diode ZD101 can overheat, resulting in improper voltage at the motor terminals. The defective cassette motor might have an open winding, worn brushes or a dead spot. Replace the suspected motor, when tapping the end-bell and the motor changes speeds, or the motor begins to rotate, when the motor pulley is turned by hand.

The cassette player with high/normal speed and motor control dubbing might contain several different transistors to control the speed of the cassette motor. The dubbing switch applies dc voltage from the dubbing/record switch to a dubbing control transistor (Q641). The dubbing motor control transistor controls the motor off/on transistor that applies a negative or positive voltage to the cassette motor terminals. Regulated motor control is supplied by Q642 and Q644. A normal or high speed is controlled by transistors Q643 and Q647 *(Figure 10-12)*. Often the dubbing speed is twice the normal tape speed. The dubbing feature enables the operator to duplicate a recorded tape cassette in tape deck 2 into a blank cassette placed in tape deck 1.

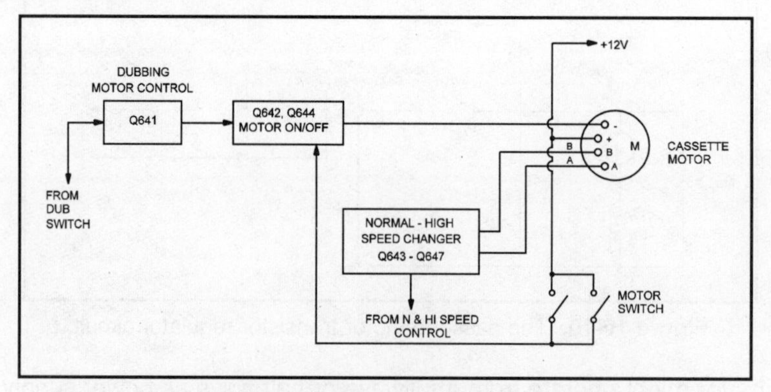

Figure 10-12. A block diagram of a dubbing control circuit with normal or higher dubbing speed circuits.

When the motor will not change speeds, check voltage on Q643, Q647 and the cassette motor. Suspect a dirty motor switch for erratic or intermittent rotation of cassette motor. Test motor control transistors Q642, Q644 and Q641 for no dubbing action. Most motor control circuit defects are caused by leaky transistors and dirty switching contacts.

AUTO RADIO-CASSETTE REVERSE CIRCUITS

In the early auto cassette-radio chassis, a motor reverse circuit would reverse the motor to play the other side of the tape. When the music came to the end of the tape, the motor would stop, then start up in the reverse direction. The auto reverse circuit might contain two or three transistors, fixed diodes and a commutator. The commutator can have wire-like tongs, built on top of the take up reel. In other models, a magnetic switch underneath a rotating magnet opens and closes, triggering the automatic-sensing circuit. The magnet is mounted on the bottom of the turntable reel.

When the turntable stops or does not rotate, the switch remains in a closed position and the relay energizes, causing the motor to change directions rapidly *(Figure 10-13)*. If the auto-cassette keeps changing direction without playing, suspect dirty tongs and commutator rings. Clean up with alcohol and a cleaning stick. Make sure the commutator is rotating and the tongs are seated properly.

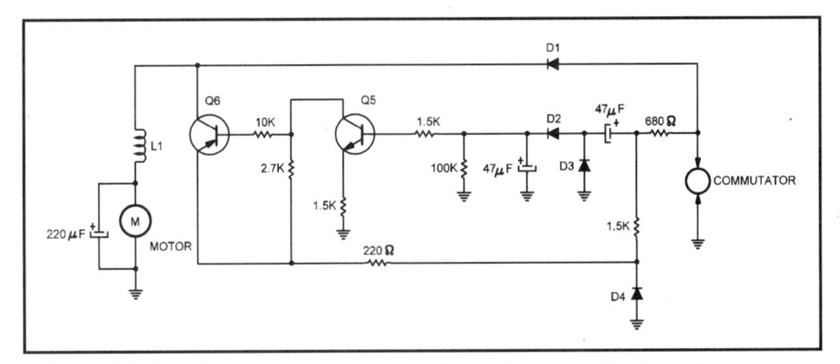

Figure 10-13. The auto-reverse cassette motor circuit in the auto receiver.

Test all transistors with in-circuit tests of the transistor tester, when the motor will not reverse directions or rotate. Check the polarity of the voltage across the motor terminals when motor switching occurs. If there is no voltage at the motor, suspect a defective Q6 transistor. Remove the positive wire from the motor terminals. Inject a 12 volt source across the motor terminals and notice if the motor rotates. Reverse the motor polarity injected voltage and notice if the motor reverses direction. Replace the defective motor, if it will not rotate, with original part number.

TROUBLESHOOTING CD PLAYER MUTE CIRCUITS

The mute circuits within the CD player are usually located in the audio line output jacks. The mute circuits operate when the power switching is applied and when any noise might be created by switching or changing of different CD operations. The mute circuits consist of transistors that ground out the audio of the line output jacks. A separate mute transistor is found in each line output circuit. The mute transistors are controlled by a system control IC or micro computer *(Figure 10-14)*.

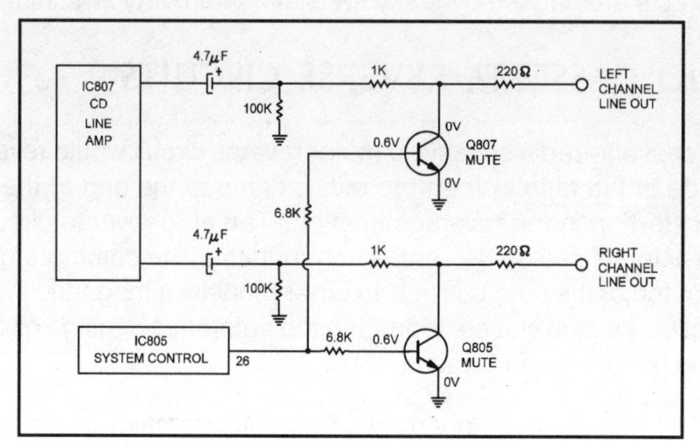

Figure 10-14. IC805 controls the muting operation of Q807 and Q805 in a CD player.

Scope the muted signal at the base of each mute transistor when the CD player begins to operate. If the voltage at the base of each mute transistor does not change, suspect IC805, Q807 and Q805. Suspect the system control IC when noise is heard in the speakers during operation. Check each transistor with in-circuit tests.

No noise or sound should be heard in startup and during switching operations with a normal mute system. Determine if the audio is dead at the line output amp IC or at the mute transistors. Remove the excess solder from the collector terminal on the PCB to remove the mute transistor from the circuit. Repair the defective mute circuit if the sound is now heard.

SERVICING THE CASSETTE PLAYER MUTE CIRCUITS

Most cassette players do not have mute circuits at the power output stages; except large and expensive models might contain certain mute circuits within the record amp and play circuits. In playback mode, the muting simultaneously controls the input and output of the decoder and encoder IC603. The circuit also eliminates the momentary noise of the play back/stop switching for both tape decks *(Figure 10-15)*.

The record muting controls the output equalization amplifier so that the signal is applied to the head of tape deck 1, only at the time of the recording. The muting takes place between the recording equalizer amp and record tape heads. A mute transistor might be found in each left and right channel tape head circuit. Q607 and Q608 mute the playback amplifier circuits.

The amplifier muting is in effect when signals other than tape signals are recorded. The amplifier muting controls the audio signal before the recording buffer amplifier so as not to record noises caused by switching of the function selector buttons and the power switch. The record input left and right signals are muted with transistors before entering the decoder and encoder IC. Q609 and Q610 mute the record/play tape head signals.

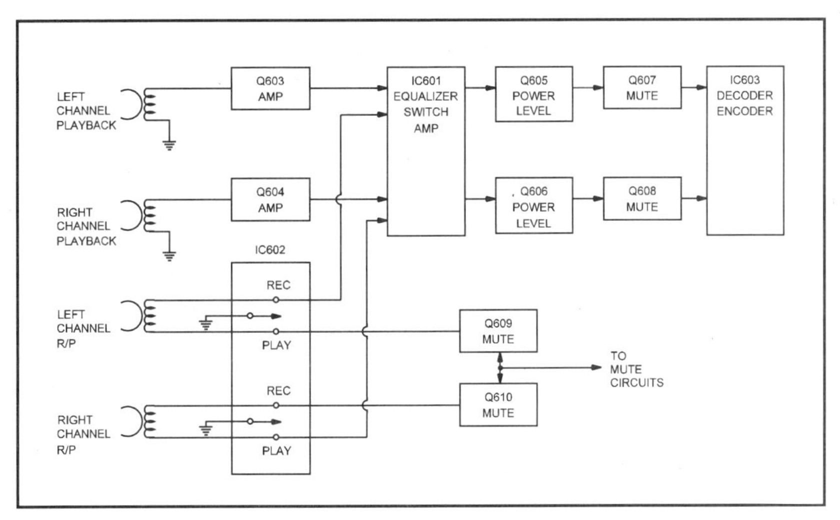

Figure 10-15. In the deluxe cassette player, the mute circuits are controlled at the input and output of a two deck head circuit.

Determine what muting circuits are not functioning. Test transistors Q607 and Q608 when the signal is dead or no muting action occurs in playback mode. Likewise, test transistors Q609 and Q610 when muting does not occur in the record/play modes. Scope the signal from each mute controller IC or transistors. Take critical voltage measurements at each transistor. If the right channel muting system is not operating, compare the voltages and signal in the left muting system that is normal.

Check for leaky 2.2 to 4.7 µF electrolytic capacitors within the muting circuits with no audio muting symptom. Suspect leaky mute transistors when one channel is dead. Check the system control IC or transistor with no muting during switching modes. Suspect a leaky mute transistor when one channel will not record and the other channel is normal in the recording mode of the deluxe cassette players.

RECEIVER MUTE CIRCUITS

Besides relay protection circuits, the large integrated stereo component system and high powered amplifiers might contain several mute circuits. The sound can be muted after the volume control or within the preamp circuits. Some high powered amp circuits might mute the power output before the speaker relay switching. Each right and left channel might have a separate mute circuit.

The system control IC or microcomputer controls each muted circuit *(Figure 10-16)*. The amplifier input circuits are muted so as not to hear the operation of power or function switch and recorded noises. Q501 and Q503 mute the incoming signal from entering the high power amp IC501. The headphone muting transistors (Q509 & Q511) are controlled by the speaker relay control transistors.

The muting circuits can be signal traced with an external audio amp or scope. Monitor the audio signal at pins 5 and 9 when the system control IC is functioning. The base voltage on Q501 and Q503 are a -3.4 volts in normal operation. Both transistors act as a switch when the system control IC is in the muting mode. The audio on both left and right channels is grounded between the two 4.7 µF electrolytic capacitors with transistors Q501 and Q503 *(Figure 10-17)*.

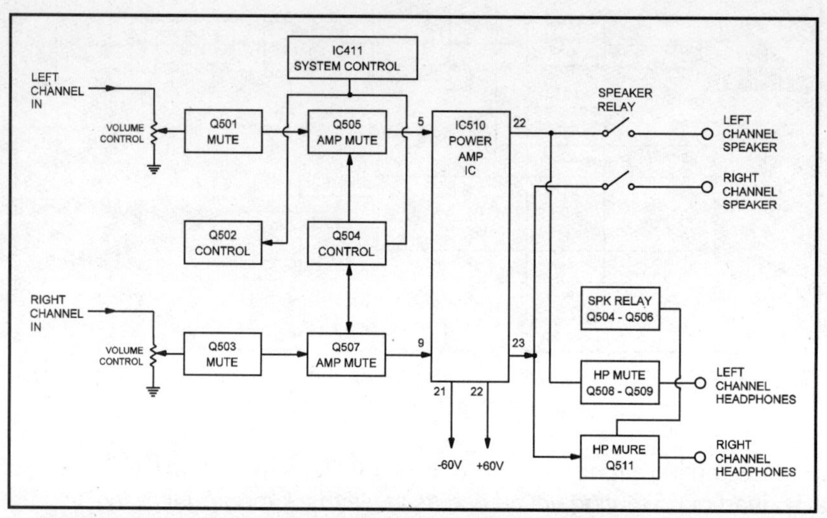

Figure 10-16. System control IC411 controls the mute audio input and headphone output circuits.

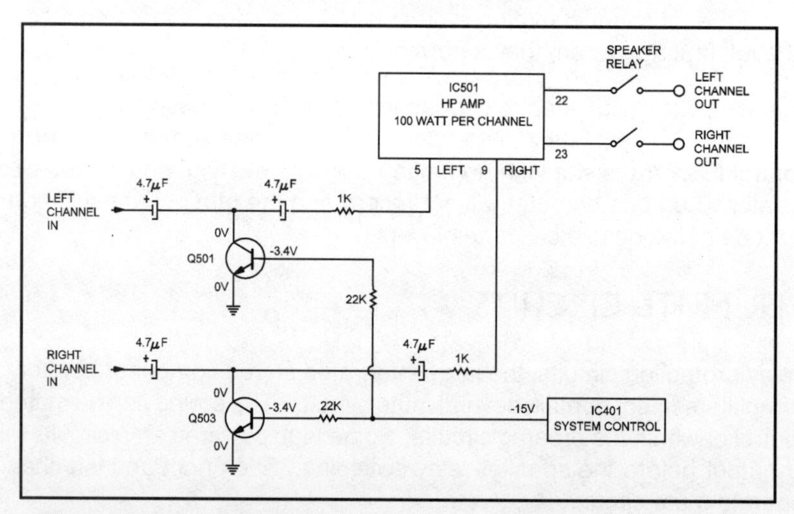

Figure 10-17. System control IC401 controls the input circuits to the high-powered output amp (IC501).

When the muting takes place, no sound should be heard at pins 5 or 9 of power amplifier IC501. Most audio signal tracing methods and critical voltage tests can locate the defective component in the muting circuits.

SERVICING THE CENTER POWER AMP CIRCUITS

The surround receiver or amplifier might have a center amplifier circuit identical to the left and right channels. A center power amplifier might consist of transistors and IC components. The most powerful center amp might contain 10 or more transistors in one channel. The center amp circuits might contain directly-coupled transistors with speaker relay and overload circuits in the output circuits.

First, locate the center power amp circuits on the chassis. If a schematic is not available, trace the center amp output IC directly tied to the center speaker terminals. Connect a load resistor across the left and right channels with a large speaker connected to the center amp terminals to monitor the audio. Determine if the defective center amp terminals contain a dc voltage. If so, connect a 100 watt load resistor across the center amp speaker terminals, for troubleshooting the audio circuits. Now repair the defective center amp circuits.

Remember, the center amp circuits are usually identical to the left and right stereo channels. If a schematic is not available, take critical voltage measurements on the normal stages and compare them to the center amp transistors.

For extreme distortion within the center amp circuits, go directly to the power output transistors or ICs. Check the voltage on each power output transistor. Sometimes you will find a leaky or shorted output and an open transistor connected to it. The directly-coupled driver transistor might also be open or shorted.

If a dc voltage is found at the center power amp output transistors and the dc detector overload circuits shut down the amp circuits, disconnect the center amp output circuits from the overload circuits. Remove one end of a coupling resistor (10K ohms in this circuit), defeating the overload circuits tied to the center amp speaker terminals *(Figure 10-18)*.

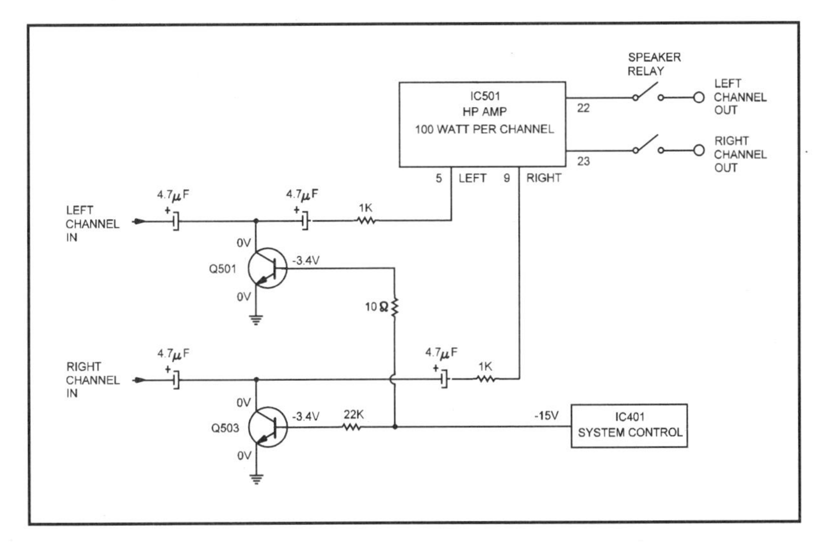

Figure 10-18. To determine if overload protection circuits are defective, remove the 10 Kohm resistor.

Test all transistors in circuit for open or leaky conditions. When one or more transistors are leaky or shorted, replace the output transistors, directly-coupled driver transistor and directly-coupled preamp transistors. If in doubt, replace all of the output transistors in that channel. Check for correct resistance of all bias resistors. Replace them if they are overheated or cracked. Remove one end of the bias diode and test using the diode-test of the DMM.

After all defective transistors have been replaced, take a resistance measurement from each collector terminal to common ground and compare the reading to a normal channel *(Figure 10-19)*. Check the resistance to common ground at the emitter terminals of each output transistor and compare to the normal left and right channels.

Figure 10-19. Compare the resistance to ground off all output transistors to determine if defective component has been replaced.

Now fire up the chassis and take critical voltages on each output transistor. Sometimes the voltage is listed on the schematic and sometimes not. Compare the voltage measurements with the normal identical output circuits.

SPEAKER RELAY PROTECTION CIRCUITS

The speaker protection circuits might consist of two or more transistors that control the speaker relay, disconnecting the speaker from the amplifier output terminals. When a fault or dc voltage is found at the speaker terminals, it is detected by the protection circuits. A protection signal is sent to the protection control transistors to switch off the speaker relay and protect the amplifier and speakers. The power output IC might have a built-in output protection circuit in some amplifiers *(Figure 10-20)*.

The speaker terminals are usually connected directly to the power output transistors or ICs. If a power transistor, IC or component breaks down in the power output circuits, a dc voltage might be found at the speaker terminals, damaging the speaker voice coil. With speaker protection circuits, the speaker relay contacts open up or disconnect the speakers from the amplifier output terminals.

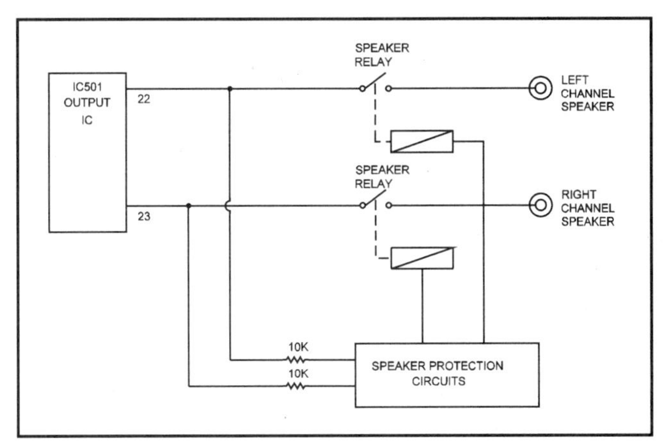

Figure 10-20. IC501 might control the speaker protection circuits.

A defective protection system might keep the relay solenoid energized and the speakers connected to the speaker terminals. Some protection circuits shut down the whole chassis, when a leaky component occurs in the output circuits.

Remove the protection circuit from the power output IC or transistors by removing one end of the resistor or capacitor from the circuit. Check the speaker circuit for a low dc voltage. If no voltage is found at the speaker terminals and the relay is continuity energized, check the defective protection circuits. Take critical voltage and resistance measurements on each transistor. Test each transistor for open or leaky conditions in the protection circuits.

SERVICING THE AC-DC TUBE RADIO AMP CIRCUITS

The "All American 5" tube radio utilized 12BA6, 12BE6, 12AV6, 50L6 and 35Z5 tubes. The 50C5 and 35W4 tubes were found in later models instead of the 50L6 and 35Z5. The 50L6 and 12AV6 are in the audio circuits.

The heater elements of all tubes were wired in series and connected directly across the ac power line. When the heater of one tube opened up, the entire bunch of tubes remained dark; a dead radio.

The audio circuits begin when a radio IF signal is detected by a diode detector element inside the 12AV6 audio amplifier tube. A 500K ohm volume control applies the audio to the input grid terminal of a triode section in the 12AV6 tube. Here the audio is amplified and capacity-coupled by C116 (0.02 µF) to the grid of the power amplifier tube (50L6GT). The audio is again amplified and coupled to the pm speaker through a matching output transformer to a 3.2 ohm speaker *(Figure 10-21)*.

Check the 12AV6 and 50L6GT tubes for weak or distorted audio. A leaky or gassy amplifier tube, leaky coupling capacitor, and burned bias resistor can cause distortion in the speaker. A weak amplifier tube or open coupling capacitor can cause weak audio reception. Check the amplifier tubes for weak or leaky conditions in a tube tester or substitute a new tube.

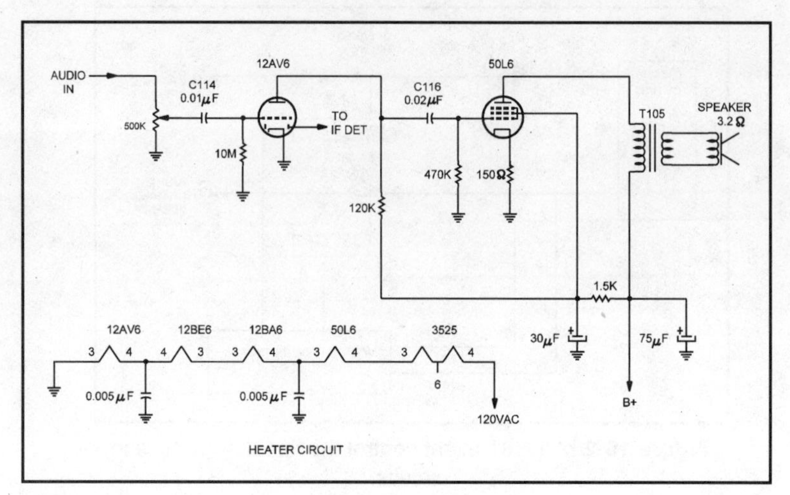

Figure 10-21. The heater and audio amp circuits found in the early "All American 5" radio.

When the output tube (50L6GT) becomes shorted, suspect a burned cathode resistor. If the shorted output tube is left to operate too long, the primary winding of the output transformer might appear overheated and burned, changing the resistance of the winding. Of course, a weak sound is heard after replacing the leaky output tube. Shunt each electrolytic capacitor for a loud hum in the speaker.

Any tube within the series heater circuit can become intermittent, causing the radio to come on and off and shut down. The thermal heater will cool down, then make contact and the whole string will heat up again. After becoming very hot, the intermittent heater element opens up again, until the whole set cools down.

The intermittent thermal heater circuit can be located by checking the ac voltage across each heater terminal. Set the voltage meter to 200 volts ac and place both meter leads across the heater terminals of each tube. You have located the intermittent tube heater or filament when the entire 120 ac volts is found across the heater terminals. Replace the intermittent tube at once.

THE TUBE AMPLIFIER CIRCUITS

There are many different tubes found as audio amplifiers within the audio circuits of a radio receiver, musical instrument amp, or audio amplifier. The triode tube has three elements called a plate, grid and cathode, besides the heater or filament elements. A tetrode tube has four elements, a cathode, control grid, screen grid and plate electrodes. A pentode tube has five electrodes, a cathode, control grid, screen grid, suppressor grid, and plate elements. The suppressor grid is added for electrons that are bombarded against the plate element and leak off to common ground.

The cathode element is connected to ground through a voltage bias resistor and has the lowest voltage on any tube element. A suppressor element is at ground potential or tied internally to the cathode element inside the tube envelope. The control grid operates at a negative voltage to control the amount of electrons applied to the plate terminal. The screen grid has a high potential voltage to help pull the electrons to the plate circuit. The outside or plate element attracts the electrons emitted from the cathode element and has the highest voltage on a tube terminal.

Tube problems result from weak, gassy or shorted elements. A weak tube occurs when very few electrons are emitted from the cathode and reach the plate element. This is usually caused after the tube has had many years of operation. The tube becomes gassy (a soft tube) and might cause distortion. A shorted tube might result from the cathode emitting material flaking off and lodging between the grid elements. The microphonic tube might have loose elements that vibrate or produce ringing noises when the tube is tapped or with sounds of music. Check the tube using the tube tester for shorted or weak conditions. Substitute a new tube when a tube tester is not handy.

TUBE BIAS CIRCUITS

In the early or musical amplifiers, the audio power output circuits contained negative bias voltage developed from the low voltage power supply. A separate tap off of the high voltage winding of the power transformer is rectified and fed to a bias control or resistor network. This negative voltage is applied to the grid resistors of the high-powered amplifier tubes *(Figure 10-22)*.

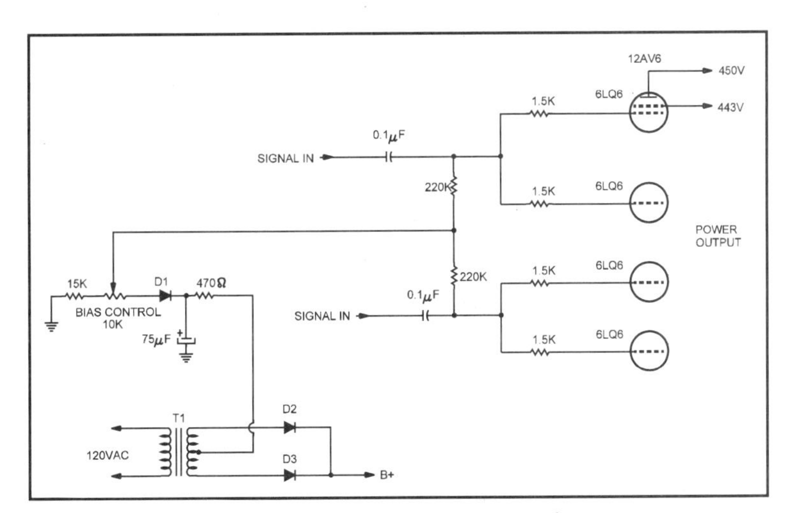

Figure 10-22. The tube bias supply is taken from a separate transformer winding in the high-wattage tube amplifier.

The negative bias voltage is fed to the cathode of a fixed silicon diode with the anode dc voltage supplied to the bias control. The bias control can apply the correct negative bias voltage to the grid circuits of the 6L6G tubes. The dc bias voltage is filtered with a 75 μF electrolytic capacitor. Improper adjustment or dirty contacts on the bias control (10K) can cause distortion within the high-powered audio output circuits. Sometimes cleaning up the control with cleaning fluid and readjustment can remove the distorted or noisy music from the speakers.

SERVICING TUBE MUSICAL AMP CIRCUITS

The musical instrument amplifier might contain several dual-purpose tubes within the preamp and AF circuits. The driver tube might operate in directly-driven triode circuits and capacity coupled to four pentode output tubes. The power output tubes might operate in parallel and push-pull, high-powered output circuits. The output tubes are transformer coupled to the external speaker jacks. Extreme high voltage is found on the plate and screen grid elements of the power output tubes.

The dead amplifier might be caused by a dead or non-lighting tube heater element and improper voltage from the low voltage power supply. First, check the high voltage across the largest filter capacitor. This voltage might be from 550 to 600 volts dc. The low voltage rectifier circuits might contain a tube or silicon diodes. Several electrolytic capacitors might be connected in a parallel and series hookup to provide adequate filtering. A blown line fuse might be caused by a leaky filter capacitor, rectifier tube or silicon diodes *(Figure 10-23)*.

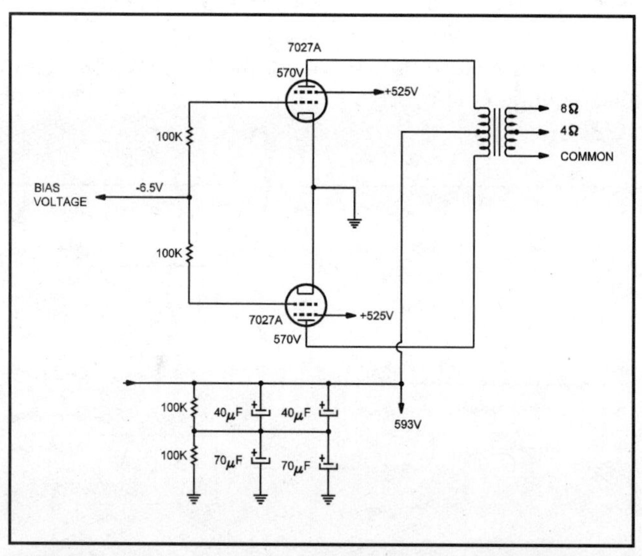

Figure 10-23. Shorted output tubes, filter capacitors or silicon diodes can blow the fuse in high-voltage transformer circuits.

A distorted amplifier might result from weak or leaky output tubes, leaky coupling capacitors, improper bias resistors and improperly applied voltage from the low voltage power supply. Replace or test all tubes.

Check all voltages at each output tube terminal and compare to the schematic. A positive or a very low negative voltage at the grid terminals of output tubes might be caused from a poor bias supply or a change in grid resistors. Measure the voltage on both sides of the coupling capacitors for possible leakage, changing the bias voltage on the AF, driver or output tube grid element.

The weak amplifier circuits can be caused by weak or gassy tubes, open coupling capacitors, leaky bypass capacitors and an improper voltage source. Check each tube or sub another new one. Signal trace the audio circuits by inserting a 1 kHz signal from a function generator to the amplifier input jack. Scope or signal trace the audio with an external audio amp until the weak stage is located. Take critical voltage and resistance measurements to locate the defective component. Shunt coupling and electrolytic capacitors with known ones to locate a defective coupling or bypass capacitor. Remember, the voltages within the tube amplifier are quite high compared to solid-state circuits and should be treated with extreme care, using proper test equipment.

HIGH-VOLTAGE TRANSFORMER PROBLEMS

The power transformer located in the tube's high-powered amplifier are physically very large, have a high-voltage secondary winding and a 6.3 or 12.6 volts ac winding to light up the heaters for 6 or 8 vacuum tubes. The low voltage power supply might have either tubes or silicon diodes as rectifiers with a 500 to 600 volt ac applied voltage. A rectified 550 to 625 volts dc is applied to the audio output transformer secondary tap to the output tubes in push-pull operation. The negative bias voltage might be taken from one side of the secondary transformer winding.

When one or more silicon diodes or tube rectifiers become leaky or shorted, either the line fuse blows or if left on too long, the power transformer becomes hot and is damaged. Sometimes leaky or shorted filter capacitors break down and provide a heavy load, destroying the power transformer. The transformer can be damaged by replacing the line fuse with a larger amp fuse than required when the silicon rectifiers break down. The defective transformer runs red hot and has a burning tar odor. Sometimes the shorted transformer begins to smoke and has an overload-humming sound *(Figure 10-24)*.

Figure 10-24. Remove all leads from suspected amp transformer for accurate tests when the transformer runs hot and smokes.

Usually the amplifier pilot lights dim with a heavy overload when the ac switch is thrown without audio. Remove all wires from the smoking transformer to the rectifiers and heater circuits. Mark down the color code of each removed wire if a schematic is not handy. With wires detached on the secondary windings, fire up the transformer and notice if it immediately becomes warm. The hot transformer might groan and hum if shorted windings are found inside the transformer.

If the transformer does not heat up, let it run for one half hour with the primary winding connected to the ac power line. Check the ac voltage on each secondary winding. A normal power transformer runs quite cool without a load on the secondary. Repair the leaky low voltage components before attaching the secondary transformer leads.

DOLBY CIRCUITS

A Dolby system is an electronic method or circuit for improving audio quality of the cassette player. For low-level sounds, the gain of the amplifier is increased during record modes and the low-level sounds are reduced in playback modes. The Dolby A system might have four frequency ranges while the Dolby B has only one band for reducing noise in the amplifying circuits.

In some cassette players, Dolby noise-reduction circuits are included to reduce the level background noise normally found during recording modes. Dolby B noise reduction occurs only in the middle and upper portions of the audio spectrum. With the Dolby circuits, noise (NR) has been reduced to a minimum in portions of the recorded program. Dolby reduction does not change the frequency response of the audio signal. At high levels, the noise is suppressed. At low input-audio levels, the signal-to-noise (S/N) decreases and the noise is heard. The amount of boost alteration depends on the level and frequency of the signal.

In the early cassette player several transistors were found in the Dolby audio circuits. In the recent cassette players, the Dolby noise reduction circuit ICs are generally located between the preamp and output IC circuits *(Figure 10-25)*.

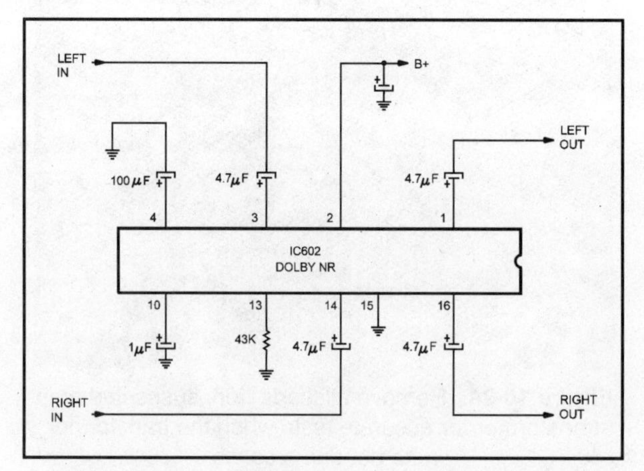

Figure 10-25. The Dolby noise reduction circuits are found after preamp circuits in the recent cassette players.

Chapter 11

Troubleshooting Consumer Electronics Audio Circuits Without a Schematic

After many years of electronic experience, the electronic technician repairs thousands of electronic products each year without a schematic. The electronic TV technician knows where to look, how to test, and how to correctly replace the defective component with a given symptom. The TV technician automatically goes to the horizontal output section with a no-high voltage symptom or a dead TV chassis. A quick horizontal waveform and voltage test indicates a defective horizontal output transistor.

Another waveform test points out a leaky count down drive IC.

Likewise, the audio technician specialist knows where to look for an extremely weak and distorted sound symptom *(Figure 11-1)*. He or she takes critical voltage and component tests in the audio output circuits, because most distorted problems occur in the output stages. Besides years of experience in electronic troubleshooting, the audio technician can rely on many short cuts and tips on how to quickly service the audio amplifier circuits. The experienced audio technician knows where to look, make critical tests, and to put their senses to work without a schematic.

Figure 11-1. Look the amplifier chassis over to locate the audio output transistors upon a large heat sink.

DOCTOR WHO?

Besides electronic knowledge and correct test equipment, the electronic technician can rely on sight, smell and sound to help locate the defective audio component. The electronic technician can tell if a power transformer or audio output transistors have become overheated with a strange odor drifting up from the audio chassis. The sweet odor from the low voltage circuits might indicate a defective selenium rectifier or silicon diode. You can smell an overheated resistor or transistor. Overheated components in the audio amplifier circuits can always be a source of trouble.

Weak or distorted music from the speaker can be traced to the audio output circuits *(Figure 11-2)*. No sound from the speaker might indicate problems within the low voltage power supply.

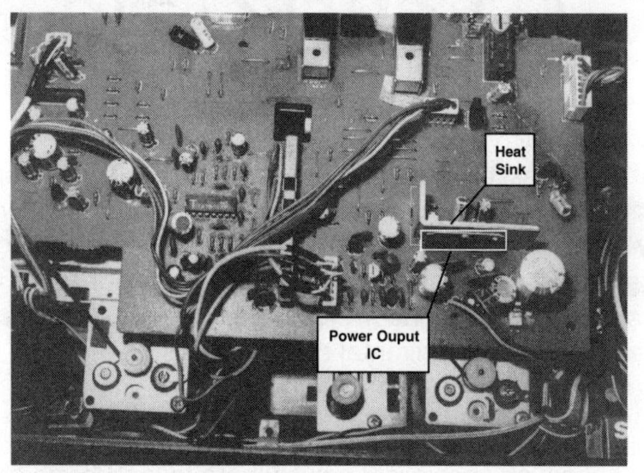

Figure 11-2. The audio output IC is found upon a heat sink within the cassette player.

A low hum symptom might be caused by a poor ground or decoupling electrolytic capacitor. A really loud hum in the speaker indicates a poor filter action in the low voltage power supply. A vibrating or mushy sound might be caused by a loose speaker cone or a dropped cone against the magnet pole. A loud groaning noise with a dim pilot light might indicate a shorted power transformer or leaky silicon rectifiers.

You can see if the power output transistor and IC is leaky or shorted with burned marks on the body area. A defective electrolytic capacitor should be replaced when white or black substance is oozing from the bottom of the capacitor. A quick peek at the audio chassis might spot a burned resistor, choke or zener diode. The overheated connection on the pc board appears as a dark section indicating a poorly soldered joint or overheated component.

When the pilot light does not come on and the chassis remains dead, this indicates a blown fuse or problems in the low voltage source, and on it goes. Simply use three senses to help locate the defective part within the audio chassis.

CHECK-UP TIME

A symptom, defined by Webster's Dictionary, is any circumstance, event, or condition that accompanies something and indicates its existence or occurrence; sign; indication. The symptom of weak sound from the speaker might indicate to the electronic or audio technician where to locate the defective component. An extremely distorted sound in the speaker might indicate a leaky or shorted output transistor or power IC. No sound from the speaker and no pilot light might result from a defective low voltage power supply.

The inoperative speaker relay system with no audio in the speaker system can indicate a defective relay, improper voltage, and defective relay protection circuits. The erratic buzz and intermittent sound symptom might indicate a defective power IC or loose mounting screws *(Figure 11-3)*.

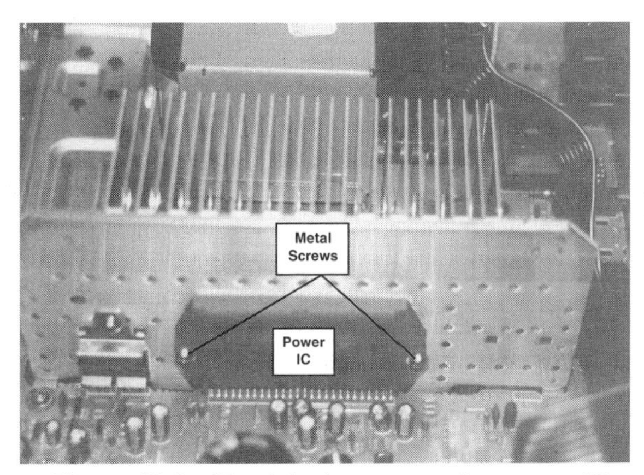

Figure 11-3. Check for loose mounting power IC screws for erratic and intermittent audio.

A distorted speaker system in the left channel might indicate leaky or open transistor or power IC in the left channel audio circuits. One channel dead with a blown fuse symptom might point out a leaky power output transistor or IC. The dead as a door-nail symptom can be caused by a defective power switch. Remember, the symptom in the audio circuits might come from the speaker (sound), smoke drifting up from an overheated component (sight), and odor from a burned part (smell).

THE DRIVING FORCE

Go directly to the power source in the low voltage power supply circuits and check the B+ voltage. Take a quick voltage measurement across the largest filter capacitor, the first thing *(Figure 11-4)*. All audio circuits must have adequate voltage or no audio operation from the circuit. If the voltage is normal in the power supply circuits, check the voltage at the defective stage or component for correct voltage. Suspect a regulator transistor or IC when the voltage is low at the defective stage or component.

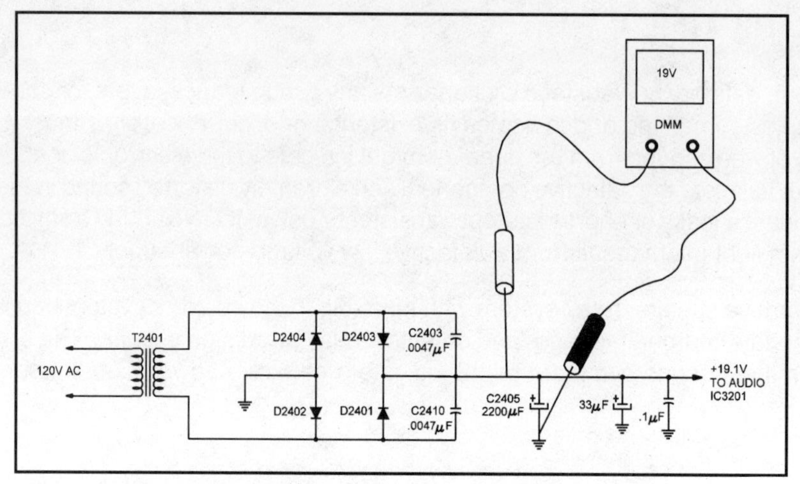

Figure 11-4. Take a quick voltage test across the large filter capacitor in the low voltage circuits in a Quasar AEDC148 chassis.

Do not overlook a leaky or shorted component pulling down the voltage source. An overloaded part can lower the voltage source. Remove the suspected overloaded circuit from the low voltage source by cutting out a section of PC wiring. This can easily be done with a pocket knife or sharp cutting tool. Now check the voltage at the voltage source.

Repair the overloaded circuit when the voltage returns at the voltage source. Do not forget to place a bare piece of hookup wire across the cut-break of PC wiring to restore the circuit.

The most frequent problems in the low voltage power supply are leaky or shorted silicon diodes, dried-up filter capacitors, and open or leaky regulator transistors. A shorted silicon diode can destroy the primary winding of the step-down power transformer or blow the main fuse. Excessive hum in the speaker might result from a defective filter capacitor in the low voltage power supply. The open regulator transistor results in a no-output voltage source and a leaky regulator transistor might lower the voltage source with poor regulation. For the first service operation, remember to check the voltage within the low voltage power supply when a dead or weak symptom exists.

QUICK AND CRITICAL VOLTAGE TESTS

Take a quick voltage measurement across the large filter capacitor to determine if the supply circuits are normal without a schematic. The voltage should be a few volts lower than the working voltage rating marked on the filter capacitor. If the working voltage at the largest filter capacitor is 25 volts, the actual measured voltage should be between 15 to 20 volts across the capacitor *(Figure 11-5)*. Likewise, check the decoupling electrolytic capacitor working voltage of 15 volts, and the measured voltage might be 9 to 12 dc volts.

Although these voltages might not be true in every case, you might find one with a lower voltage. By quickly checking the voltage across the various electrolytic capacitors, you can determine whether the voltage sources are normal or improper without a schematic.

Figure 11-5. The working voltage upon a filter capacitor should be 5 or 7 volts lower than marked upon the electrolytic capacitor.

Quick voltage tests on a transistor will indicate if the transistor is an NPN or PNP type. The collector terminal of an NPN transistor is always the highest positive voltage (10v) with a low positive voltage (2v) on the base and very low (1.7v), or the least positive on the emitter terminal.

A PNP transistor has the highest negative or the lowest positive voltage (0.69v) on the collector terminal with a more positive or low negative voltage (7.3v) on the emitter and a very low negative or more positive voltage (7.06v) on the base terminal.

The collector voltage is the highest positive voltage on the NPN transistor while the emitter is the most positive voltage in a PNP transistor circuit. Just compare the suspected transistor terminal voltages to the same transistor in the normal audio stereo channel without a handy schematic.

Locate the transistor terminals with C, E and B stamped on the pc board wiring when no schematic is available. Take transistor in circuit tests with a transistor tester or diode-junction tests of a DMM, when the terminals are not marked on chassis and no schematic is available. Locate the base terminal with common measurements on collector and emitter terminals. The highest voltage measurement, either positive (NPN) or less negative (PNP), will indicate a collector terminal. Check from chassis to emitter terminal with a low resistance measurement, since most audio emitter resistors are connected to common ground; the terminal remaining is the base terminal.

The voltages found on audio transistors can be checked with another similar model or chassis amplifier schematic *(Figure 11-6)*. Compare these voltages with the defective chassis. Another method is to compare voltages on the identical transistor found in the normal stereo channel. For instance, if the left channel is dead and really low voltages are found upon the AF transistor, compare these voltages to the normal right stereo AF transistor.

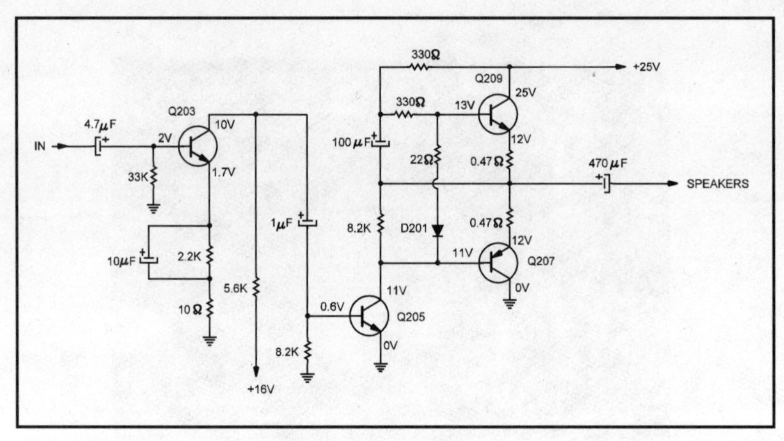

Figure 11-6. Compare the voltages found upon another similar circuit of another audio chassis.

A quick in circuit voltage test on a suspected transistor can indicate if the transistor is open or leaky. A higher than normal voltage upon the collector terminal might indicated the transistor is open. A very low voltage on the collector or on all three terminals, that are about the same voltage, indicates a leaky transistor.

RESISTANCE RUNS DEEP

A quick resistance junction-diode test between any two transistor terminals can indicate if the transistor is open or leaky *(Figure 11-7)*. Take a resistance test on transistors with the diode-test of a DMM. A normal and comparable resistance test between base and collector, base and emitter should be quite close, with the red probe at the base terminal. A low resistance in both directions indicates a leaky transistor. The leaky transistor often has a low resistance measurement between the collector and emitter terminals.

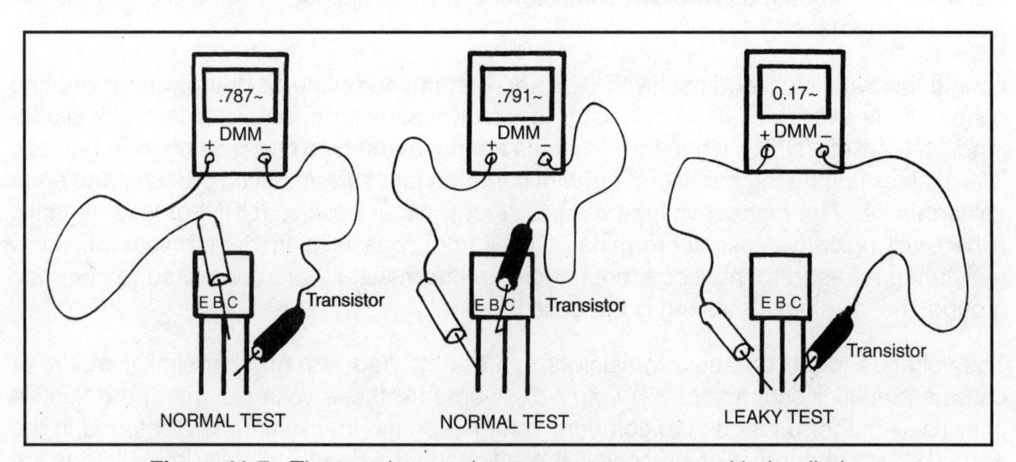

Figure 11-7. The transistor resistance measurements with the diode-test of DMM.

Resistance tests within the audio circuits might have an improper measurement, if a diode or coil is found in the base or collector circuit *(Figure 11-8)*. If the resistance measurement is quite low from base to emitter, remove the base terminal from the PCB wiring with a solder wick and iron. Now take another measurement between the base and emitter terminals. Replace the leaky transistor if a low resistance measurement is found with the base terminal removed from the circuit.

Figure 11-8. A quick resistance test with the diode-test of DMM can locate a defective transistor.

A quick resistance measurement across a capacitor, diode, and resistor can indicate if the component is leaky or open. Remove one end of the diode or resistor from the circuit for a correct measurement. Sometimes another component is shunted across the part to be measured and results in an improper reading. When locating a leaky or shorted transistor in the audio circuit, always check the emitter bias resistor for correct resistance. Sometimes the resistor can become overheated, burn and have a change in resistance. A normal silicon diode should have a resistance measurement in only one direction.

A low resistance measurement between a fixed capacitor terminal indicates a leaky capacitor. The normal capacitor should have an infinite resistance measurement. A leaky capacitor will show a low measurement in both directions. The electrolytic capacitor should charge up and down with a normal component on the ohmmeter. Remove one end of the capacitor from the circuit for an accurate resistance measurement. Now check the suspected capacitor with the capacitor tester, if handy.

EASY TRANSISTOR TESTS

You can quickly locate a leaky or open transistor with an in-circuit voltage, resistance and diode-junction test with the DMM. Locate the suspected transistor with signal tests. Take a quick voltage test on the collector terminal and common ground. Now take a voltage bias test between the base and emitter terminals. A really low collector voltage might indicate a leaky transistor. A higher than normal voltage on the collector terminal indicates the transistor is open *(Figure 11-9)*. No voltage measurement on the emitter terminal results in an open transistor or emitter resistor. To determine if the emitter resistor is open, take a resistance measurement between the emitter terminal and common ground.

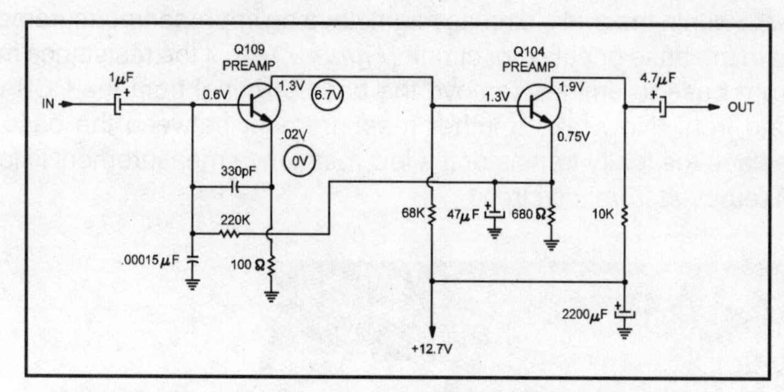

Figure 11-9. Zero voltage measured on the emitter terminal indicates the transistor or emitter resistor is open.

Take a quick resistance test with the power off, between the collector and emitter terminals. Locate the collector terminal and take a resistance measurement to common ground. A low resistance measurement indicates a leaky or shorted transistor. Check the resistance between collector and base, base and emitter, and collector to emitter. If the resistance is low between any two or all three elements, replace the leaky transistor. Remove the suspected transistor and test it out of the circuit. Sometimes the solid-state component will test normal after heat is applied and defective before removing it from the circuit; then, just replace it.

DAY IN AND DAY OUT

The quickest method to locate a defective IC component is with audio signal in and out tests. Locate the input terminal by tracing the audio signal from the volume control, through a coupling capacitor or resistor to the IC input terminal. The output terminal is generally capacity coupled from the IC output terminal to the speaker terminals. Check for a 220, 470, or 1000 µF electrolytic capacitor on the chassis, close to the power IC, to locate the output terminal.

Another method of locating the various IC preamp or power output terminals is in the universal semiconductor manual. Take the part number stamped on top of the IC and look up the universal replacement. Now, check the input, output and supply voltage terminals on the universal IC drawing *(Figure 11-10)*. Besides finding the correct terminals, the required operating voltage might be found upon the supply voltage terminal (Vcc).

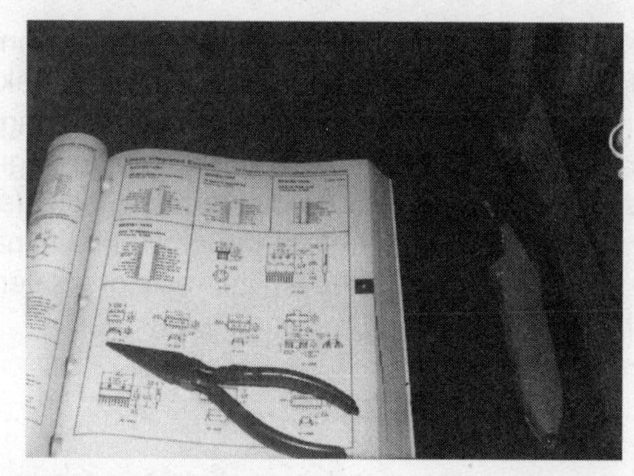

Figure 11-10. Check the part number of the IC against the universal replacement outline for correct terminal connections.

Check the supply voltage applied to the IC component. This voltage is always the highest and is called the supply voltage. Suspect a leaky or shorted IC if the supply pin voltage is quite low. Feel the body of the IC and notice if it is running quite warm. Remove the supply pin terminal from the pc wiring with solder wick and iron. Now take another voltage measurement on the supply pin pad and common ground.

Suspect a leaky IC when the supply voltage rises higher than the normal supply voltage. Double check by taking a resistance measurement on an unsoldered supply pin and common ground. Replace the leaky IC if the resistance is below 100 ohms.

Signal in and out tests with the audio signal generator can locate defective transistors, ICs, and coupling capacitors. If one stereo channel is weak or dead, signal trace the audio circuits with the function generator. Check the signal in and out on the suspected IC. Suspect a defective transistor when audio signal is applied to the base terminal and no or weak audio on the collector terminal.

Check the suspected coupling capacitor by injecting audio on one side and then the other side. When one side of the capacitor will not respond in the speaker, check the coupling capacitor for open conditions. Sometimes the intermittent capacitor might pop on with a loud signal tone in the speaker; replace it.

FROM HERE TO THERE

A quick continuity or resistance test across any electronic component might determine if the part is defective. Rotate the DMM to the 2K ohm scale and place the test prods on each side of a coil, diode and capacitor for a quick continuity measurement *(Figure 11-11)*. The meter reading can indicate if the component is open or leaky. The continuity measurement of dial lights, LEDs, fuses and transformer windings can indicate if the part is open.

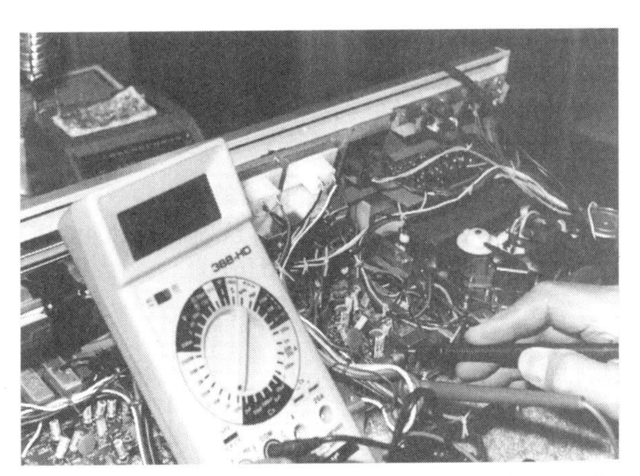

Figure 11-11. An open coil, defective resistor and poor connection can be located with a quick low ohm continuity measurement.

When locating a defective component in the low voltage power supply, a quick continuity measurement of the primary and secondary windings of a power transformer indicates if the winding is open or has continuity. A continuity measurement across the silicon diode can quickly locate a shorted or leaky diode. The shorted filter capacitor can be located with continuity measurements across the positive and negative terminals.

A leaky regulator transistor can be located with a low continuity measurement. Although, the continuity test can quickly locate a defective part, remove one end of the suspected component and take accurate resistance, transistor and diode tests.

NO LIVE ACTION

Of all the troubles that might exist in the audio amplifier, receiver, CD and cassette player, the quick and easy problem to locate is the dead symptom. Nothing usually operates in a dead chassis. Most dead chassis problems originate in the low voltage power supply and power output circuits of the audio amplifier. One stereo channel of a receiver might be dead and the other normal.

The dead audio amplifier symptom might be caused by no sound in the speaker of either channel and no pilot lights light up. The problem might indicate a blown power fuse or defective power transformer. Check for leaky silicon diodes after replacing the line fuse. The dead amplifier might result from an open regulator transistor and power resistors. A leaky power output transistor and IC can cause a dead symptom. The dead chassis might be caused by a defective power switch. A defective speaker relay might cause a dead chassis symptom with no sound in the speakers.

The no left channel symptom in the deluxe AM/FM/MPX amp receiver might be dead from an open AF, driver or output transistors. You might not be able to hear even a slight hum in the speaker. Both audio channels might be dead with a defective dual-power output IC. A dead receiver might result from a defective voltage regulator transistor in the power supply.

Both speakers might be dead with a defective speaker relay or relay circuit. The right channel might be dead with a blown speaker fuse. The receiver chassis might be dead with a blown power fuse that was caused by leaky and shorted output transistors. A dead symptom with arcing noise might result in a defective power push-on switch.

A no sound symptom with pilot lights on, might be caused by a leaky zener diode, open regulator transistor, or filter capacitor in the power supply circuits *(Figure 11-12)*. Do not overlook a leaky or shorted output IC for no sound from the speakers. The audio speakers were dead in a Fisher CA270 amplifier with defective Q114, Q115, Q104, D412, R165 and C153 components.

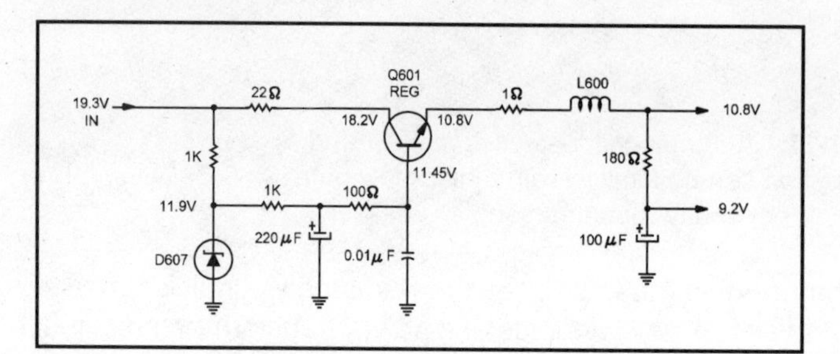

Figure 11-12. Suspect an open regulator transistor for a dead receiver system.

The dead symptom in the auto-stereo receiver might be caused by a blown fuse and leaky output transistors. Locate the suspected output transistor or IC on a heat sink. A shorted output IC might result in no audio in the speakers. Suspect burned "A" lead wires and pc wiring with a dead auto receiver. The defective on/off switch can cause a dead receiver symptom. Check for burned bias resistors after replacing leaky output transistors. The dead auto speaker might result from an open voice coil or broken speaker wires.

KEEPS BLOWING FUSES

When the main power line fuse keeps blowing, a short or leaky component is loading down the low voltage power supply. A blown speaker fuse might be caused by a dc voltage placed on the speaker terminals or too much volume applied to the speaker. Although you can easily see a blown fuse, test the continuity of the fuse with the ohmmeter. The black internal glass area of the fuse indicates a directly-shorted component. When only the center area of the fuse opens up, a power line outage condition might occur or too much power may be applied to the speakers. Test run the amplifier or receiver for several hours and notice if the fuse opens.

Suspect a leaky or shorted component within the low voltage power supply when the fuse keeps blowing. Check for shorted or leaky silicon diodes. The dead chassis with the blown fuse might be caused by leaky power output transistors in the audio amplifier. You may have to replace both output transistors. When the speaker fuse blows each time the chassis is fired up, check for dc voltage on the speaker terminals caused by a shorted transistor or IC. The fuse did blow in a Pioneer SX-1000TW amplifier with leaky 2SC897 output transistors.

The power line fuse might keep blowing in the large receiver with a dead chassis symptom. Check the output transistors and power output IC for leaky conditions. One channel can be out with a blown speaker fuse in a Fisher CA65 receiver. A quick voltage test on the speaker terminals was around +65 volts. An open Q111 (2SC15011) and R140 (78 ohms) were replaced to restore the dead audio channel.

DISTORTED MUSIC

Extreme distortion symptoms are generally located in the power output stage of the audio amplifier or receiver. The left channel was extremely distorted at high volume with an open R713 in a Sony STRAV1020 receiver. Readjustment of the bias control helped solve the distorted left channel.

After the receiver warms up, both channels might contain strong distorted music with a leaky voltage regulator and audio driver IC. Two or more leaky output transistors and a dual-output IC can cause extreme distortion in both channels. Look for the power IC on a large heat sink. Check leaky output electrolytic capacitors to the speaker terminals when distortion exists.

After several hours of warm-up, a Magnavox receiver (AS305SM3701) contained extreme distortion with a leaky regulator Q7300 (IC7520). Remember, that extreme distortion in the audio channels of a receiver can result from leaky transistors and IC components. Do not overlook an improper voltage source. Check for a change or burned bias resistors for distortion. These bias resistors are tied to the emitter and base terminals of the power output transistors.

The extreme distortion found in the audio circuits of a TV chassis can be caused by a shorted output IC. Extreme distortion might be caused by a leaky SIF/Amp IC in the TV chassis. Improper alignment of the discriminator coil can cause distortion in the speaker. A leaky Multi-Sound IF Decoder and balance sound IC can cause extreme distortion in the recent TV chassis. Replace L201, C208 (91 pF) and C209 (7.2 pF) for extreme distortion in the RCA CTC110 chassis *(Figure 11-13)*.

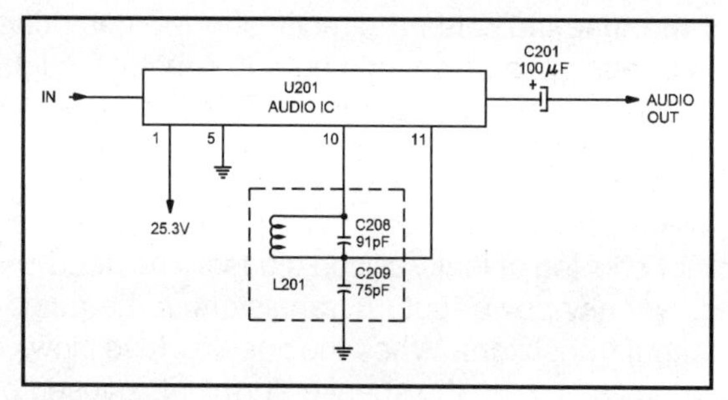

Figure 11-13. Replace C208, C209, and L201 for extreme distortion in the sound circuits of an RCA CTC110 chassis.

CORRECT PARTS

Exact replacement components are easy to replace for they were designed for a certain chassis and circuit. So try to obtain the original parts when possible. Replace high powered dual-IC components with exact replacements. Replace those high-powered output transistors found in the high-wattage auto amplifiers with originals. Obtain the original part number from the manufacturer or replacement depot, and from electronic mail order firms.

Low signal and general purpose transistors and ICs can be replaced with universal parts. Simply cross-reference the semiconductor to the universal semiconductor manual. Most resistors and capacitors can be replaced with regular replacements. Silicon and zener diodes can be replaced with parts found at the local electronic distributor.

The power and special resistors should be obtained from the manufacturer. The dual-volume controls should be ordered from the parts depot. Replace the stereo volume control and switch found in the auto receiver with the exact part number *(Figure 11-14)*. Special type coils, chokes and transformers should be ordered from the manufacturer. Replace the special correct size and correct voice coil impedance speaker with original part number. Special function and power switches should be replaced with original components.

Figure 11-14. Replace the defective volume control and switch in the auto receiver with the original part number.

Filter capacitors can be placed in series or parallel when the original is not available. In the high-voltage tube amplifier chassis, 450 volt electrolytics can be placed in series to operate at a high working voltage and shunted to add capacity. Most bypass and coupling capacitors can be obtained from the local electronic distributor. Test all new components before replacing them in the audio circuits.

YEARS BEFORE AND AFTER

By keeping a record of the unusual or tough dog repairs, the electronic technician can use a case history on another chassis to quickly solve the same electronic symptom. The different audio case histories can be recorded using a card file system, the computer, or right on the schematic. By circling the defective component and writing the symptom on the edge of schematic, the service problem is there when you need it.

Case histories can save alot of valuable service time. It seems the same electronic trouble in one model can occur again or in another chassis. Sometimes the same trouble happens again and again. The most common service problem might be remembered, while the unusual one might be forgotten in this busy world of servicing. Just take a minute and mark it down for a hurry up and rush-critical repair down the road.

There are many case histories sold today in literature form and on computer disk. The case histories might be repairs on TVs, Projection TVs, camcorders, CD players, and audio equipment. Check the classified adds in the magazine, Electronic Servicing & Technology for the different listings. Here are a few service firms that have case histories for sale:

Mike's Repair Service
P.O. Box 217
Aberdeen Proving Ground, MD 21005
410-272-4984

Electronic Software Developers, Inc.
826 S. Main St.So.
Farmingdale, NY 11735
1-800-621-8477

KD-TV
514 3rd St.
Aurora, IN 47001
1-888-KD-Stips

Higher Intelligence Software
60 Farmington Lane
Melville, NY 11747
1-800-215-5081

Tv-Man Tech-Tips
2082 Augusta
Weston, Florida 33326
1-800-474-3588

Magazine:
Electronic Servicing and Technology
76 N. Broadway
Hicksville, NY 11801
516-681-2922

HOT CIRCUITS-HOT PARTS

The dead or excessively distorted chassis symptom usually consists of a hot audio output transistor or IC component. A shorted or leaky output transistor not only provides a dead or loud-hum symptom, but might be running red hot. Some normal audio output transistors and power ICs operate warm on the large heat sink. The red hot transistor or IC has a dark gray body appearance and is too hot to touch. Be very careful when touching the body of a suspected power output component; you might end up with burned fingers.

Remove the overheated part from the audio chassis. Test the output IC or transistor for leakage *(Figure 11-15)*. The red hot output transistors can have a direct short between collector and emitter terminals. A leaky power IC might have a very low resistance between the voltage supply pin (Vcc) and ground terminal. Do not install the red hot component even if it tests normal; discard it.

Figure 11-15. Remove the hot output transistor from the heat sink and test out of the circuit.

Before replacing the red hot output transistors, test the driver or AF and all directly-coupled transistors in the output circuits for leaky or open conditions. Sometimes it is best to replace all output and driver transistors within the audio output circuits, when one or more is found to be leaky or shorted. Double check all bias resistors and diodes in the output circuits. Likewise, check for leaky or open electrolytic capacitors within the output circuits. Replace all burned resistors or those with a change in resistance when connected to the red hot IC.

LOCATING THE WEAK COMPONENT IN THE CASSETTE PLAYER

Clean the tape heads for a weak or distorted sound within the cassette player. Make sure all packed tape oxide is removed from the tape head. Sometimes a flat plastic dowel helps to remove the packed oxide from the shiny surface. Make certain that the magnetic area of the tape head is free and clear of oxide dust *(Figure 11-16)*.

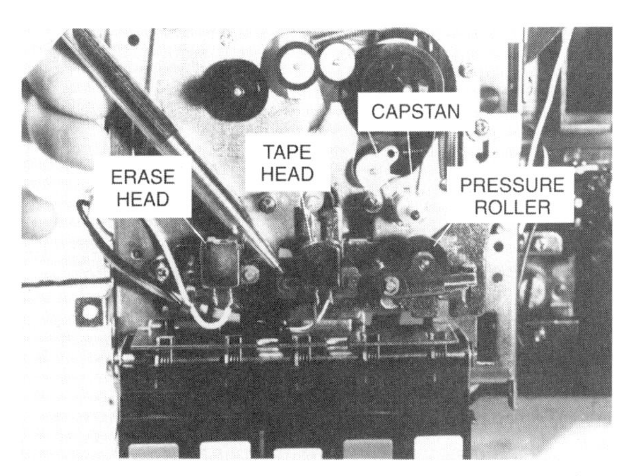

Figure 11-16. Make sure the tape head is clean in the cassette player for weak audio symptom.

A weak audio channel is generally caused by a defective transistor, IC, coupling capacitor and improper voltage source. Measure the voltage at the large filter capacitor. Check the voltage at the power output transistor or IC. Signal trace the audio circuits if everything appears normal at this point.

Insert a test cassette and check the signal at the volume control and ground with the scope. Check the weak signal against the normal channel at the volume control. If the signal is weak in one channel at the volume control, proceed toward the preamp transistors. When the signal at the volume control is normal or at the same audio level, signal trace the audio output circuits.

Another method used to locate a weak stage is to signal trace with an external audio amplifier. Plug the cassette player into the isolation transformer. Check the signal at the stereo tape heads. Proceed from the base to the next coupling capacitor and base of the preamp, to base of AF amp transistor or IC. When the signal becomes very weak, you have located the defective circuit. Locate the defective component with voltage, transistor and resistance test methods.

In a Philco 11-75198-1 chassis, the audio was weak in the left stereo speakers. Since the audio was also weak in the AM/FM/MPX receiver and phonograph, the defective component must be in the audio output circuits. A local radio station was used as the signal source. The volume on both left and right channels were normal at the volume controls.

When signal tracing with the external audio amp, a weak signal was noticed on the base terminal of the Voltage Amp (Q502). Again the signal was checked at the volume control in the left channel. C504 (1 µF) electrolytic was found to be defective just off of the left volume control terminal *(Figure 11-17)*.

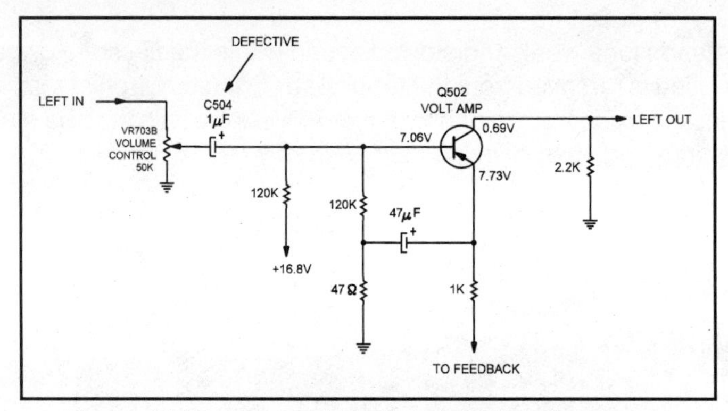

Figure 11-17. The weak audio in a Philco chassis was caused by defective C504 (1 µF) electrolytic.

TROUBLESHOOTING INTERMITTENT AUDIO AMPLIFIER CIRCUITS

Intermittent conditions within any electronic product are very difficult to locate. Most components producing intermittent audio are transistors, ICs, coupling capacitors, and poor terminal board connections. Try to divide the audio circuits in half by monitoring the audio at the volume control. Proceed towards the preamp or AF circuits, when the defective channel cuts up and down.

Sometimes a suspected transistor or IC can be sprayed with coolant to make it act up. It's best to let the chassis operate until it acts up and then signal trace the intermittent channel.

Monitor the right intermittent stereo channel at the driver transistor base terminal, when the signal is normal at the volume control. Attach the external audio amp probe to the base terminal of the audio output transistor or input terminal of power output IC. Notice if the sound pops up or cuts down. If the sound quits, the defective part is located ahead of the power output stage. Scope the driver and AF circuits. Sometimes when the intermittent transistor, IC or coupling capacitor are touched with the scope or external amp probe, the audio snaps back in.

The sound might appear intermittent when simply touching or moving a part with the probe tip. Monitor the input and output of a suspected component until you are sure the part is intermittent. Apply heat from a hair dryer or apply coolant from the spray can to make the component act up.

The sound would cut up and down in a Marantz 2230 audio amp's right channel. Sometimes the sound would cut in and become distorted. Signal tracing the audio circuits with the external audio amp located an intermittent driver transistor. The driver transistor (2SC959) was replaced with an ECG-128 universal replacement.

REPAIRING DEAD RECEIVER CIRCUITS

The dead chassis has no sign of life or even a hum in the speaker. Most dead receiver symptoms are caused by leaky power output transistors or ICs. Check for a blown line or speaker fuse. Both audio channels might be dead with an open power line fuse; only one channel is dead with an open speaker fuse.

Next, check the low voltage power supply at the main filter capacitor terminals. If the pilot lights are on in the latest receiver chassis, usually the low voltage power supply is okay. If a low hum is heard in the speaker, voltage is applied to the audio output circuits.

Touch the center terminal of the volume control with a test probe or metal screwdriver and the audio circuits are usually normal, if a loud hum is heard. Proceed toward the AF and preamp circuits if the receiver and input signals are dead. Signal trace the input stages until you have located the dead stage.

Look the chassis over to possibly locate a burned, cracked or damaged component. In a J.C. Penney 3233 model, several pieces of blown electrolytic capacitor paper and foil covered the low voltage circuits. Several resistors were burned around a filter regular transistor Q601 *(Figure 11-18)*. Since several of these models had been worked on, the low voltage power supply circuits were well known. The C602 capacitor had blown into pieces, R601 and R604 were burned. Replacing D201, R601, C602, R604 and Q601 solved the dead chassis.

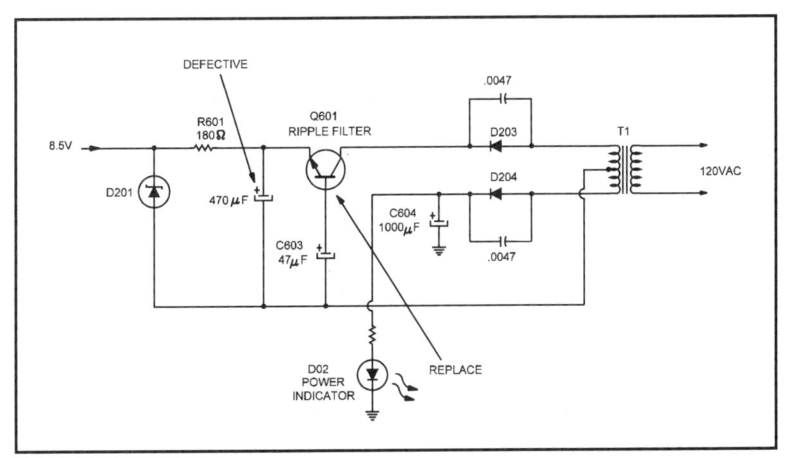

Figure 11-18. C602 was blown open and Q601 was replaced with a dead symptom in a J. C. Penney 3233.

SERVICING NOISY AUDIO TV CIRCUITS

Extreme cracking and frying noises in TV audio circuits might result from a defective transistor, IC or electrolytic capacitor. A popping or cracking noise might be caused by a defective power output IC. The frying noise, hum and distortion in the speaker can be cause by electrolytics, ICs and transistors. You might find several different components with noise in the audio circuits. Remember you cannot locate the defective noisy component with regular transistor or IC tests.

In the RCA CTC140 TV chassis, intermittent popping was cause by a defective power IC U1900. A high-pitched sound in both channels resulted in a defective C1807 (4.7 µF), off of pin 7 of U1800 *(Figure 11-19)*. A no sound symptom with static in the audio was solved by readjustment of the discriminator coil (L2306). A distorted and noisy sound in the main chassis resulted from a leaky C1600 and C1695, off of pin 12 of U1601. These actual noisy problems were taken from case histories listed on the CTC140 RCA schematic.

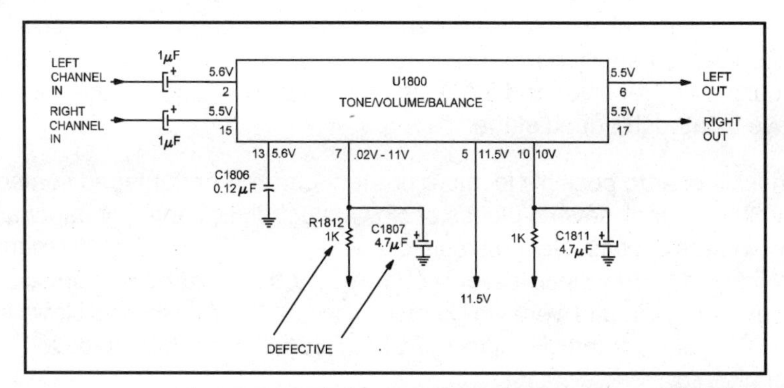

Figure 11-19. Replace C1807 for a high-pitched squeal in both audio channels on RCA CTC140 TV chassis.

When trying to locate defective components within the audio circuits, check the layout of parts on the chassis. Most stereo left and right channels are located to the left and right side of the chassis; sometimes they are lined up in a row. Most audio components are located close to and around the power output transistors and IC parts located on a large heat sink. If in doubt, simply trace the audio circuits back from the speaker terminals to the output transistors or ICs; the left and right channel components are found in the same manner.

Locate the low voltage power supply with large filter capacitors and silicon diodes mounted close together. Locate audio transistors with a body part number and look them up in the universal replacement manual. Compare the audio section with another product or schematic when the exact diagram is not available. Remember, the highest voltage is found on the collector terminal of a transistor and IC power supply terminal.

SERVICING REMOTE CONTROL CIRCUITS

Besides the TV, a camcorder, VCR, and AM/FM/MPX receiver might have a remote to control the various functions. The hand-held remote contains an infrared transmitter that triggers a infrared receiver inside the product to be controlled. The AM/FM receiver remote transmitter might also control a TV, VCR, tuner, tape and CD player in one remote package *(Figure 11-20)*. The TV remote might control functions in the TV, VCR, audio amp and DSS cable disc operations.

Figure 11-20. The infrared remote control transmitter might control the TV, VCR, tuner, tape, CD player, and receiver functions.

The separate receiver remote transmitter might control the on/off switch, volume up and down, and mute the audio at any time. If the receiver has a built-in CD player, the remote controls the disc/deck, right or left skip, fast forward skip and what disc to select. All of these functions are controlled by two self-contained AAA batteries.

When the remote will not trigger any functions of the electronic products, check the small batteries. Remove the batteries and insert new ones. Clean the battery terminals with cleaning fluid. Rub the ends of the batteries on a towel or cloth to help clean the contacts.

The intermittent remote control operation might result from loose batteries or dirty contacts. Clean battery contacts and spread out the battery terminals to make a tight connection. Since remote controls are rather inexpensive, it is best to replace the remote control instead of replacing components inside the transmitter. Try another remote before determining the infrared receiver is not functioning.

The infrared remote receiver is located inside the electronic product and consists of a photo-transistor pickup that connects to a preamp circuit. The new infrared receivers have an IC component as amplifier within the TV chassis. The infrared signal is then fed to a microcomputer or system control IC. The control IC controls the many functions within the TV.

When the infrared signal strikes the infrared element CR3401, the signal is amplified by IR amp (U3401) in an RCA CTC157 chassis *(Figure 11-21)*. The IR signal is fed to pin 36 of the microcomputer AIU (U3300) IC. Here, the volume is controlled out of pin 37 to the volume control IC (U1801). Try to check another similar schematic of the control system when the schematic is not available, since there are many pin terminals on the system control IC.

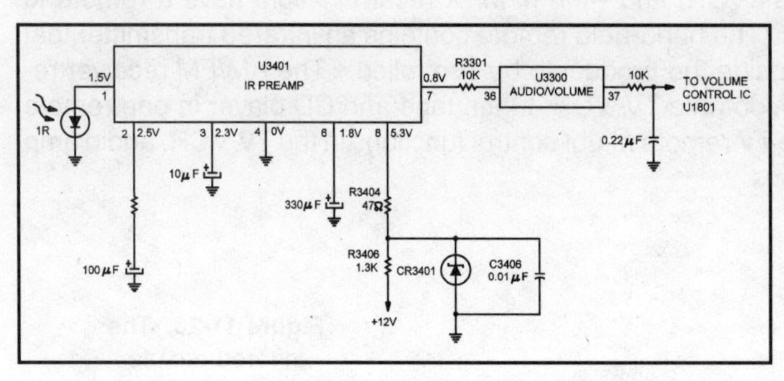

Figure 11-21. The IR signal is amplified by an IR preamp IC and fed to the AIU IC of an RCA CTC157 chassis to control the volume in the audio circuits.

Although the infrared receiver produces very few service problems, the IR preamp IC has been known to break down. Check the supply voltage at pin 8 (5.2v) and all other pin terminal voltages. Suspect a leaky IR preamp IC or zener diode (CR3401), when the voltage is low on pin 8.

Do not overlook the various electrolytics connected to the IC pin terminals. Take a quick resistance test from each pin terminal to common ground, to locate a leaky component. When other remote functions seem to operate and there is no volume control action, suspect the AIU IC (U3300) or volume control IC (U1801) in the TV chassis.

Chapter 12

Important Audio Tests And Adjustments

Signal tracing the defective audio circuits by injecting a signal from the audio signal or function generator can help to locate a defective audio component. The external audio amplifier can locate the weak or distorted circuit in the audio stages. Using the oscilloscope as indicator can quickly locate a defective preamp or audio output circuit with waveforms.

Correct adjustments within the audio circuits or components can improve the sound of the audio amplifier *(Figure 12-1)*. In this chapter, you will learn how to make your own test catridges and cassettes for use in signal tracing the signal from input to output terminals of the audio amplifier.

Figure 12-1. The function generator, power supply and frequency counter can quickly locate a defective stage in the audio amplifier.

SIGNAL INJECTION

Signal injection is a method of quick audio or RF troubleshooting in which a signal injector or generator is used to inject a signal into the audio circuits. The signal injection frequency might be 1 kHz, 3 kHz, and 10 kHz. Signal injection can be accomplished with the audio generator, audio oscillator, function and noise generator test instruments. The generator can inject a sine or square waveform into the audio circuits. The audio signal can be injected at the input or at any point in the audio circuits. Signal injection can quickly locate a defective stage or component, weak or distorted, and a dead audio circuit.

The audio signal must have an indicator that produces the audio tone, waveform, or analog meter movement. The oscilloscope can quickly identify a dead, distorted, clipped, or weak audio circuit with a signal input waveform. An external audio amplifier can locate a distorted, weak or dead audio signal in any audio stage or circuit.

A frequency counter can be used as an audio indicator and also in audio alignment procedures. The VTVM, FET-VOM and AC voltmeter can be used as a signal indicator and in alignment of balance and head azimuth adjustments. Of course, the audio indicator can also be the test speaker connected at the amplifier output terminals.

The audio, function and noise generator can be connected at the input terminals of the audio amplifier and the audio tone can be checked from stage to stage with the scope or external amplifier as an indicator. Simply check the signal from base to base of each transistor in the preamp, AF, driver and audio output circuits. If the preamp and audio output circuits are comprised of IC components, check the signal at the input and output terminals of every IC *(Figure 12-2)*. When the signal is lost or diminished (weaker), you have located the defective circuit.

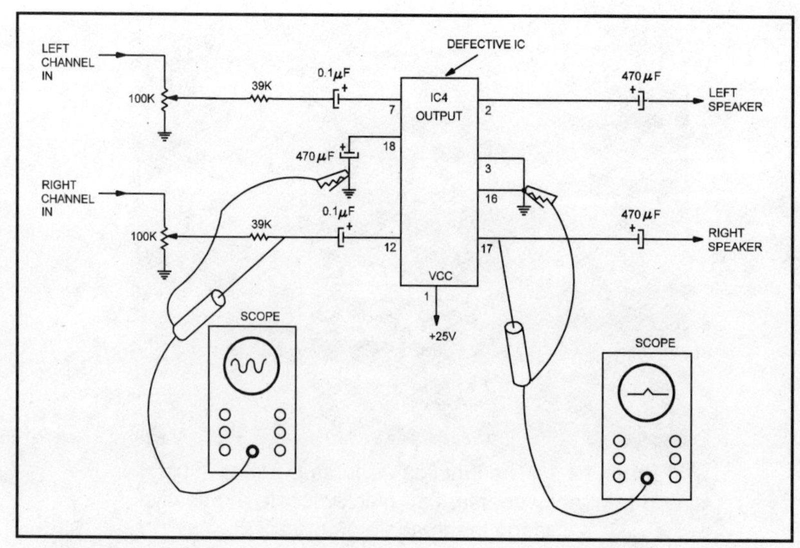

Figure 12-2. Scope the signal in and out of the suspected audio IC.

The audio signal from the audio, function and noise generator can be injected at any point in the audio circuit to locate a defective component. Instead of moving the indicator at different points within the audio circuit, simply inject the audio signal. Likewise, go from base to base terminal of the audio transistors and input to output terminals of IC components. The most common indicator of the audio signal is the connected PM speaker *(Figure 12-3)*. The sound of the signal generator is heard in the speaker until the defective circuit is located. The noise generator can be used in RF and IF circuits, as well as in audio injection troubleshooting.

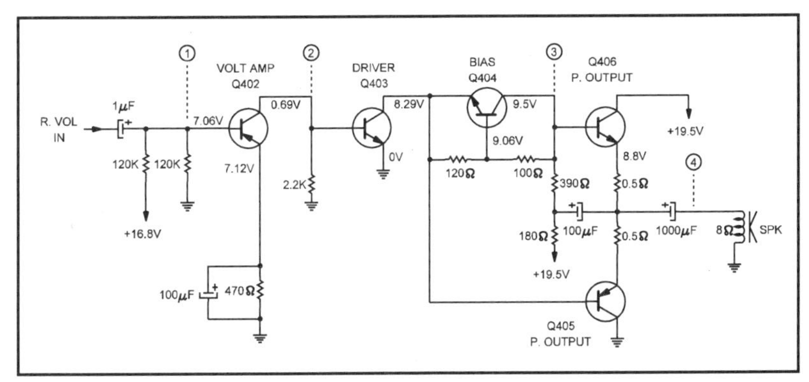

Figure 12-3. Injecting the audio signal at the transistor base terminal can quickly locate the defective circuit with the speaker as an indicator.

Break the audio circuits down by starting signal injection at the volume control. Inject the audio 1 kHz signal into the left and right stereo circuits. In the RCA CTC146 TV chassis, the sound was weak and sometimes the sound level could not be controlled. A 1 kHz audio signal was injected at the base terminal of the first audio amp (Q1201) and the sound appeared normal. Since the IF/SIF IC component fed into the audio circuits, U1001 was suspected of causing the weak sound.

All voltages appeared normal on U1001. A further check of the audio circuits indicated the sound was controlled by the Analog Interface Unit (U3300) at pin 30 of U1001. Perhaps, the audio symptom existed in the audio control circuits. The control volume was traced back to pin 37 of U3300. Here, diode CR3306 was found to be leaky and was replaced with a universal ECG-177 silicon switching diode. In another RCA CTC146 chassis, C3314 off of the leg of CR3306, produced no control of the volume. Replace both the CR3306 and C3314 (0.22 µF) capacitor when either one is found to be defective *(Figure 12-4)*. Check outside of the sound output circuits for possible sound problems.

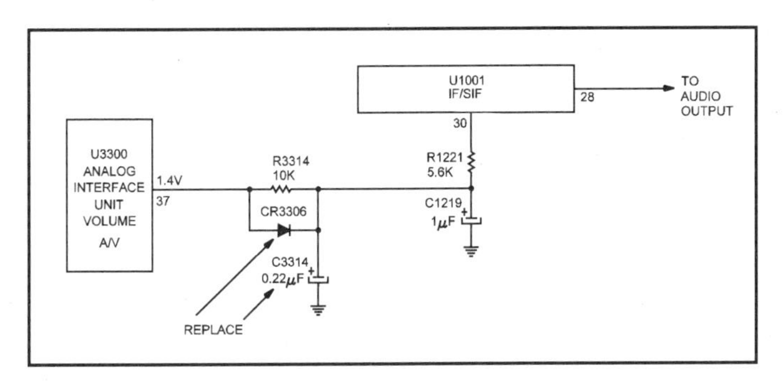

Figure 12-4. The weak sound symptom in an RCA CTC146 chassis was caused by CR3306 and C3314 off of pin 37 of AIU (U3300).

AUDIO SIGNAL TRACING

The external audio amp is used in signal tracing the distorted, weak, dead, or intermittent audio within the audio amplifier. The audio amplifier can have a PM speaker or headphones as an indicator at the output of the audio signal tracer. The audio signal can be traced from the preamp to the audio output terminals with the external audio amp. Again, the audio is signal traced starting at the volume control and proceeding either way to locate the defective stage or circuit.

Insert the electronic products ac cord into an isolation transformer for test equipment and product protection. When the audio is weak in the AM radio, tune in a local radio station as the signal source. Insert a test cassette when signal tracing the distorted sound in the cassette player. A test disc can be used when signal tracing the audio output circuits in the CD player.

Clip an audio or function generator to the input terminals of the high-wattage amplifier for a signal source and signal trace the audio circuits with the external audio amp. Rotate the signal generator to 1 kHz frequency for audio tests.

Go from base to base of each succeeding transistor from preamp to output circuits, to check the audio signal with the external amp *(Figure 12-5)*. Clip the ground terminal to common ground and place a test probe on each base terminal. Check the audio signal at the volume control. When the defective stage is in the output circuits, began at the volume control and check each output transistor. Test for audio signal at the input and output terminals of a suspected IC with the external amp.

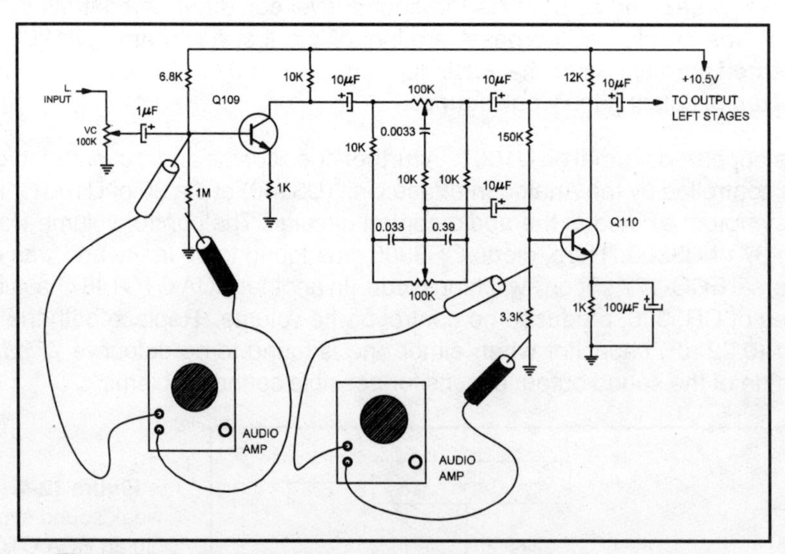

Figure 12-5. Troubleshooting the transistor audio circuits with the external audio amplifier.

When the audio signal becomes weak, intermittent, or distorted within the external audio amp, you have located the defective circuit. Remember, the audio signal should become stronger as you proceed toward the speaker terminals. Simply turn down the audio at the external audio amplifier. The external amplifier can monitor the defective stage for weak and distorted reception. You can pin-point the weak audio component with the external audio amp. Take critical voltage and resistance measurements at the defective stage or circuit. Then, test each suspected transistor with in-circuit transistor tests.

The audio was distorted in the PM speaker of an RCA CTC145 TV chassis. Sometimes a touch-up of the quadrature sound coil (L1201) off of IF/SIF IC (U1001) can cure the extremely distorted audio. The audio signal within the external audio amp was normal on output pin 28 of U1001 and base of Q1201 *(Figure 12-6)*. The audio was normal on the base of Q1202 and Q1203, but distorted at the speaker terminal. The audio was distorted on both sides of C1207 (100 µF). No doubt, Q1202 and Q1203 were defective. Q1202 tested open with a in-circuit transistor test. The collector voltage had increased to 19.7 volts. Both output transistors were replaced, Q1202 with a SK3854, and Q1203 with a SK3867A universal replacement transistor.

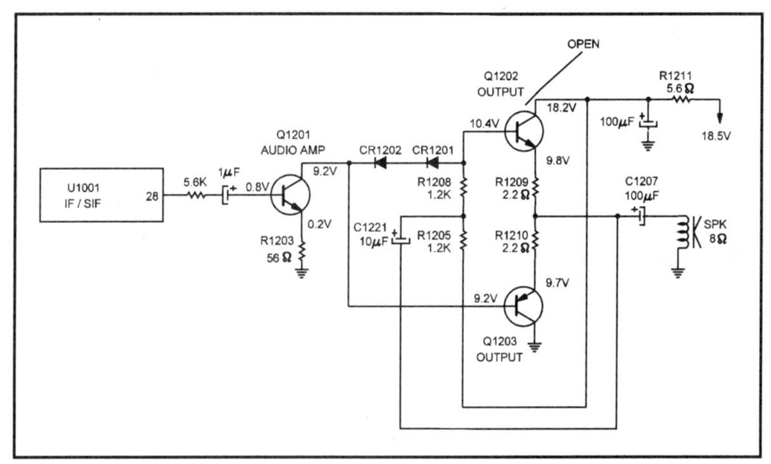

Figure 12-6. Open Q1202 caused extreme distortion in the RCA CTC145 audio output circuits.

1 IC AUDIO AMP

The external audio amplifier for chasing down the lost, distorted or weak audio in the audio circuits, does not have to be a complicated or high-powered audio circuit. The audio amp can be constructed around a LM386-1 audio amp IC. To acquire the highest wattage output, choose either a LM386-1 or LM386-3 IC *(Figure 12-7)*. Turn the volume down (R1) as the external amp tests each AF or driver transistor. Mount C4 as close to pin 4 as possible to prevent oscillations in the speaker. The external amp components are listed in the part lists:

IC1 LM386-1 or LM386-3 audio IC
C1 0.1 µF 250 volt capacitor
C2 1 µF 35 volt electrolytic
C3 10 µF 35 volt electrolytic
C4 220 µF 35 volt electrolytic

C5, C6	0.1 µF 50 volt ceramic capacitor
C7	100 µF 35 volt electrolytic
R1	10K ohm audio taper control
R2	560 ohm 1/4 watt carbon resistor
R3	4.5K ohm 1/4 watt carbon resistor
R4	10 ohm 1/4 watt carbon resistor
Spk	4 inch 8 ohm PM speaker
Batt	9 volt energizer battery
Sw1	SPST on rear of R1
Misc	PC board, 8 prong IC socket, hookup wire, cabinet, solder, etc.

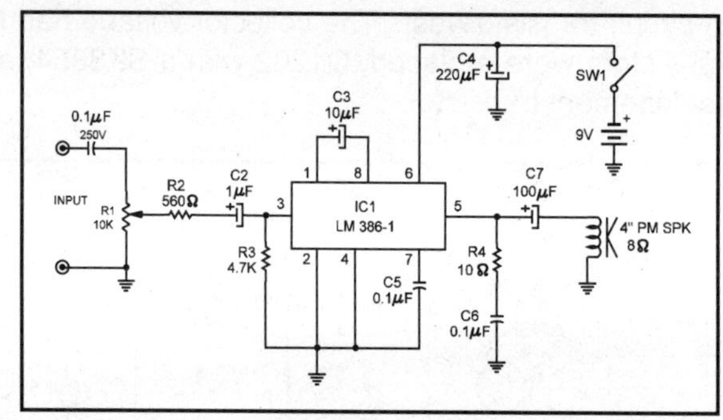

Figure 12-7. The small external audio amp was built around the LM386-1 or LM386-3 low-powered amplifier.

TEST CASSETTES

There are several different test cassettes and discs used in cassette and CD players for proper adjustments. The azimuth adjustment of the tape head should either be a 6.3 kHz or 10 kHz test cassette for proper alignment. The scope, ac voltmeter and frequency counter can be used as the indicator across a 4 or 8 ohm load at the speaker terminals. The bias, record gain and record/playback sensitivity adjustments should be made with a 1 kHz test cassette.

The tape speed adjustment can be checked with a 1 kHz or 3 kHz test cassette, with the frequency counter as an indicator. The playback sensitivity adjustment, Dolby NR level adjustment, and bias current adjustment should be made according to the manufacturer's service literature. The flutter meter might use a 3 kHz and 3.15 kHz test cassette for wow or flutter tests.

ROLL YOUR OWN TEST CASSETTE

Test cassettes with different frequencies can be used in troubleshooting the audio circuits, cassette azimuth and cassette speed adjustments. The most common test cassettes used are the 1 kHz, 3 kHz and 10 kHz. You can make your own test tapes with the audio signal generator and cassette recorder.

Simply inject the 1 kHz audio signal into the input jack of cassette recorder and record the 1 kHz signal on a 1/2 hour cassette. Keep the output control of the generator at a normal volume level. Likewise, record the 3 kHz and 10 kHz audio signals on a separate blank cassette. These audio cassette tapes can be used in troubleshooting and adjustments on the cassette player.

The commercial microcassette alignment tape might have a 15 minute alignment cassette of a 1 kHz audio tone recorded at -4dB level. The commercial reference level cassette might have a frequency of 6.3 kHz, 8 kHz, and 10 kHz. A commercial cassette speed tape might operate at 50 Hz, 3.15 kHz, and 3 kHz. The commercial frequency response cassette might have a wide range of frequencies between 315 Hz, 1 kHz, 6.3 kHz, 10 kHz, 14 kHz, and 18 kHz.

TROUBLESHOOTING WITH TEST CASSETTES

The different test cassettes can be used to signal trace the audio circuits within the cassette player. When the cassette player has intermittent, weak and distorted sound problems, insert a 3 kHz or 10 kHz test cassette into the cassette recorder. Use a scope or external audio amplifier as an indicator *(Figure 12-8)*. Start at the tape head terminals and trace the audio signal through the preamp and AF circuits, if the audio defects are not heard at the volume control terminals. The audio cassette signal can be traced clear through to the audio speaker terminals.

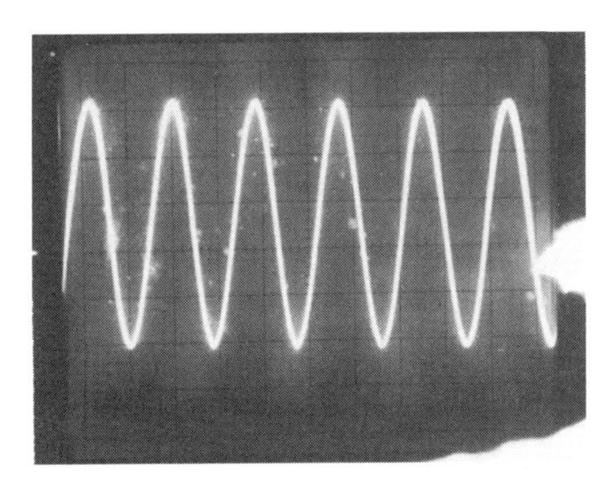

Figure 12-8. Insert a test cassette and check the various waveforms throughout with the scope as the indicator.

When the cassette signal is missing, weak or distorted, take critical voltage and resistance measurements on the components in that stage. Suspect the AF transistor when the signal is weak at the collector and normal at the base terminal. Poor component soldered terminals on diodes or regulator transistors can produce an intermittent or dead voltage source.

The cassette audio signal and scope can locate an open or leaky transistor and IC in the audio output circuits. A defective speaker relay circuit can be located with the test cassette and scope as indicator.

VOLTAGE INJECTION

When the voltage source is derived from another source in a battery operated audio amplifier, radio, auto receiver, TV and special audio circuits, an external working voltage can be injected from a separate power supply *(Figure 12-9)*. For instance, in the TV chassis, most sound circuits are powered by a voltage source developed in the flyback transformer windings. Thus, the horizontal circuits must operate before any secondary voltage is developed in the horizontal input transformer. A defective component in the sound circuits might shutdown the TV horizontal circuits.

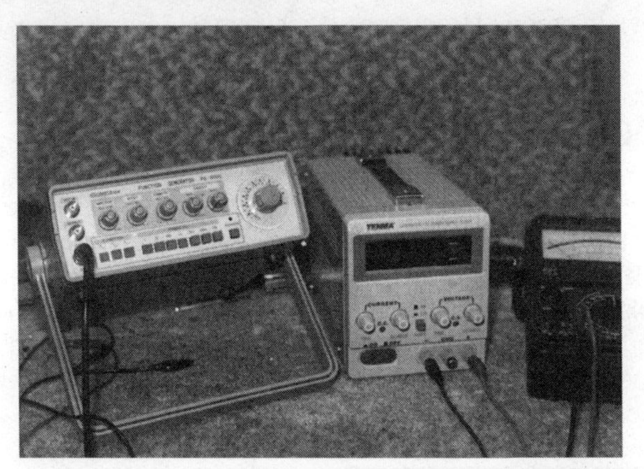

Figure 12-9. The external power supply can provide a voltage source for a defective audio circuit.

The sound circuits can be tested by injecting an external voltage from the power supply. Check the schematic for the supply voltage terminal source operating the output IC or transistors; this is usually the highest working voltage applied to the component. In a Panasonic model CTL-1030R TV, the supply voltage from the flyback circuits was a +15 volts. By injecting the supply source at R210 (10 ohm) resistor on pin 9 of IC201, the audio circuits should now be alive.

Inject the 1 kHz audio signal from the signal tracer, signal or function generator at the C209 (4.7 µF) electrolytic at input terminal 2 of IC1201 *(Figure 12-10)*. With a voltage source injected at pin 7 of IC201, the audio should be heard in the loud speaker. If no sound, inject the audio signal at output pin 8 of IC201. A low audio tone should be heard in the speaker, indicating IC201 or components tied to it are defective.

Take a voltage and resistance test on each pin terminal of IC201 to locate a leaky electrolytic capacitor. Replace defective IC201 when the voltage and resistance measurements were quite normal to those on the schematic.

CRITICAL WAVEFORMS

The audio stereo circuits within the TV chassis or high-powered amplifier can locate the weak and distorted audio stage or circuits. By injecting a sine or square waveform audio signal into the input of both left and right stereo circuits, the oscilloscope probe can quickly locate the defective stage. Connect a speaker or dummy load to both stereo output speaker terminals. Besides the speaker load, a PM speaker can be connected to each audio channel to indicate what channel is defective. A weak stereo channel does have a lower audio sound than the normal channel.

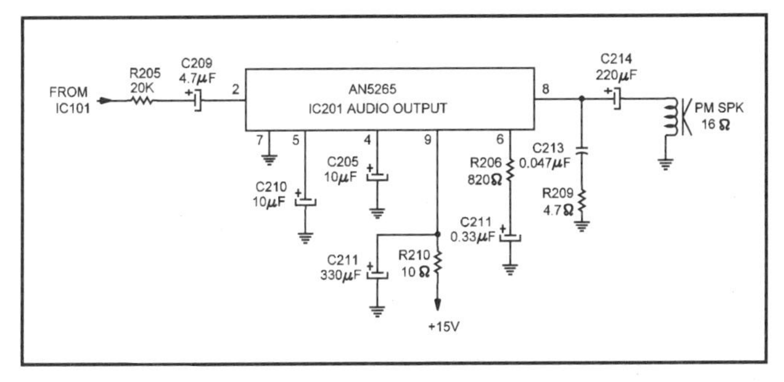

Figure 12-10. A dc voltage from external power supply located a leaky IC1201 in the Panasonic TV sound output circuits.

Keep the input audio signal as low as possible to prevent clipping or over-driving the audio circuits. Use the dual-trace scope to test both audio channels at the same time. Clip the scope probes to both stereo balance or volume controls. Notice if one channel is weaker than the other. Proceed to the AF or driver circuits if the sine waveform at the controls have the same waveform and amplitude *(Figure 12-11)*. Check the waveform of the defective channel against the good audio channel.

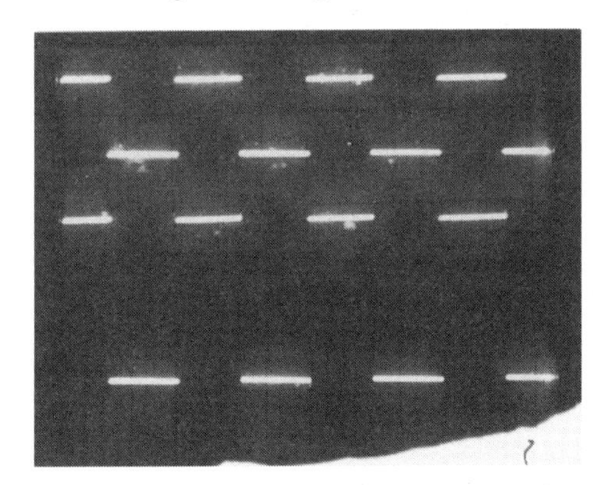

Figure 12-11. The scope waveforms in the right channel were quite weak with a square wave in the audio amplifier.

The stereo sound can be checked stage by stage in each stereo channel. Proceed slowly and deliberately to trace both audio channels. When the audio waveform changes in form and amplitude, the defective part is close at hand. The defective waveform might be weak (lower amplitude), have rounded corners, or a different shape than the normal audio channel. Make sure each scope test is on the same point in each stereo circuit.

Take critical voltage measurements on a transistor or IC component that indicates an improper waveform. Compare these voltage measurements with the same spot in the normal channel. If a schematic is not available, check the voltages in the normal circuit and compare with the defective measurements. The audio channels in the stereo amplifier are much easier and quick to service, since comparison voltages, resistance and waveforms can be measured without a schematic.

TAPE HEAD PROBLEMS

The defective tape head might have an open winding, poor internal solder connections, or torn wires from the tape head terminals. When the volume is wide open with only a rushing noise in the speaker, suspect an open head or torn wires from the head terminals. The cassette audio might be intermittent with poor internal connections. Sometimes moving the tape head while the player is operating will cause the sound to cut in and out. The sound can be intermittent within the auto cassette player in reverse mode with broken head wire terminal leads.

The tape head might not have any sound in record or playback modes when the tape head is moved backwards or out of the line of the tape path. Sometimes the mounting screws on the tape head can work loose and let the tape head swing out of line. In players where the R/P head is pivoted downward, the head may not reach its full position in play mode; thus, no sound from the tape. At other times, the tape head might come loose from the welded support causing no audio in the speaker *(Figure 12-12)*.

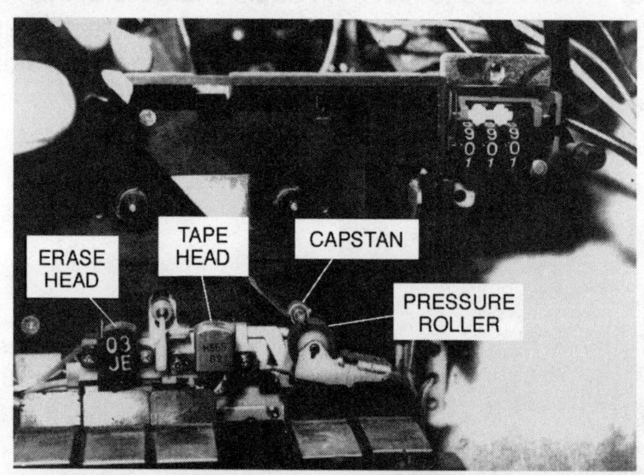

Figure 12-12. Check the tape head terminals for poor connection and to see if they have moved out of the line of rotating tape.

Check the face of the tape head when the recordings become tinny or contain a high-pitched music sound. Look for a worn front area of the tape head. This worn condition usually occurs after many years of operation.

The tape head can be magnetized with slight distortion and a loss of high frequency response in the speakers; demagnetize the tape head. The typical tape head resistance might be from 200 to 850 ohms. The erase head resistance might be from 200 to 1000 ohms. Both stereo tape head windings should be quite similar in a resistance measurement.

DEMAGNETIZE THE TAPE HEAD

When servicing the cassette player audio circuits, do not use magnetized tools near the tape head, as the head can easily be magnetized. With normal use, the head will retain a small amount of residual magnetism, resulting in increased noise and distortion, and loss of high frequency response. Long hours of tape playing can magnetize the tape heads, causing high-end frequency response and increased distortion, resulting in overall poor performance.

There are many different kinds of demagnetizer cassettes and tools on the market. A modern cassette demagnetizer automatically demagnetizes the record and playback heads of all standard decks and restores the best performance. This system quickly and conveniently restores audio performance, displaying a red light when the system is operating *(Figure 12-13)*.

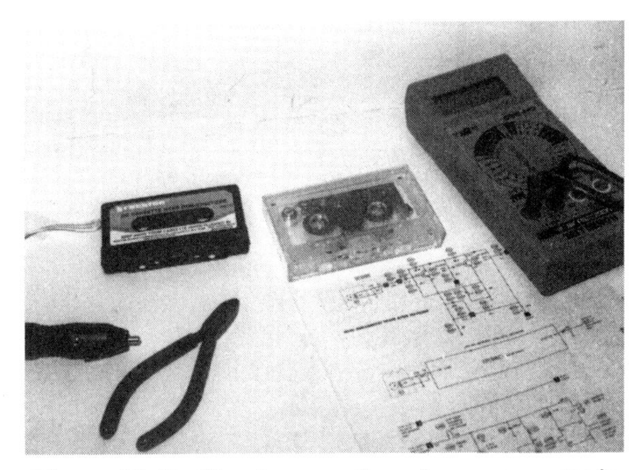

Figure 12-13. The demagnetizer cleaner cassette is self-powered while the demagnetizing cassette receiver power is from the cigarette lighter.

Insert the demagnetizer like a regular cassette. Be sure the front side faces up. Set the cassette player to "play" mode. Notice if the red light is on. Eject the demagnetizer cassette after 10 seconds of operation. The demagnetizer should be used after 20 to 30 hours of playing time. Clean up the tape heads with an audio cassette cleaner or alcohol and cleaning stick. Keep all test cassettes away from the demagnetizer cassette when operating.

A cassette demagnetizer might operate directly from the cigarette lighter on the car dash board. Insert the cassette demagnetizer into the player. Place cassette deck into "play" mode. Plug the cord of the demagnetizer into the cigarette lighter. Withdraw the demagnetizer from the player after 10 seconds of operation. Place recorded tapes at least 3 feet away from demagnetizing operation to avoid accidental erasure of the recordings. Always clean the erase and tape heads after repairs, and then demagnetize the tape heads.

HEAD AZIMUTH ADJUSTMENT

Make sure the tape head is clean. Insert a music tape into the cassette player and select a side of the cassette which contains piano or violin strings. Rotate the tone control to the high position. Check the side of the tape head which has the adjustment screw with a spring on one side of the tape head *(Figure 12-14)*. Insert a screwdriver blade to rotate or turn the azimuth adjustment screw. Sometimes a hole is found in the top cover of the auto-cassette player for azimuth adjustment.

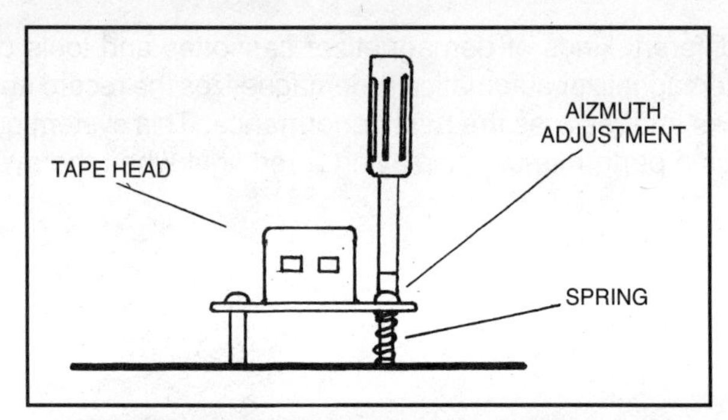

Figure 12-14. Adjust the azimuth screw with a tension spring for maximum sound in the speaker or on AC meter.

Turn the screwdriver left or right until the high frequency reproduction is the clearest. If readjusting the azimuth does not reproduce the high frequencies, check for a worn magnetic head, defective component in the audio circuits and a magnetized magnetic tape head. Demagnetize the tape head and try again.

The head azimuth screw can be accurately adjusted by inserting a 6.3 kHz or 10 kHz test cassette into the player. Connect a 4 or 8 ohm dummy resistance load across the headphone jack, line output, or the speaker terminals *(Figure 12-15)*. Connect a high sensitivity AC voltmeter or FET-VOM across the dummy load. Playback the test tape with the volume at the proper level. Adjust the azimuth screw for maximum output reading on the meter. After completing the alignment, lock the screw with a screw lock, glue or paint.

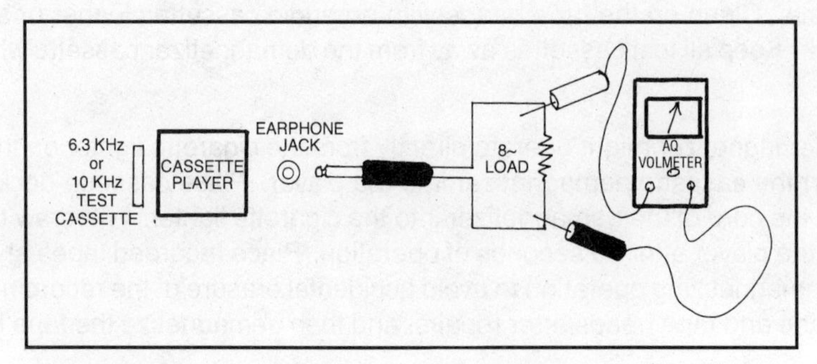

Figure 12-15. Insert a 10 kHz test cassette with 8 ohm load across headphone jack and ac voltmeter for correct azimuth alignment.

SPEED TESTS

The tape speed in the cassette player might run slow or too fast. The cassette motor might contain a wow or flutter sound in the speaker with oil spots on the belt. Slow speeds can be caused by a stretched or loose motor drive belt. When the motor speed changes, the cassette player results in poor music reproduction. Clean up all belts, pulleys, dry bearings and capstan. If the motor is still running slow, make a tape speed adjustment.

A cassette motor might have a separate speed adjustment within a speed regulator circuit or on the rear bell of the motor. Play a standard test tape (3 kHz) and adjust the variable resistor on the motor or speed circuit for correct speed. Connect a frequency counter to the headphone jack, line output, or speaker terminals with a 4 or 8 ohm load resistor. Adjust the motor tape speed VR control to obtain an output of 3 kHz with a + or -10 Hz on the frequency counter *(Figure 12-16)*. The speed regulator circuit might have a normal and a high speed adjustment. Use a 3 kHz cassette for normal speed and the 6.3 kHz cassette for the high speed adjustment.

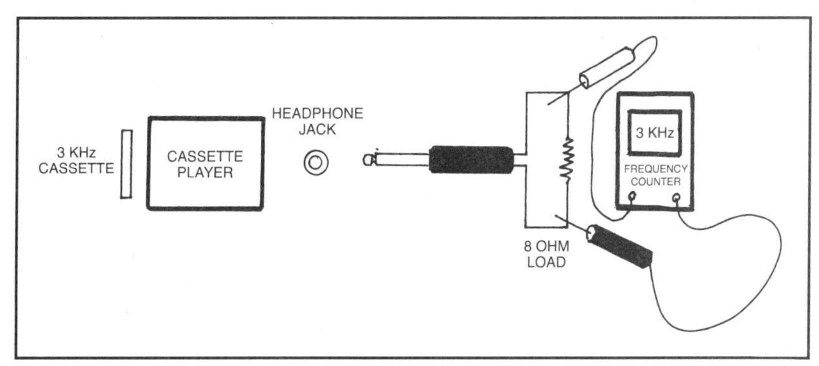

Figure 12-16. Check the tape motor speed with 3 kHz cassette and frequency meter connected to 8 ohm load resistor.

PHONO SPEED TESTS

The phonograph speed can be checked by a speed-strobe disk that is placed upon the turntable of the record player. The different 33, 45 and 78 RPM speeds will stand still with a strobe, neon or fluorescent light above the turntable. The electronic technician's fluorescent bench light is ideal. When the speed stripes or bars stand still at any given speed, the turntable speed is normal. When the stripes or bars slowly goes backwards, the turntable is running slow. Likewise when the speed is too fast, the bars and stripes are slowly moving ahead. Check all three speeds with the strobe disk.

Adjust the motor speed control for the bars and stripes to stand still under a strobe or fluorescent light *(Figure 12-17)*. When no speed adjustments are found in the early models, the slow speed indicates a worn idler wheel, dry motor bearings or slippage of the belt on the rim of the turntable. A good cleanup of the idler, motor pulley and bearings might solve the slow speed problem. If not, replace the idler wheel, rubber tire or motor drive belt. Resurfacing the turntable rim with liquid rosin might prevent idler slippage.

Figure 12-17. Adjust the speed control until the lines or bars standstill with overhead fluorescent light.

A commercial speed-strobe digital turntable speed readout uses digital graphics instead of hypnotic bars or stripes. The optical disc can check any strobe set speed with unsurpassed accuracy. The digital speed readout can check the 16, 35, 45 and 78 RPM turntable speed (model # 32-10345). Order from :

MCM Electronics
650 Congress Park Dr
Centerville, OH 45459-4072
1-800-543-4330

PLAYBACK CASSETTE SENSITIVITY ADJUSTMENT

Always follow the manufacturer's alignment procedures when making playback sensitivity, line-out Dolby NR level, bias current, playback gain and record gain adjustments. The playback sensitivity adjustment might also be called playback gain adjustment. These are usually located with preset controls after the preamp tape head transistors or IC components.

A Dolby NR calibration tape (400 Hz), 200 pwb m/m), (MTT-150 test cassette) might be used with a VTVM or FET-VOM and oscilloscope connected to the Dolby output test points. Adjust V304 and V404 for a certain manufacturer 580 to 590 mVs *(Figure 12-18)*. In a Sanyo MX720K cassette player the Dolby output adjustment was 580 mV. This playback adjustment actually selects a certain volume of sound to be placed in the Dolby and output circuits.

TAPE HEAD BIAS ADJUSTMENT

Although the small tape head within the portable cassette player might not have a separate control to adjust the bias current on the tape head, insert a 100 ohm resistor in series with the grounded head terminal. To make sure the bias oscillator is operating, take a waveform upon the ungrounded tape head terminal in record mode. A current test at the tape head can be made by taking a voltage measurement across the 100 ohm resistor.

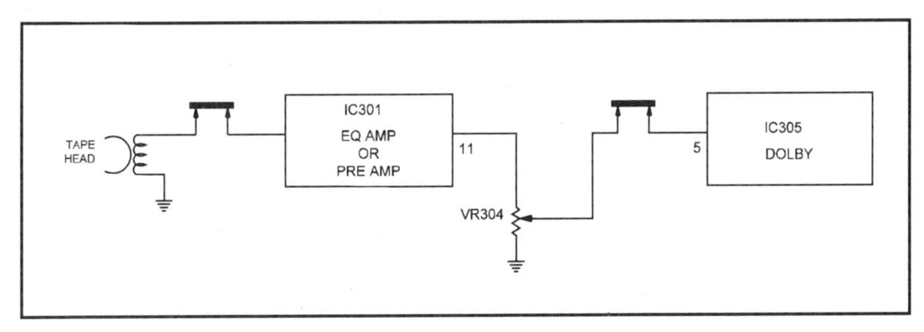

Figure 12-18. Block diagram of the playback sensitivity or gain adjustment (VR304) in a Sanyo MX720K cassette portable.

Place the cassette player in record mode. Measure the voltage across the 100 ohm resistor with a VTVM, FET-VOM or DMM. The tape head current should be between 5 to 65 mVs. Some players have test points at the tape head terminals for these type of tests. Simply adjust the bias controls from the bias oscillator for correct manufacturers mV measurement.

In larger cassette players that have Dolby circuits, connect the audio signal generator to the AUX input terminal. Connect the AC voltmeter to the output test points of Dolby IC. Load the deck with a normal tape and set the player in the recording mode. Adjust the input level with attenuator control until the meter reads an average 30 mV with a 1 kHz and 12.5 kHz signal from the generator. Record both audio signals on the tape *(Figure 12-19)*.

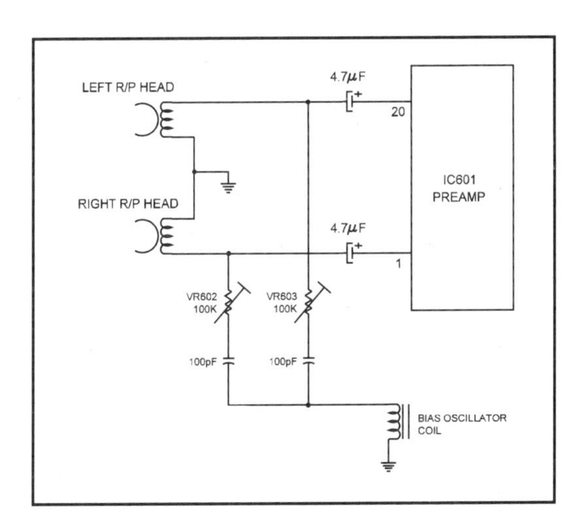

Figure 12-19. Adjust for correct head bias in each stereo channel with VR602 and VR603.

Playback the recorded signals and adjust VR602 for the left channel, so that the difference in output of the 1 kHz and 12.5 kHz signals become +1 dB - 0 dB. Adjust control VR602 in the same manner for the right channel. Follow the manufacturer's exact head bias procedures found in the service literature.

Follow the manufacturer's alignment procedures for cassette amp alignment, output level, record/playback (R/PB) and the various mechanical alignments. The other special alignment procedures are level meter, metal and chrome tape, Dolby level, normal recording levels, Dolby noise reduction check and erase current test alignments. You may find more required adjustments within the deluxe cassette deck than the common cassette player used everyday.

BIAS CONTROL ADJUSTMENTS

The high-powered solid-state or tube amplifier might have correct bias adjustments within the audio output circuits. The bias adjustment on the push-pull output tubes is to adjust the negative bias from the low voltage power supply applied to the grid circuits of the output tubes. An improper adjustment of the bias control can develop noise and distortion in the output circuits.

For instance, the correct bias adjustment of two EL-34 tubes in push-pull operation is -32 volts on pin 5, with a plate voltage on pin 3 of 415 dc volts. A 6550 tube with a +530 dc plate voltage should have a bias of -66 volts for proper operation.

The voltage measurement on the grid terminals should be made with the VTVM or FET-VOM. Improper bias adjustment can cause extreme distortion. These screwdriver type bias controls can produce a noisy or erratic sound in the speaker with dirty wiping contacts. A weak-distorted sound might result from a poor wiping contact of the bias control. Sometimes, just spraying the control with cleaning fluid can remove the noisy condition. Follow each manufacturers correct bias and level control adjustment procedures.

In the early high-priced solid-state directly-coupled audio output circuits, the bias adjustment control was found in the base of the driver amp. The VTVM or FET-VOM was connected across the emitter resistor and output terminal of TR714. VR704 was adjusted for a low mV measurement on the voltmeter *(Figure 12-20)*. The Panasonic SA-6200 (100 watt) amplifier was adjusted for 3.7 mV on the meter. Follow the manufacturer's bias adjustment procedures for correct bias and level control adjustments.

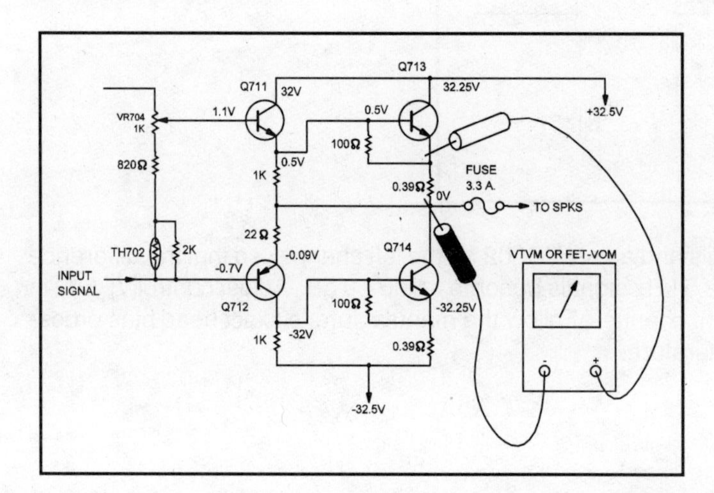

Figure 12-20. Adjust VR704 to provide the manufacturer's audio bias adjustment (MV) upon the emitter resistor of Q713 in the early high-powered amplifier.

TYPICAL FREQUENCY RESPONSE TESTS

The frequency response of the deluxe high-powered amplifier might be 20 Hz to 20 kHz. The typical frequency response of any amplifier can be checked by connecting the audio signal generator to the input and ground terminal. Set the volume and tone controls at mid-range. Plug the amplifier into the isolation power transformer. Begin the audio generator at 20 Hz and rotate through the 20 kHz range.

If the audio tapers off at either end of the frequency sound spectrum, in the speaker, either your ears cannot respond to the sound or the amplifier has reached the end of the frequency test. Do not overdrive the generator output to show clipping on the oscilloscope. A dual-channel AC or audio voltmeter can be used as an indicator in stereo circuits and response tests.

Clip an 8 ohm 100 watt load resistor to the left speaker terminals and check the stereo amplifier frequency response of the left channel. Connect the signal generator to the input terminals of the left channel. Now clip a frequency counter across the load resistor *(Figure 12-21)*. Start at the lower frequency and notice where the meter hand begins a roll off reading.

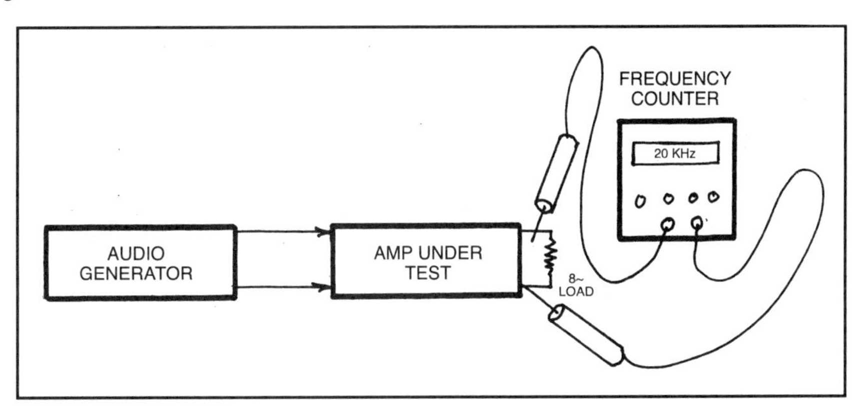

Figure 12-21. Typical hookup for a frequency response test with frequency counter, ac multimeter or scope as an indicator.

Lets say that the lower frequency begins to measure at 40 Hz. Rotate the audio generator up past the 20 kHz frequency and notice where the audio signal tapers or rolls off. If the meter hand falls rapidly at 18 kHz, the typical frequency response measurement is 40 Hz to 18 kHz. Of course, most men cannot hear above 11 kHz, while women can hear above 15 kHz. Follow the manufacturer's frequency response test if one is found in the service literature.

WOW AND FLUTTER TESTS

The wow and flutter conditions of a cassette player might occur during playback or record modes. The short change of sound might be called flutter, while the longer duration of sound might be a wow condition. The wow and flutter sounds might not be heard by some technicians, while those with better hearing can hear these changes of sound.

Wow and flutter conditions within the tape deck can be located with a test tape and a wow and flutter test instrument. The frequency counter can be used to monitor the test tape during playback modes. Insert a 3 kHz test tape and connect the frequency counter or audio meter to a fixed load at the speaker terminals. Simply notice the deviation in frequency of the counter or meter as the tone test tape is played on the cassette player. A greater deviation in frequency (3 to 10 kHz) of the played test tape indicates a wow or flutter condition.

The wow and flutter test instrument is required in a service establishment that repairs cassette and VCR machines. A crystal-controlled oscillator inside the test instrument provides the most accurate 3 kHz and 3.15 kHz frequency for wow and flutter tests. A prerecorded tone test tape or a recorded signal from the wow and flutter test instrument can be used in making wow and flutter tests.

Connect the output of the cassette player to the input terminals of the wow and flutter meter *(Figure 12-22)*. The oscillator output of wow and flutter meter is connected to the record input terminals of the cassette player. Follow the test instrument and manufacturer's operation procedures to check the wow and flutter measurements of the cassette player.

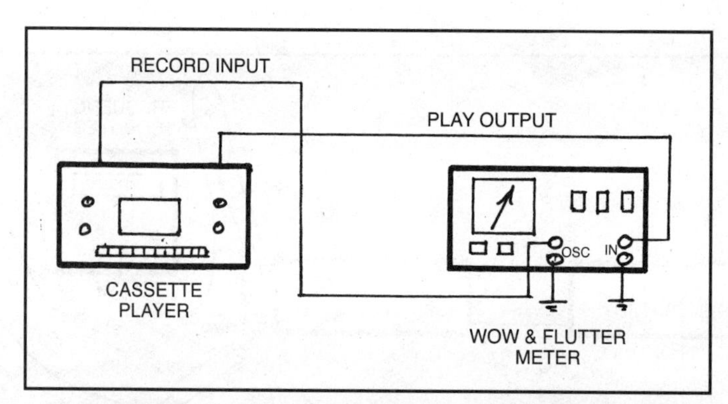

Figure 12-22. Connect the oscillator output terminals to the record input and output of cassette player to input on wow and flutter meter.

LEVEL ADJUSTMENTS

Audio level adjustments should be made in the amplifier sections that do not have a balance control in the stereo circuits. Insert a 1 kHz cassette and adjust the VU meters to 0 VU, if a VU meter is included within the cassette player or amplifier. Adjust the right and left level adjustments at the same voltage. Follow the manufacturer's level procedure if a schematic is handy.

The stereo amplifier with left and right line output jacks can be adjusted for the same voltage reading. Insert a 1 kHz cassette or connect a 1 kHz audio signal to the audio input terminals. Connect a VTVM, DMM or ac meter to the line output jacks. Adjust each level control to read according to the manufacturer's specifications. Most line output level voltage adjustments are from 400 mV to 1 volt.

Glossary
AUDIO TERMS

Amplitude The height of a waveform, either positive or negative.

Analog Filter A filter system in the CD player that reduces and cancels out noise.

Acoustic Suspension (AS) or air suspension speakers with low frequency reproduction enclosed in a solid box to produce natural, low-distortion bass output.

Air Suspension Another name for an acoustic-suspension speaker.

Ambience Simulation Refers to the sound environment of a given space in a room.

Amp Abbreviation for amplifier.

ANRS A noise-reduction system operating similar to the Dolby systems.

Auto Eject The tape is automatically ejected at the end of the playing time in the cassette player.

Automatic Shutoff A switching arrangement that automatically shuts off a device, product or circuit.

Auto Record Level The automatic control of the recording level.

Auto Reverse The cassette player automatically reverses direction to play the other side of the tape in the auto tape player.

Azimuth The angle of the tape head. A low high frequency response is noted with improper tape head alignment.

Baffle The board on which the speakers aremounted.

Balance A balance control equalizes the left and right channel audio output.

Bass Reflex To improve the bass response, the sound waves are vented through a tuned vent or port.

Bias A high-frequency signal switched to the tape head winding to prevent low distortion and noise on the tape during recording.

Block Diagram A boxed diagram showing the various electronic circuits of a cassette and CD player, radio and amplifier.

Booster Amplifier A separate amplifier connected between the main unit and the various speakers in the auto HI-FI system.

Bridging Combining both stereo speaker outputs to produce a monaural signal to almost double the normal car stereo amplifier.

Capstan The shaft that rotates the tape at constant speed past the tape heads in a cassette player or VCR.

Cartridge A component that holds the stylus or needle in the record player. The cartridge develops the voltage from movement on the record to the audio amplifier.

Cassette Radio A combination of an AM/FM tuner, amplifier and cassette player in one cabinet.

CH The abbreviation for channel. The stereo amplifier has a right and left channel output.

Chip Devices Many different chip SMD (surface-mounted devices) such as thick-film resistors, chip ICs, chip transistors, multi layer ceramic chip capacitors, and minimould chip diodes are found in todays electronic chassis.

Clipping Clipping can occur if a stage is distorted or with too much volume is applied in the amplifier. Clipping can be seen on the oscilloscope.

Coaxial Speaker A speaker with two speakers mounted in one frame. Usually, the tweeter is mounted ahead of the woofer speaker in the car radio system.

Compact Disc (CD) The compact disc player plays a small disc of digitally encoded music. Mount the rainbow-like surface on the disc holder. Place the disc label upwards.

CPU A computer-type processor (central processing unit), used in micro, master or mechanism circuits of electronic products.

Crossover A filter network that applies certain audio frequencies to each speaker. The highs to the tweeter and the low frequencies to the woofer speakers.

Crosstalk Crosstalk is leakage of one channel into the other caused by improper tape head alignment.

D/A Converter The stage that separates the digital from the analog or audio signal.

DBX Abbreviation of decibels in a noise-reduction system in which the program is compressed before being recorded and expanded in playback.

DC Direct current found in batteries or the low-voltage power supplies.

Decibel (dB) A measure of gain; the ratio of output power or voltage with respect to the input.

Digital Information expressed in binary terms.

Digital Filter A low-pass filtering network.

Direct-Access Volume A control feature that lets you simply touch a place in the volume scale to set it.

Dispersion The angle in which the speaker radiates the sound.

Distortion Distortion might appear as harmonics or multiples of the input frequency. Clipping of the audio waveform is a form of distortion. The deformation of an audio signal waveform.

DNR (Dynamic noise reduction) A noise-reduction system that reduces the high frequencies when the signal is at a low level.

Dolby Noise Reduction A type of noise reduction that by increasing the high-frequency sounds during recording and decreasing them during playback, returns the audio signal to the original level and eliminates hiss tape and other recorded noises.

Driver In a speaker system, each separate speaker might be called a driver.

Dual Capstan Dual capstans and flywheels are found in auto-reverse cassette players.

Dynamic A dynamic speaker has a voice coil that carries the signal current in a pm magnetic field. A pair of headphones operate in the same manner.

Dynamic Range The ratio between the maximum signal level and minimum level expressed in decibels (dB).

Efficiency A percentage of electrical input power to a given speaker that is connected to audio energy.

Electrostatic A electrostatic speaker, headphone, or meter that uses a thin diaphragm having voltage applied to it. The electrostatic field is created by the signal voltage, which moves the diaphragm to create sound.

Equalization (EQ) Alteration of the frequency response so that the frequency balance of the output equals the frequency balance of the input. Equalization is also used to correct response deficiencies in speakers and tape player circuits.

Equalizer A device to change the volume of certain frequency in relation to the rest of the frequency range.

Erase Head The erase head with applied voltage or current removes the previous recording on the tape. The erase head is mounted ahead of the play/record tape heads.

Fader A control in auto radios or cassette players used to control music between front and rear speakers.

Fast Forward (FF) The motor or the large idler pulley is engaged to speed up the forward rotation of tape in the cassette player or VCR.

Filter A circuit that attenuates certain frequencies, but not others. The large capacitor in the low voltage power supply might be called a filter capacitor.

Flutter A change in fast speed of a tape transport. A low-speed change might be called wow.

Folded Horn Speaker A system that forces the sound of the driver to take a difficult path to the listener.

Four Track Four separate parallel magnetic tracks can be recorded on regular tape width found in quad-tape players.

Frequency Response The range of frequencies the amplifier can pass to the listener. The frequency response of a given amplifier might be 20 Hz to 20 kHz. The average listener can hear from 35 to 15-kHz.

Full-Range-Speaker A speaker system with only one driver that reproduces the normal frequency range.

Gain The amplification of a given signal in decibels (dB).

Gain Control A control to adjust the volume or boost amount of signal.

Gap The distance between the pole pieces of a tape head. A gap with excessive tape oxide might result in a weak and distorted sound.

Graphic Equalizer An equalizer with a series of sliders that provides a graphic display.

Ground A common point of zero return for components within electronic circuits. The common ground can be a metal chassis in the amplifier or receiver.

Harmonics A series of multiples of the fundamental frequency.

Harmonic Distortion Harmonics is indicated by the amount of harmonic distortion. A tape player should have less than 1% distortion.

Head A magnetic monaural or stereo tape head that picks up the signal from the moving magnetic tape.

Hertz (Hz) The unit of frequency in cycles per second (cps).

Hiss The annoying background noise in tape and cassette players. Defective transistors and IC components can produce a hiss or frying sound in the speakers.

Hum Hum noise originates from the power line. Pickup hum might result in poor solder input or ground terminals of audio components.

Idler A wheel found in the cassette player to determine the speed of the capstan/ flywheel.

IDM (inter modulation distortion) Distortion at frequencies that are the sum and differences of multiples of the input frequency.

Impedance The symbol is Z and the unit in ohms. The impedance of a speaker might be 4, 8, 10, 20, and 32 ohms. The impedance of a yaggi antenna is 300 ohms.

Infinite Baffle A completely sealed box that encloses speakers.

Integrated Amp A single component containing all amplifier components in one envelope.

Integrated Circuit (IC) A single component with many internal parts. ICs are used throughout the stereo amplifier in the TV set.

Inter modulation Distortion (IMD) The presence of unwanted frequencies that are the sum and differences of test signals.

IPS (inches per second) The measurement of tape speed.

Jack A female part of a plug and receptacle.

kHz (kilohertz) 1000 Hz or 1000 cycles per second.

LED (light emitting diodes) These are found as optical readouts and displays or indicators within the electronic product.

Level The strength of a signal. The alignment of a tape head.

Line Line input and output jacks are found in the amplifier, cassette deck and CD players.

Loading Motor The motor in the CD, VCR, camcorder or cassette player.

Loading Tray The tray that holds the compact disc during loading of the CD player.

Long Play (LP) A speed of the VCR that provides four hours of recording on a 120-minute VHS cassette.

Loudness The volume of sound. Loudness is controlled by the volume control.

Loudness Compensation A switch that boosts a low-level to compensate for the natural loss of sound at the human ear.

LSI (Large scale integration) Many electronic components built inside one large chip with many terminals. ICs, processors and CPUs might be called LSI components.

Magnetic Metal Attraction The magnetic pickup, tape head, VOM or VTVM have magnetic components.

Microprocessor A large IC chip with many functions and terminals.

Megahertz 1 MHz equals 1,000 kHz or one million cycles per second.

Metal Tape The high-frequency response and maximum output level are greatly improved with metal tape.

Monitor The scope and external audio speaker can monitor the intermittent audio circuit. To compare signals.

Monophonic One audio channel. Stereo sound has a right and left audio channel.

MOSFET Metal-oxide semiconductor field-effect transistor.

Multiplex A demodular or decoder circuit that converts a single carrier signal into two audio stereo channels.

Mute Switch A switch that turns off the sound in the TV, CD player or large receiver.

Noise Any unwanted signal found in the reproduction of sound.

Noise suppressor A filter to reduce background noise.

NR Abbreviation for noise reduction.

Output Power The output power of an amplifier, rated in watts.

Oxide The magnetic coating compound on the magnetic tape that rubs off and becomes packed on the cassette tape head.

Passive Radiator A second woofer cone that is added without a voice coil.

Pause Control The pause control stops the tape movement from the magnetic tape head in the cassette player or VCR.

PBX The noise-reduction system in which the program is compressed before being recorded and expanded in playback.

Peak The level of power or signal.

Phase Sound waves in sync with one another that connect the speakers in phase with one other.

Piezoelectric Speaker A ceramic element that expands or bends under applied signal voltage in a speaker.

Playback Head The only head found in a playback only cassette player. The playback head might also be used as a P/R head in the cassette player.

PLL (phase-locked-loop) A variable control oscillator (VCO) tied to the digital control processor.

Port An opening in a speaker enclosure for back-bass radiation.

Power The output power of an amplifier is given in watts.

Preamplifier The preamp is the first stage in the amplifier and is connected to the tape head winding within the cassette player.

Rated Power Bandwidth The frequency range over which the amplifier supplies a certain minimum power factor (20 to 20,000 Hz).

Recording Power Meter A meter that indicates how much audio signal is recorded on the tape.

Reject Lever A lever that rejects or deletes a given track in a cassette player.

Remote Control A means of operating a TV, VCR, compact disc, cassette player or receiver from a distance.

Ribbon Speaker A high-frequency driver or tweeter that uses a ribbon material suspended in a magnetic field to generate audio sound.

Rumble Low frequency vibrations that are transmitted or picked up by the audio system.

Self-Erase A partial erasure of information on the magnetic tape.

Self-Powered Speaker A speaker with a built-in amplifier.

Sensitivity The sensitivity of a speaker is the measured output of the speaker in dB compared to the input.

Separation The separation of two stereo audio channels.

Signal A form of music or voice carried in electronic form.

Signal Processing Converting the laser beam from digital to audio in the CD player.

Signal-to-Noise Ratio (SN) The higher the signal-to-noise ratio, the better the sound. The ratio of the loudest signal to that of hiss or noise.

SMD Surface mounted devices. Tiny electronic components that mount directly on the pc wiring.

Solenoid A coil with an iron-core that switches in a cassette, CD player, or TV set. The for adjustment and troubleshooting procedures.

Speaker Enclosure The cabinet or box in which the speakers are mounted.

Standard Play The speed that a two-hour (T120) VHS cassette plays on a VCR.

Standing Waves A wave created of bouncing or reflected sound back to the original wave. Standing waves can cause distortion.

Subwoofer A speaker designed to handle frequencies below 150 Hz.

Test Cassette Recorded signals on a test cassette used for adjestment and troubleshooting procedures.

Tone Control A control circuit designed to increase or decrease the amplification of a specific frequency range.

Total Harmonic Distortion (THD) The amplifier total distortion is to feed a signal in and measure the harmonic distortion at the output terminals.

Transducer A device that converts energy into another form. The microphone converts sound into electrical energy.

Tweeter A high-frequency driver speaker.

Vented Speaker System A speaker cabinet with a port to let the back waves of the speaker to escape.

Voice Coil A coil of copper wire attached to the cone of a speaker that converts electrical signal to movement of cone to create audible sound.

Watts The measurement of power.

W/CH Watts per channel.

Woofer The low frequency driver or largest speaker in the audio system.

Wow A slow change in fluctuation or speed of tape. The fast speed variation is called flutter.

Index

Symbols

A

B

C

D

E

H

I

J

K

L

M

N

O

P

Q

R

T

U

V

W

X

Z

ES&T Presents TV Troubleshooting & Repair

TV set servicing has never been easy. The service manager, service technician, and electronics hobbyist need timely, insightful information in order to locate the correct service literature, make a quick diagnosis, obtain the correct replacement components, complete the repair, and get the TV back to the owner.

ES&T Presents TV Troubleshooting & Repair presents information that will make it possible for technicians and electronics hobbyists to service TVs faster, more efficiently, and more economically, thus making it more likely that customers will choose not to discard their faulty products, but to have them restored to service by a trained, competent professional.

Originally published in *Electronic Servicing & Technology*, the chapters in this book are articles written by professional technicians, most of whom service TV sets every day. These chapters provide general descriptions of television circuit operation, detailed service procedures, and diagnostic hints.

Troubleshooting & Repair
226 pages • paperback • 6 x 9"
ISBN: 0-7906-1086-8 • Sams: 61086
$24.95

ES&T Presents Computer Troubleshooting & Repair

ES&T is the nation's most popular magazine for professionals who service consumer electronics equipment. PROMPT® Publications, a rising star in the technical publishing business, is combining its publishing expertise with the experience and knowledge of *ES&T's* best writers to produce a new line of troubleshooting and repair books for the electronics market.

Compiled from articles and prefaced by the editor in chief, Nils Conrad Persson, these books provide valuable, hands-on information for anyone interested in electronics and product repair. *Computer Troubleshooting & Repair* is the second book in the series and features information on repairing Macintosh computers, a CD-ROM primer, and a color monitor. Also included are hard drive troubleshooting and repair tips, computer diagnostic software, preventative maintenance for computers, upgrading, and much more.

Troubleshooting & Repair
256 pages • paperback • 6 x 9"
ISBN: 0-7906-1087-6 • Sams: 61087
$24.95

To order your copy today or locate your nearest Prompt® Publications distributor : 1-800-428-7267 or www.hwsams.com

Prices subject to change.

Speakers For Your Home and Automobile, *Second Edition*

Gordon McComb, Alvis Evans, Eric Evans

The cleanest CD sound or the clearest FM signal are useless without a fine speaker system. *Speakers for Your Home and Automobile* will show you the hows and whys of building quality speaker systems for home or auto installation. With easy-to-understand instructions and clearly-illustrated examples, this book is a must-have for anyone interested in improving their sound systems.

The comprehensive coverage includes:

Speakers, Enclosures, and Finishing Touches
Construction Techniques
Wiring Speakers
Automotive Sound Systems and Installation
Home Theater Applications

Audio
192 pages • paperback • 6 x 9"
ISBN: 0-7906-1119-8 • Sams: 61119
$24.95

Complete Guide to Audio

John J. Adams

Complete Guide to Audio was written for the consumer who wants to know more about sound systems. With comprehensive, simple explanations, it answers questions you may have asked salespeople in the past but were unable to get answers for.

In addition, this book explains some common problems that you may run into while setting up your home entertainment center. The information in the book will help you make successful purchasing decisions and demystify the jungle of wires and connections that come with your audio system.

Topics include: Audio Basics, Sound, Stereophonics, Home Theater, Amplifiers and Preamplifiers, Receivers and Surround-Sound, Cassette and CD Decks, DVD, MiniDisc and Phonographs, Speakers, Headphones and Microphones, Computer Sound, Brands and Choices, Hookups and Accessories.

Audio
163 pages • paperback • 7-1/4 x 9-3/8"
ISBN: 0-7906-1128-7 • Sams: 61128
$29.95

To order your copy today or locate your nearest Prompt® Publications distributor : 1-800-428-7267 or www.hwsams.com

Prices subject to change.